ANATOMICAL KINESIOLOGY

Michael Gross, DA

Associate Professor
Department of Kinesiology
College of Education
University of North Georgia - Gainesville Campus
Oakwood, Georgia

JONES & BARTLETT
LEARNING

World Headquarters
Jones & Bartlett Learning
5 Wall Street
Burlington, MA 01803
978-443-5000
info@jblearning.com
www.jblearning.com

Jones & Bartlett Learning books and products are available through most bookstores and online booksellers. To contact Jones & Bartlett Learning directly, call 800-832-0034, fax 978-443-8000, or visit our website, www.jblearning.com.

17616-2

Production Credits
VP, Product Management: Amanda Martin
Director of Product Management: Cathy L. Esperti
Product Manager: Sean Fabery
Product Assistant: Andrew LaBelle
Project Manager: Jessica deMartin
Project Specialist: Kristine Janssens
Digital Products Manager: Jordan McKenzie
Senior Digital Project Specialist: Angela Dooley
Digital Project Specialist: Rachel Reyes
Director of Marketing: Andrea DeFronzo
VP, Manufacturing and Inventory Control: Therese Connell
Composition: Exela Technologies
Project Management: Exela Technologies
Cover Design: Michael O'Donnell
Text Design: Michael O'Donnell
Senior Media Development Editor: Troy Liston
Rights Specialist: Rebecca Damon
Cover Image (Title Page, Part Opener, Chapter Opener): © BraunS/E+/Getty Images
Printing and Binding: LSC Communications
Cover Printing: LSC Communications

Library of Congress Cataloging-in-Publication Data
Names: Gross, Michael R., author.
Title: Anatomical kinesiology / Michael R. Gross.
Description: Burlington, MA : Jones & Bartlett Learning, [2021] | Includes bibliographical references and index.
Identifiers: LCCN 2019032348 | ISBN 9781284175646 (paperback) | ISBN 9781284176162 (other)
Subjects: LCSH: Kinesiology. | Human anatomy.
Classification: LCC QP303 .G758 2021 | DDC 613.7–dc23
LC record available at https://lccn.loc.gov/2019032348

6048

Printed in the United States of America
24 23 22 21 20 10 9 8 7 6 5 4 3 2 1

Brief Contents

Contents

Preface

Welcome to *Anatomical Kinesiology*! This text is intended for undergraduate students in a variety of academic disciplines such as biomechanics, exercise physiology, athletic training, sports medicine, rehabilitation and health sciences, and other related areas. It lays the foundation for future study in their professional field.

Purpose

The main purpose of this text is for the student to master the muscles most responsible for body movement. To be more specific, their names, their locations, their attachments, the motions they cause, and the nerves that stimulate them to act. How can a personal trainer, strength and conditioning specialist, sport coach, physical educator, exercise scientist, or exercise physiologist design an exercise program without this knowledge? How can an athletic trainer, occupational therapist, physical therapist, or other healthcare professional design rehabilitation programs without this knowledge? The material in this text is foundational to all these professions and many more.

Organization and Approach of This Text

The text is organized into five sections and eighteen chapters. Section 1, composed of four chapters, provides foundational information relevant to anatomical kinesiology. Chapter 1, "Fundamentals of Anatomy," provides an overview of the human body and defines terminology that is essential for application throughout the rest of this text. For instance, the different motions are presented and defined in this chapter. Equipped with this, it becomes easier to grasp the application of the motions to the actions of the muscles in later chapters.

The next three chapters of Section 1 cover the skeletal, muscular, and nervous systems, respectively. These systems are the most pertinent to anatomical kinesiology. The bones are arranged about joints so that a vast array of body movements are possible. However, the bones will not move without the actions of the muscles. Additionally, these muscles will not act upon the bones unless the nervous system "tells" them to do so. Therefore, all three systems must be learned to understand their interrelated roles in body movement.

In Section 2, the focus is turned toward the bones and their landmarks. Upon completion of this section, the reader will have mastered the name, location, and significant landmarks of each bone. The reader will thus be prepared to apply this knowledge in later chapters that examine the structure of joints and attachment sites of muscles. This section consists of three chapters to facilitate this outcome. Chapter 5 addresses the bones of the axial skeleton. Chapters 6 and 7 discuss the bones of the appendicular skeleton: respectively, the upper extremities (arms) and then the lower extremities (legs).

The final three sections are the "heart" of the text and its pedagogical purpose. Everything has led up to this. Section 3 begins at the bottom of the body. The lower extremities are presented first because they are easier for students to learn in my experience. Once the student becomes more knowledgeable in this area, they will have more confidence to "tackle" the rest of the body. The subsequent sections systematically move through the axial skeleton and conclude with the upper extremities.

Each chapter within the final three sections is dedicated to a region of the body and the muscles that produce the motions that occur within that region. The muscles are grouped by the common motion they produce (i.e., the knee flexors), but they are also examined individually. The names, locations, attachment sites, motions caused, and nerve innervations of the main muscles for body movement are presented.

What Makes This Text Unique

Table of Contents

The table of contents is arranged in a logical order, beginning at the most fundamental first step and proceeding without skipping any steps and returning to them later. A clear example of this is the order of the muscle chapters. The muscle chapters begin in the lower extremities and move up the body; other sources do the opposite. This is no accident. Because the lower extremities are easier to learn than the upper extremities, beginning with the easier body regions results in more student success and confidence to tackle the more difficult areas later in the text.

Organization of Chapter Content

The order of information presented in each chapter is also designed to aid student learning. As stated above, the muscle attachments, motions produced, and nerves are presented individually, but are also grouped in a meaningful order. A good example of this are the muscles of the ankle. The first muscle group covered is in the front of the lower leg. Then, the chapter proceeds to the muscles on the lateral side and finally the two posterior groups. Each of the four groups causes one of four ankle motions. For the first three groups explained, it's relatively easy to understand the motion they cause. The fourth group is not. Therefore, applying this order results in only one choice for the last muscle group to produce the last motion.

Tables

Several tables exist throughout the text to summarize and reference information. For instance, there are several for anatomical terms, for motions, for bones, for landmarks of bones, and for muscles. Another advantage exists in the various tables for the muscles. There are tables for each individual muscle and tables for muscles groups. The tables for muscle groups summarize the entire group and reveal commonalities where they exist. Why are commonalities advantageous? It is easier to recall that all four of the knee extensors attach on the tibial tuberosity than to try to remember them separately.

Phonetic Spelling

The names of many muscles are difficult to pronounce. Therefore, a phonetic spelling has been provided immediately following those words in parentheses.

© BraunS/E+/Getty Images.

How to Use This Text

- **Learning Objectives** at the beginning of each chapter focus students on key concepts and the material they will learn.

CHAPTER 1

Fundamentals of Anatomy

CHAPTER OBJECTIVES

After completing this chapter, the student will be able to:

1. describe how the body is organized;
2. define and use anatomical terminology;
3. recall the body regions;
4. name the body planes and the axes of rotation; and
5. match the body planes and axes of rotation.

- **Key Terms** will be bolded upon their first use in the text, with definitions appearing in the end-of-text **Glossary**.

The title of this book is Anatomical Kinesiology. The word "anatomical" refers to the study of the structures of an organism. From the word kinesiology, the root word "kines" means movement and the suffix "-ology" is "the study of." Thus, **kinesiology** is the study of body movement. Therefore, **anatomical kinesiology** is the study of the structures most relevant to body movement. Additionally, **physiology**, the study of the functions of the structures of an organism, must also be learned. This chapter will start with a general overview of the anatomy and physiology of the entire body and end with the terms most relevant to anatomical kinesiology.

Organization of the Body

All organisms, including the human body, have a hierarchy of organization. This **hierarchy** refers to an increasing complexity of structural and functional units. From the smallest to the largest, the hierarchy follows this pattern:

- Cells
- Tissues
- Organs
- Organ Systems
- Organism

Cells

The cell is the smallest functional unit of life. There are smaller parts within a cell, but none are considered a "unit of life." The cells of an organism are ultimately responsible for all the functions of an organism. However, a single type of cell only performs one function. Therefore, there are a number of different types of cells to perform various functions of an organism. For instance, a skin cell and a muscle cell play completely different roles.

Some cells are less complex than what is shown in **FIGURE 1.1**, but there is much commonality among most types of cells. All cells have three main components: the *cell membrane*, the *cytoplasm*, and the *nucleus*. The **cell membrane** serves as a semipermeable barrier between its internal environment and what lies outside of it. This barrier allows for the free flow of some substances and a controlled sharing of other substances.

The Joints of the Foot **71**

FIGURE 8.17 The Abductor Digiti Minimi Muscle.

ORIGIN

lateral half of calcaneal tuberosity

INSERTION

lateral base of proximal phalange V

NERVE

Tibial

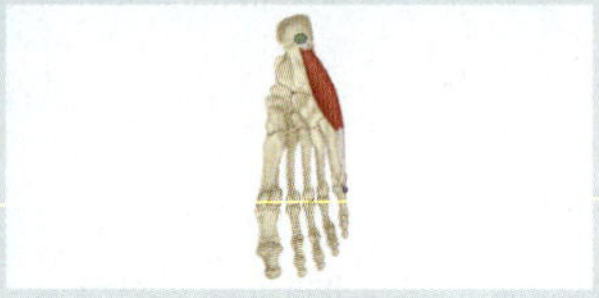

TABLE 8.6 The Muscles that Abduct the Toes

Muscle	Joint(s) Acted Upon	Other Actions	Nerve
Abductor Hallucis	1st MP		
Dorsal Interossei	2nd–4th MP	none	Tibial
Abductor Digiti Minimi	5th MP		

PAUSE TO CHECK FOR UNDERSTANDING

Table 8.6 should serve as a good review for much of the material, but you should also review all the origins and insertions.

- **Pause to Check for Understanding** boxes incorporated throughout the chapters provide students with an opportunity to reflect on important concepts and to consider what they have learned from the preceding section of the chapter.

The Toe Adductors

Only two muscles are primarily responsible for adducting the toes. They are the *adductor hallucis* and the *plantar interossei* muscles. The toe adductors are summarized at the end of this section in **TABLE 8.7**.

The **adductor hallucis** (ha-LOO-sis) is unique having two bellies that lie in the plantar aspect of the foot (**FIGURE 8.18**). The transverse belly originates from the heads (distal ends) of the third through fifth metatarsals as it passes across them. The oblique belly originates from the bases (proximal end) of the second through fourth metatarsals. The two bellies merge and insert on the lateral side of the base of the first proximal phalange.

The **plantar** (PLAN-tar) **interossei** (in-ter-ROSS-ee) are actually a group of three muscles lying between metatarsals II–V (**FIGURE 8.19**). They originate from medial side of the diaphyses of the third through fifth metatarsals. Each belly passes between their respective metatarsals and insert on the medial side of the base of the third through fifth proximal phalanges.

- A perforated **Workbook** is included at the end of the text, providing students with review questions, labeling exercises, and study material that will help them memorize and understand the function of the various bones and muscles of the body.

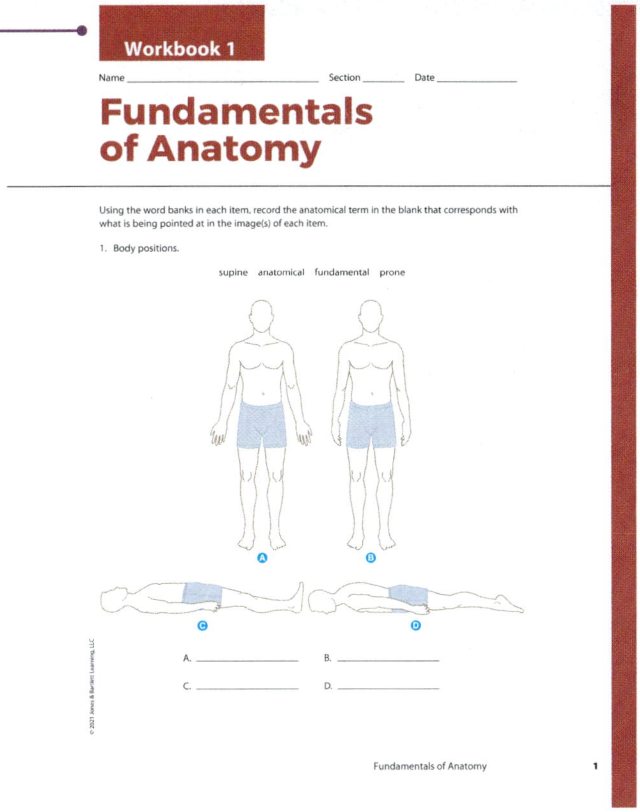
Workbook 1

Name ______ Section ______ Date ______

Fundamentals of Anatomy

Using the word banks in each item, record the anatomical term in the blank that corresponds with what is being pointed at in the image(s) of each item.

1. Body positions.

supine anatomical fundamental prone

A. ______ B. ______

C. ______ D. ______

© 2021 Jones & Bartlett Learning, LLC

Fundamentals of Anatomy 1

For the Instructor

Qualified instructors can also request access to a full suite of instructor resources, which include the following:

- Test Bank
- Slides in PowerPoint format
- Instructor's Manual
- Sample Syllabus

Access to the Anatomy & Physiology Review Module is also available online for both instructors and students.

It is my sincere hope that the reader will find this resource a must for the study of anatomical kinesiology. In addition, I hope you decide to keep it in your personal library as invaluable material to refer back to and refresh your knowledge when needed.

—*M.G.*

Reviewers

J.P. Barfield, DA
Associate Professor and Chair
Department of Health and Human Performance
College of Education and Human Development
Radford University
Radford, Virginia

Renee Borromeo, PT, DPT, MA
Program Head and Professor-in-Charge, PTA Program at the Pennsylvania State University
and
Program Coordinator, PTA Program at the Mont Alto campus
The Pennsylvania State University, Mont Alto Campus
Mont Alto, Pennsylvania

Brittany Gorres-Martens, PhD, FACSM
Assistant Professor and Chair
Exercise Science and Sport Sciences Department
Augustana University
Sioux Falls, South Dakota

Sukho Lee, PhD, FACN, CPT
Associate Professor and Associate Chair
Department of Counseling, Health, and Kinesiology
College of Education and Human Development
Texas A&M University
College Station, Texas

Chrystal McDonald, DPT
Professor and Program Coordinator
Physical Therapist Assisting Program
Blue Ridge Community and Technical College
Martinsburg, West Virginia

Keith Naugle, PhD, ATC, NSCA-CPT
Clinical Associate Professor
Department of Kinesiology
School of Health and Human Sciences
Indiana University – Purdue University Indianapolis
Indianapolis, Indiana

Ronda Sturgill, PhD, ATC, CHES
Director, Exercise and Nutrition Science Program
and
Associate Professor, Department of Health Sciences and Human Performance
College of Natural and Health Sciences
The University of Tampa
Tampa, Florida

Matthew Wagner, PhD
Associate Professor
Department of Kinesiology
College of Health Sciences
Sam Houston State University
Huntsville, Texas

Henry Wang, PhD
Associate Professor
Exercise Science Program
School of Kinesiology
College of Health
Ball State University
Muncie, Indiana

Malcolm T. Whitehead, PhD, CSCS*D, ACSM EP-C, FMS
Associate Professor
Department of Kinesiology and Health Science
Stephen F. Austin State University
Nacogdoches, Texas

© BraunS/E+/Getty Images

SECTION 1

Anatomy Relevant to Kinesiology

CHAPTER 1

Fundamentals of Anatomy

CHAPTER OBJECTIVES

After completing this chapter, the student will be able to:

1. describe how the body is organized;
2. define and use anatomical terminology;
3. recall the body regions;
4. name the body planes and the axes of rotation; and
5. match the body planes and axes of rotation.

The title of this book is Anatomical Kinesiology. The word "anatomical" refers to the study of the structures of an organism. From the word kinesiology, the root word "kines" means movement and the suffix "-ology" is "the study of." Thus, **kinesiology** is the study of body movement. Therefore, **anatomical kinesiology** is the study of the structures most relevant to body movement. Additionally, **physiology**, the study of the functions of the structures of an organism, must also be learned. This chapter will start with a general overview of the anatomy and physiology of the entire body and end with the terms most relevant to anatomical kinesiology.

Organization of the Body

All organisms, including the human body, have a hierarchy of organization. This **hierarchy** refers to an increasing complexity of structural and functional units. From the smallest to the largest, the hierarchy follows this pattern:

- Cells
- Tissues
- Organs
- Organ Systems
- Organism

Cells

The cell is the smallest functional unit of life. There are smaller parts within a cell, but none are considered a "unit of life." The cells of an organism are ultimately responsible for all the functions of an organism. However, a single type of cell only performs one function. Therefore, there are a number of different types of cells to perform various functions of an organism. For instance, a skin cell and a muscle cell play completely different roles.

Some cells are less complex than what is shown in **FIGURE 1.1**, but there is much commonality among most types of cells. All cells have three main components: the *cell membrane*, the *cytoplasm*, and the *nucleus*. The **cell membrane** serves as a semipermeable barrier between its internal environment and what lies outside of it. This barrier allows for the free flow of some substances and a controlled sharing of other substances.

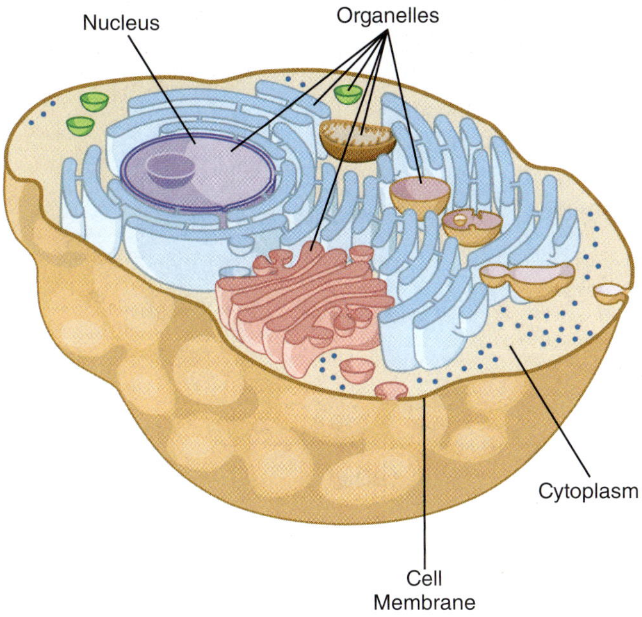

FIGURE 1.1 A Typical Cell.

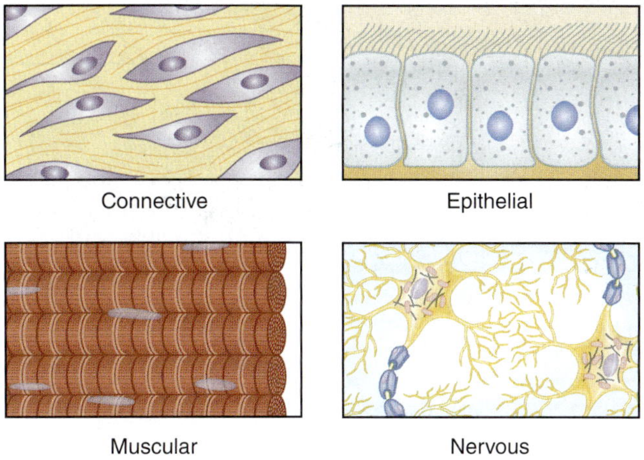

FIGURE 1.2 The Types of Tissue.

The **cytoplasm** is everything within the plasma membrane except the nucleus. The cytoplasm includes the organelles, the cytoskeleton, and the cytosol. The **organelles** are a group of structures that collectively carry out the function of the cell. The **nucleus** is also grouped with the organelles, but it is not considered part of the cytoplasm. The nucleus houses the genetic code (DNA) which controls the activities of the cell. The **cytoskeleton** is a network of protein filaments that provide structure for the cell and organize its contents. The **cytosol** is a gel-like substance that fills all the gaps within the cell, but not inside the nucleus.

Tissues

A **tissue** is a group of similar types of cells and the fluid in between them called intracellular (between cells) fluid or tissue fluid. There are four main types of tissues. They are epithelial, connective, muscular, and nervous (**FIGURE 1.2**). **Epithelial tissue** covers internal and external surfaces. The skin is a prime example of epithelial tissue as well as the membranes that line internal body cavities and organs. **Connective tissue** does exactly that; it connects things such as tendons connecting muscles to bones. Connective tissue has the most diverse varieties of any other tissue type. Some of these are bones, ligaments, cartilage, tendons, and adipose (fat) tissue.

Nervous tissue and muscular tissue are sometimes called the *excitable tissues* because they both are built to respond to stimuli. **Nervous tissue** is highly specialized and reacts to a much broader range of stimuli than muscular tissue. Basically, nervous tissue and its cells react to various stimuli to provide a communication highway between the brain and the body. When **muscular tissue** is excited, it does something that no other tissue is capable of: it contracts. Muscular contraction uses energy to perform mechanical work. In other words, muscular contraction generates a force which is used to move things.

Tissues can also be described as being in layers. The **general layers of tissue**, from superficial to deep, are 1) the **skin**, 2) the **hypodermis**, 3) the **muscular** layer, and finally 4) the **skeletal** layer (**FIGURE 1.3**). All of these general layers contain more than one type of tissue but are named so because of the tissue that is the most abundant in that layer.

Organs

Organs are groups of different tissues that perform a more complex function than any single tissue or cell. An example is the heart organ. The purpose of the heart is to apply pressure to the blood so it will flow through the blood vessels (**FIGURE 1.4**). Some of the tissues are epithelial and create an outer shell that holds the heart together. Other inner tissues provide the structure of the chambers of the heart. Still others, which are muscular tissue, squeeze the blood in the chambers. All these tissues work together to apply pressure to the blood so it will flow through the blood vessels and supply all the cells.

Organ Systems

Organ systems are groups of organs working together to perform an even more complex role than a single organ. For example, the cardiovascular system has three main sets of organs which are the blood, the heart, and the blood vessels (**FIGURE 1.5**).

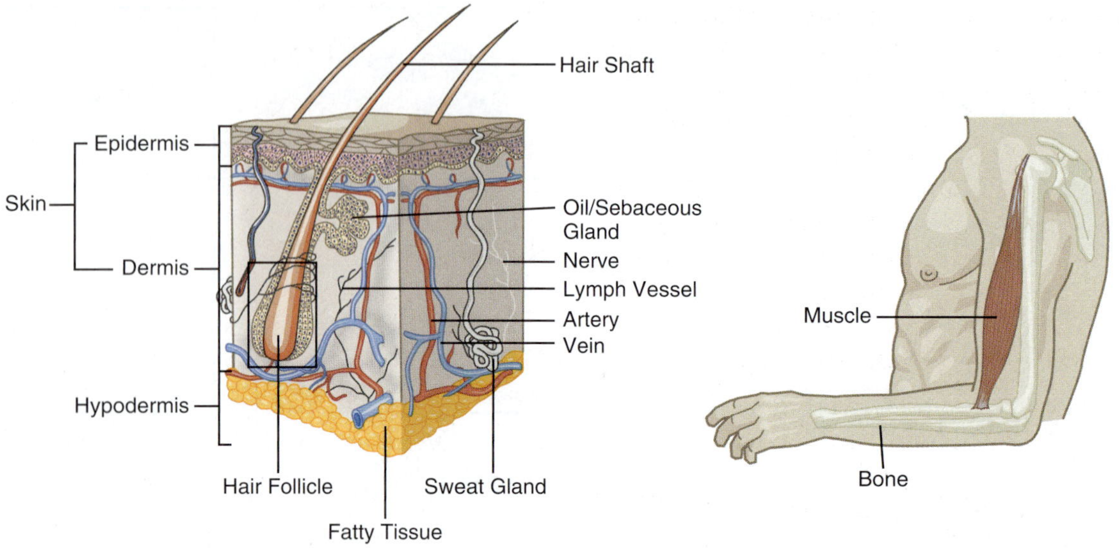

FIGURE 1.3 The Skin and Hypodermis (left) and the Muscle and Bone Layers (right).

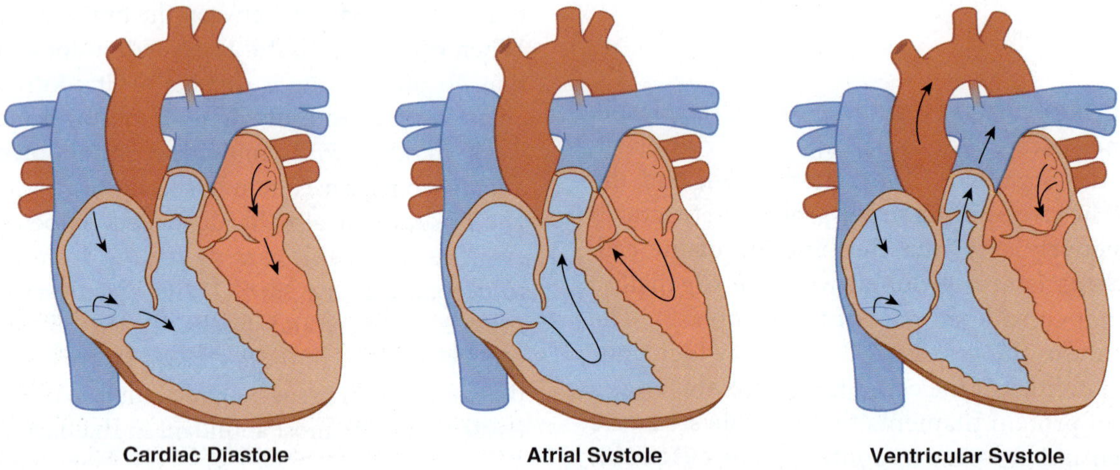

FIGURE 1.4 Contraction of the Heart.

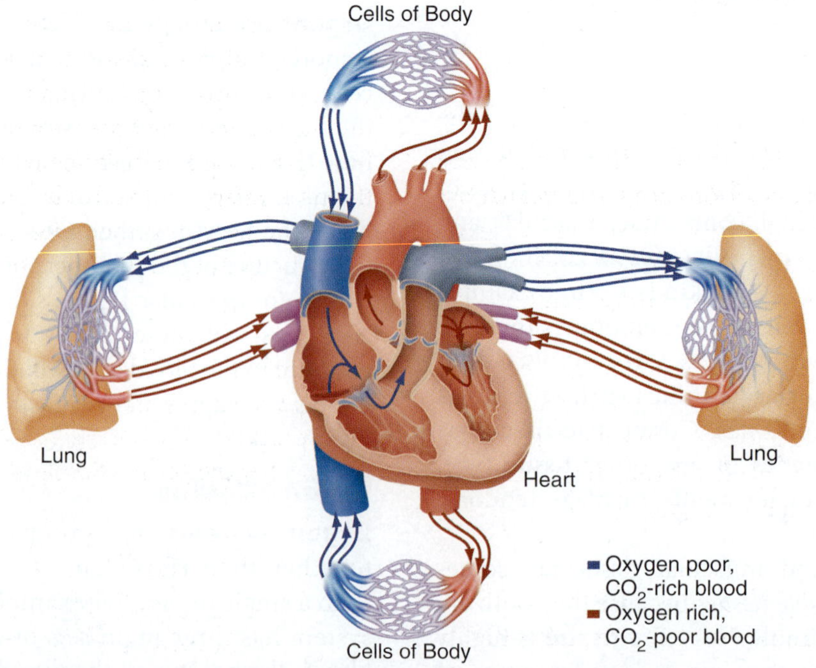

FIGURE 1.5 The Cardiovascular System.

The overall, basic purpose of the cardiovascular system is to supply the cells with needed items and transport waste from the cells for excretion. The blood is the medium through which all these things are transported. However, blood will not flow throughout the body without the heart applying pressure to it. Furthermore, the blood vessels are needed to contain and direct the blood to the cells. So, the organs of the cardiovascular system work together to perform a larger function. Finally, the collective roles of the cells, the tissues, the organs, and the organ systems result in an organism (person) having a vast and amazing array of abilities.

PAUSE TO CHECK FOR UNDERSTANDING

1. Is the word "anatomy" or "physiology" defined as the study of structures?
2. Rearrange the following terms so they are in the correct order from the most complex to the most basic:
   tissue organism cell organ organ system
3. What is the smallest, basic unit of life?
4. What are the four types of tissues and their basic function?

Anatomical Terminology

Knowledge of the terms that describe anatomy is imperative. It is necessary to label or name the body parts so there is commonality for discussion and explanation.

Body Positions

General body positions include standing, sitting, and lying down. These body positions are self-explanatory. However, these general body positions do not specifically address the position of body parts in relation to each other. *Anatomical position* solves this dilemma.

Anatomical position is the established reference point for body position (**FIGURE 1.6**). For instance, having a reference point is necessary to establish what is to be called the front, the back, the top, the bottom, the right, and the left sides. The gold standard, accepted anatomical position is standing erect, facing forward, arms straight at each side with forearms rotated so that the palms face forward, and legs straight with the feet together. Another standing position is **fundamental position**, which is the same as anatomical except the forearms are rotated so that the palms are facing the sides of the body (**FIGURE 1.6**). Fundamental position is the typical, natural stance of a person and this position is used when measuring joint motions.

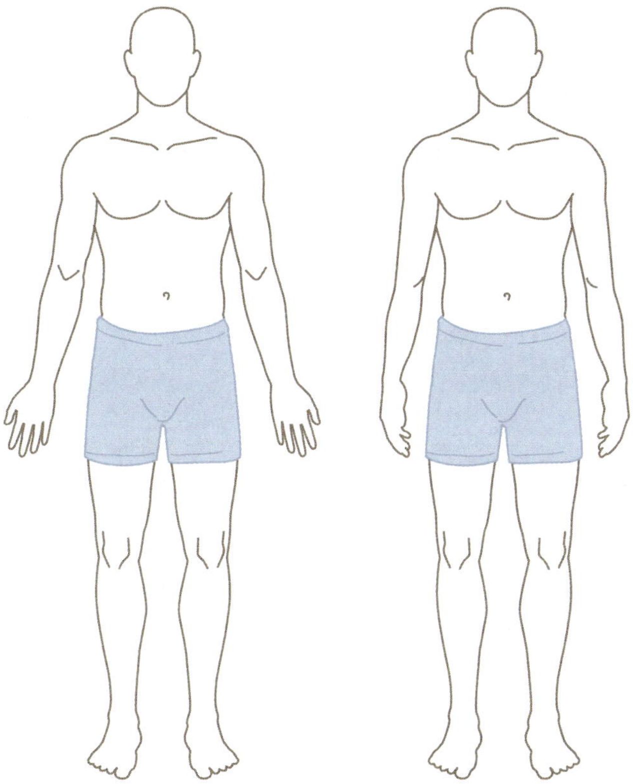

FIGURE 1.6 Anatomical Position (left) and Fundamental Position (right).

Supine and *prone* are two more general body position terms used to describe lying down. (**FIGURE 1.7**). **Supine** refers to the body lying flat on the back. A **prone** body position is the opposite with the body lying flat on the front side.

Directional Terms

There are several terms that describe the location of body parts (**FIGURE 1.8**). **TABLE 1.1** summarizes the definition of these terms. Some of the directional terms

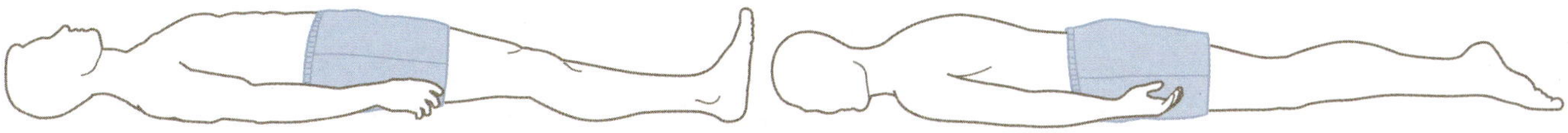

FIGURE 1.7 Supine (left) and Prone (right) Body Positions.

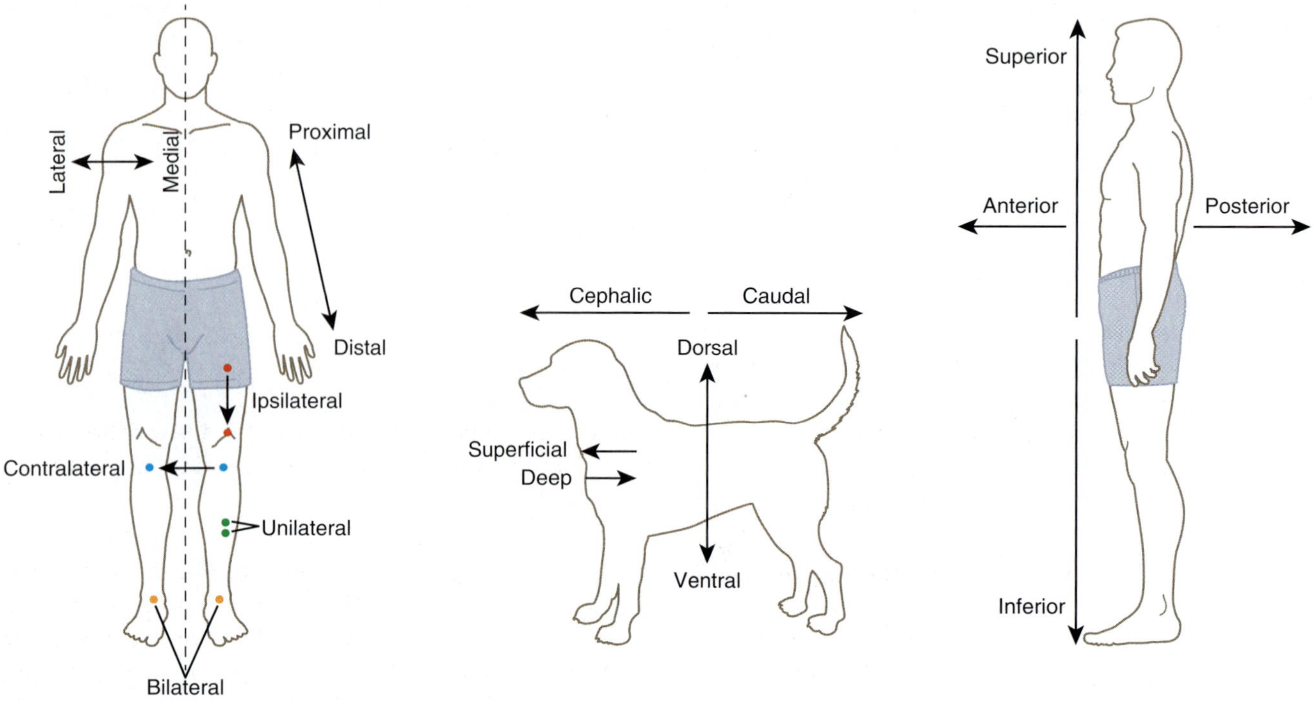

FIGURE 1.8 Directional Terms.

TABLE 1.1 Definitions of Directional Terms

Term	Definition
Anterior	towards front
Posterior	towards back
Ventral	towards belly or front
Dorsal	towards back or spine
Cephalic	towards head
Caudal	towards tail or lower end
Superior	above
Inferior	below
Proximal	nearer the midline (beginning of structure)
Distal	farther from midline (end of structure)
Medial	nearer the midline (side of structure)
Lateral	farther from midline (side of structure)
Ipsilateral	same side

Contralateral	opposite side
Unilateral	one side
Bilateral	both or two sides
Superficial	closer to body surface
Deep	farther from body surface

were derived from the study of animals. Many years ago, scientists utilized animals more so than humans because of access and the general similarities between the species.

Anterior means "towards the front" and **posterior** is the opposite meaning "towards the back." Similar to these terms are ventral and dorsal and are examples of terms originating from the study of animals. **Ventral** literally means "towards the belly."

However, it is still important to understand because it sometimes refers to the front because the belly of a human is on the front side. **Dorsal** means "towards the spine." Like ventral, dorsal sometimes refers to the back in humans because that is where our spine is. The spine in four-legged animals is on the top side of the body and is used in this manner in the human foot. The top side of the foot is the dorsal side.

Cephalic and caudal are two more terms that originated from the study of animal anatomy. **Cephalic** means "towards the head." **Caudal** means "towards the tail" and is the trailing or end part of an animal. In human anatomy, caudal means "towards the lower end" which is towards the feet. Similar to these are superior and inferior. **Superior** means "above" as in towards the head. **Inferior** means "below" and also refers to towards the feet. Superior and inferior are used mostly to indicate direction or relative position in the axial region (head, neck, chest, back, abdomen, and pelvis). Two different terms are used in a similar fashion mostly for the appendages (arms and legs).

Proximal means "nearer to the midline" and refers to being towards the beginning of a structure or the end that is closer to the midline. **Distal** means "farther from the midline" and refers to the end of a structure; the end farther away from the midline. Medial ("nearer the midline") and lateral ("farther from the midline") are two more terms that share the same root definitions as proximal and distal. However, these terms are used to indicate sides rather than ends of a structure. For instance, the medial forearm bone refers to the bone that is closest to the midline while the lateral forearm bone is on the side farthest from the midline.

Four other terms are related to sides. They are *ipsilateral*, *contralateral*, *unilateral*, and *bilateral*. **Ipsilateral** literally means "same side" and **contralateral** is "opposite side." For example, the right arm is ipsilateral to the right leg while the left leg is contralateral to both of them. These terms could also be used to indicate the location of structures on the same body part. **Unilateral** means "one side" while **bilateral** refers to both sides.

Most of the terms to this point are relevant to height and width, but what about depth? The terms superficial and deep indicate depth in both the axial and appendicular regions of the body. **Superficial** means "closer to surface of a body" while **deep** means "farther from the surface of a body." For instance, the skin is the most superficial body part while all other organs are deep to the skin.

PAUSE TO CHECK FOR UNDERSTANDING

1. Describe the differences between anatomical and fundamental position.
2. Is "prone" or "supine" lying on the back or face up?
3. Draw a three-dimensional box and label each side including the top and bottom with the appropriate directional terms.
4. To represent legs, add two more three-dimensional long boxes to the bottom of your first box. Using the rest of the directional terms, label them.

Body Regions

The body is divided into a number of regions and sub-regions. The two main regions are the axial and the appendicular regions (**FIGURE 1.9**). The **axial region**

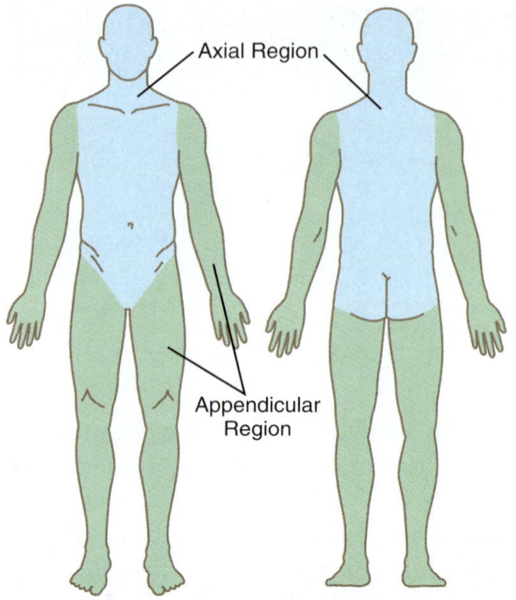

FIGURE 1.9 The Axial and Appendicular Regions of the Body.

is composed of the head, neck, and trunk (chest, back, and pelvis). The **appendicular region** includes the four appendages (or extremities): the two upper extremities (arms) and the two lower extremities (legs).

The sub-regions of the axial region are indicated in **FIGURE 1.10**. The sub-regions of the upper extremities and lower extremities are indicated in **FIGURE 1.11** and **FIGURE 1.12**, respectively.

Body Planes

Observing an organism by its surface anatomy tells us much, but limits our understanding of the inner parts. Imaginary planes are used to "slice" a body into different parts to observe these inner parts and to view them from different sides. The imaginary planes are the *sagittal*, *coronal* (frontal), and *transverse* planes (**FIGURE 1.13**).

FIGURE 1.10 The Sub-Regions of the Axial Region.

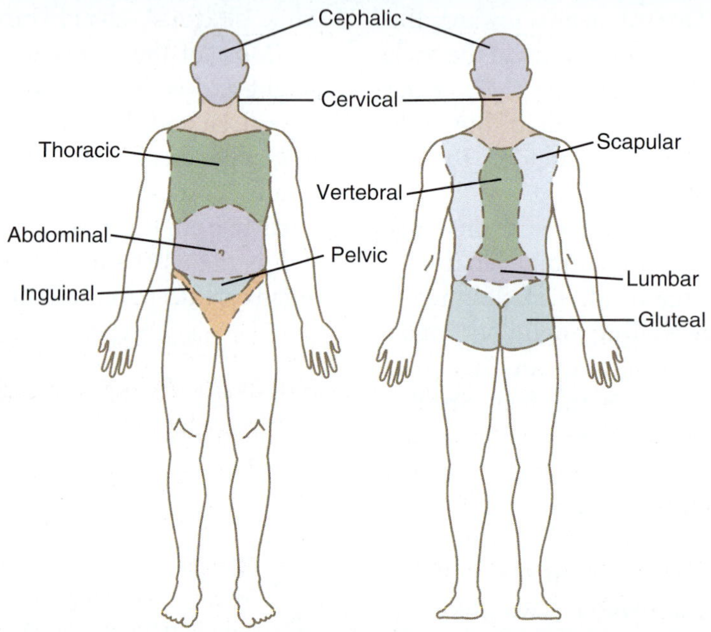

SUB-REGION	PERTAINING TO THE:	SUB-REGION	PERTAINING TO THE:
Cephalic	head	Inguinal	groin
Cervical	neck	Scapular	upper back
Thoracic	anterior chest	Vertebral	center of back
Abdominal	belly	Lumbar	low back
Pelvic	lower end of anterior trunk	Gluteal	buttocks

FIGURE 1.11 The Sub-Regions of the Upper Extremities.

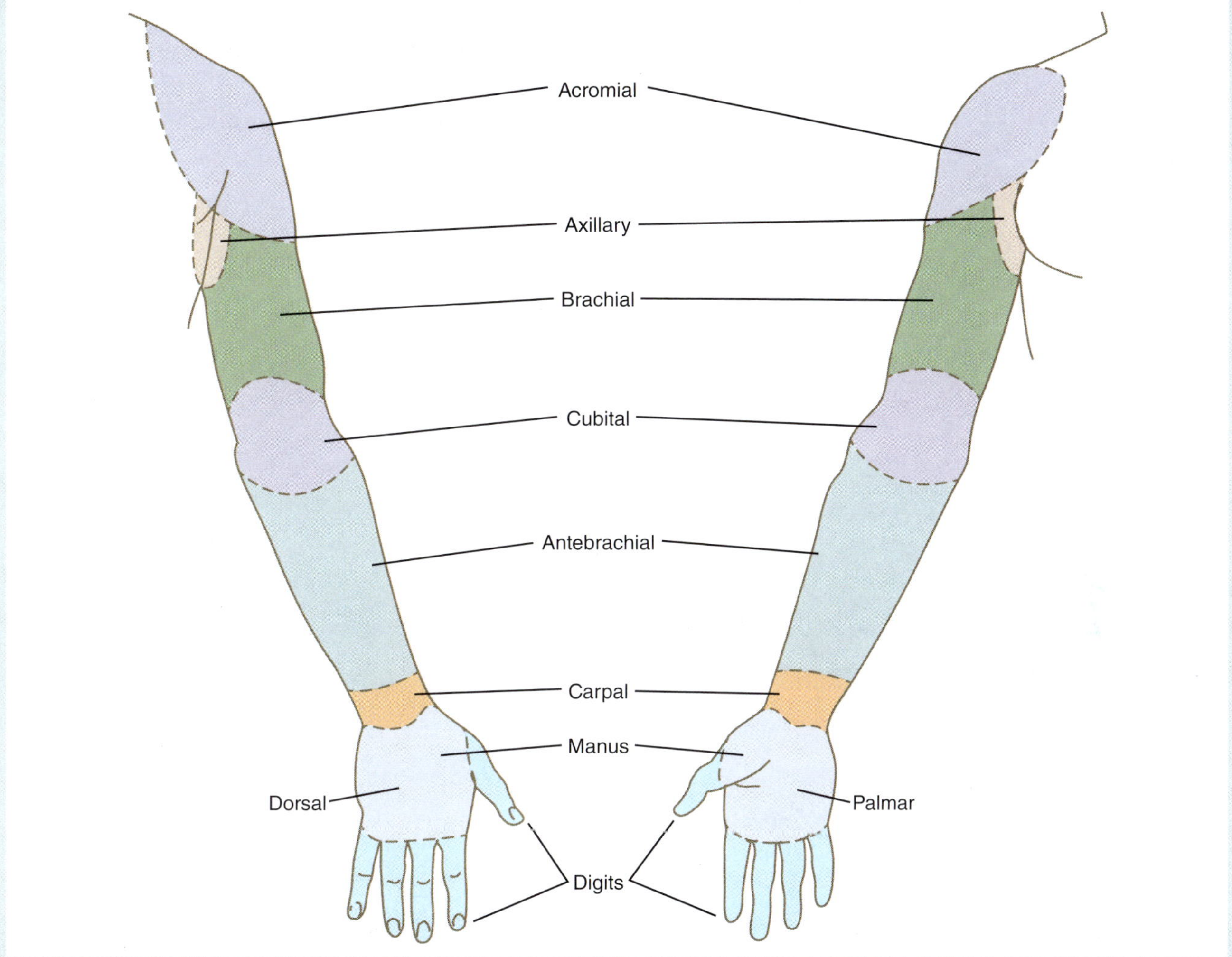

SUB-REGION	PERTAINING TO THE:	SUB-REGION	PERTAINING TO THE:
Acromial	shoulder	Carpal	wrist
Axillary	armpit	Manus	hand
Brachial	upper arm from shoulder to elbow	Palmar	anterior side of hand
Cubital	elbow	Dorsal	posterior side of hand
Antebrachial	forearm	Digits	fingers

The **sagittal plane** divides an organism into left and right portions and gives a side view of a body part. A sagittal plane used to "slice" through the very center of the body is called the **midsagittal** plane. The midsagittal plane is also called the **median** or **midline**. The midline divides the body into equal left and right halves. The **coronal plane** (also called **frontal plane**) divides an organism into anterior and posterior sections. The **transverse plane** "cuts" an organism into superior and inferior parts. All of these planes can be used to "slice" the body in any number of areas in order to realize internal structures from a different point of view (**FIGURE 1.14**).

FIGURE 1.12 Sub-Regions of the Lower Extremities.

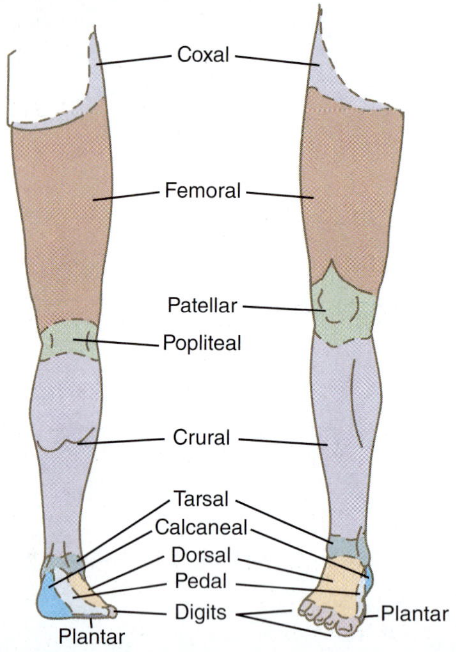

SUB-REGION	PERTAINING TO THE:	SUB-REGION	PERTAINING TO THE:
Coxal	hip	Calcaneal	heel
Femoral	thigh from hip to knee	Pedal	foot
Patellar	anterior side of knee (kneecap)	Dorsal	superior side of foot
Popliteal	posterior side of knee	Plantar	inferior side of foot
Crural	lower leg from knee to ankle	Digits	toes
Tarsal	ankle		

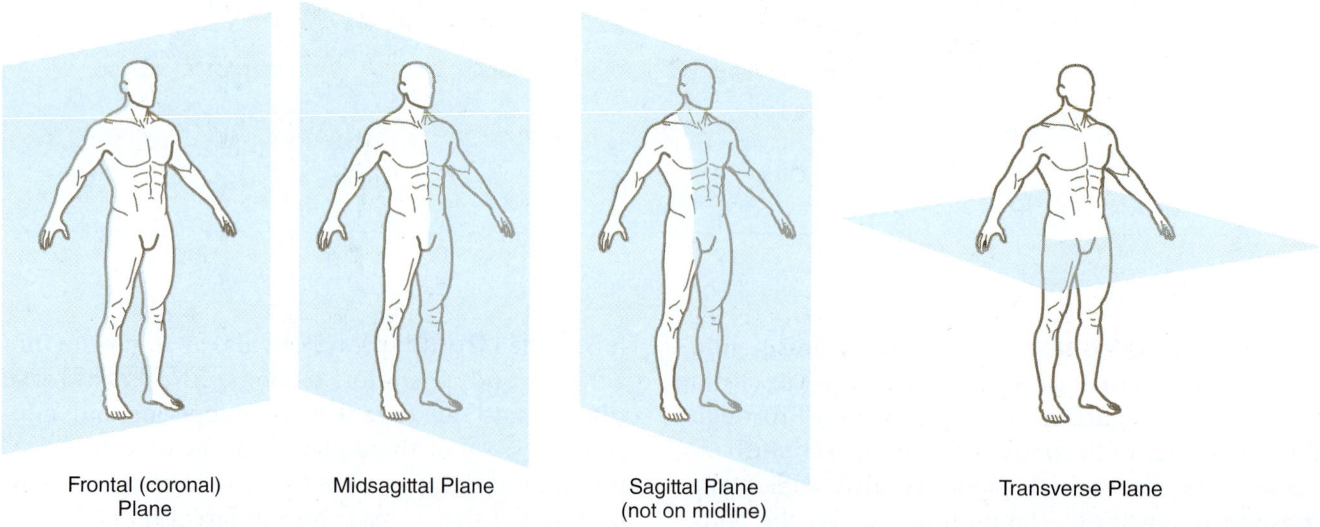

FIGURE 1.13 The Body Planes.

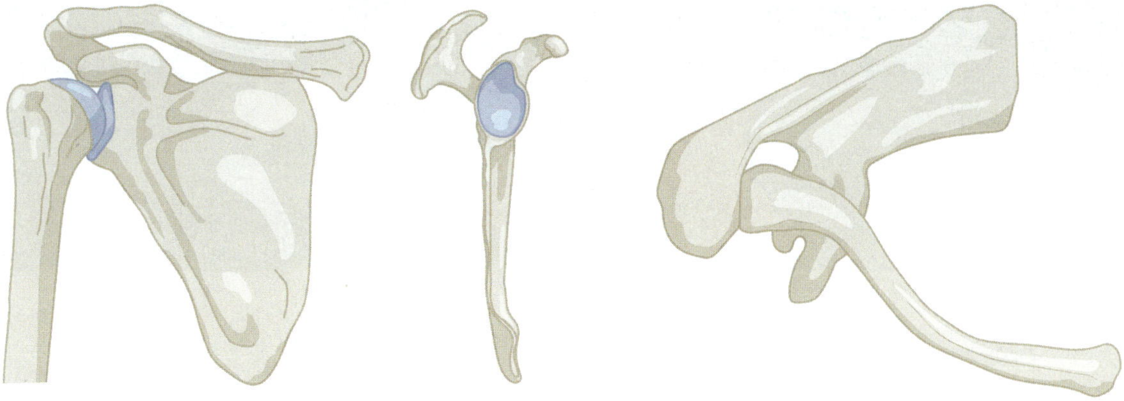

FIGURE 1.14 Three Views of the Shoulder.

Axes of Rotation

Each plane has an axis associated with it (**FIGURE 1.15**). Each of these axes is an imaginary line or rod that the different body regions rotate about. For example, when the knee is extended, it is rotating around the imaginary center line of the knee. When it rotates in the opposite direction, it is called knee flexion.

The **sagittal axis** passes from anterior to posterior through the frontal plane. The **frontal axis** passes from side to side of the body through the sagittal plane. The **longitudinal axis**, also called **vertical axis**, passes from superior to inferior through the transverse plane.

PAUSE TO CHECK FOR UNDERSTANDING

1. Draw an anterior and posterior blank body similar to Figure 1.11. Then, label all the body regions. Maybe even add lines to delineate the regions.
2. What are the body planes and how do they divide the body?
3. List the axes of rotation and tell which axis goes with each body plane.

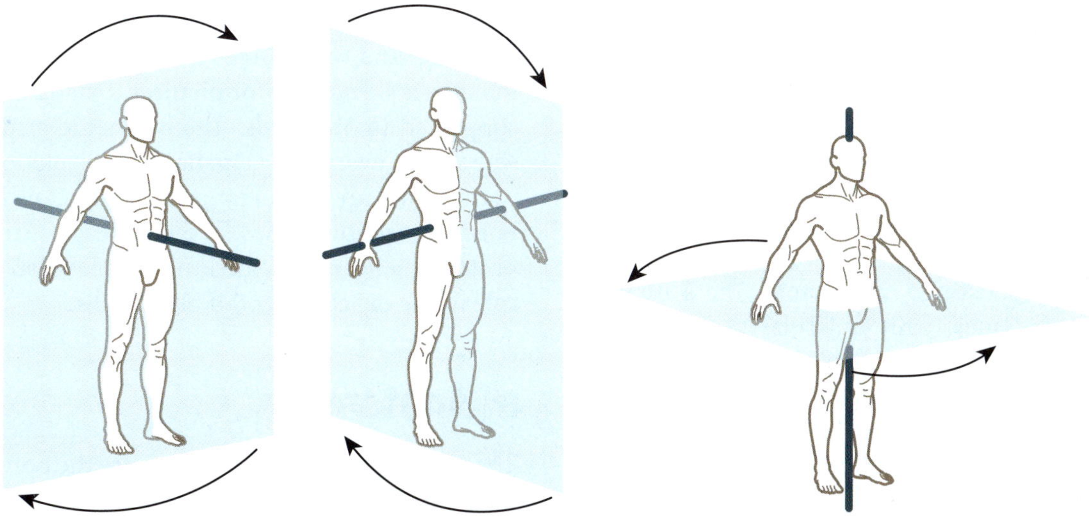

FIGURE 1.15 Axes of Rotation within Planes.

© BraunS/E+/Getty Images

CHAPTER 2

The Skeletal System

CHAPTER OBJECTIVES

After completing this chapter, the student will be able to:

1. describe general anatomy of a bone;
2. recall the types of bone classified by shape;
3. employ terminology related to anatomical features of bones;
4. realize the functional and structural classifications of joints; and
5. recognize the types of synovial joints.

The skeletal system is comprised of several organs called bones (**FIGURE 2.1**). The main functions of the skeletal system and its bones are:

1. support (framework),
2. protection, and
3. blood cell formation.

Most, if not all resources include movement as a function of the skeletal system. This text takes the position that the skeletal system is ***not*** responsible for movement. As we learned in Chapter 1, there are four main types of tissues. Only muscular tissue has the unique characteristic of contractility. That is, whatever a muscle is attached to, it applies force to that object and if the force is great enough, moves it. Bone tissue is a sub-type of connective tissue and no form of connective tissue moves anything. The numerous bones do intersect each other at joints and if not for the arrangement, body movement would be extremely restricted. However, that does not make it a function. For instance, under "normal" conditions, the muscles do not act unless the nervous system "tells" it to do so. But, movement is not normally listed as a main function of the nervous system.

The bones are arranged as a relatively solid framework which supports all the other organs directly or indirectly. It is like the steel beams in a commercial building. The walls directly attach to the beams. This makes a solid surface for pictures to hang from and be supported by the walls. The bones are arranged in a way so that everything else has a stable platform. Many of the bones are also arranged to protect other organs. For instance, the bones of the rib cage surround and protect the heart and the lungs. Bones also play a key role in the development of blood cells.

Anatomy

The organs of the skeletal system are the bones, such as the femur. And, as you have already learned, an organ is a composition of tissues and a tissue is composed of cells and matrix (extracellular material). The different tissues that comprise a bone organ are marrow, cartilage, vascular, nervous, and osseous tissue (**FIGURE 2.2**). And, different types of cells comprise these tissues. Osseous means bone. So, osseous tissue is a tissue that is composed of bone cells. This can be confusing.

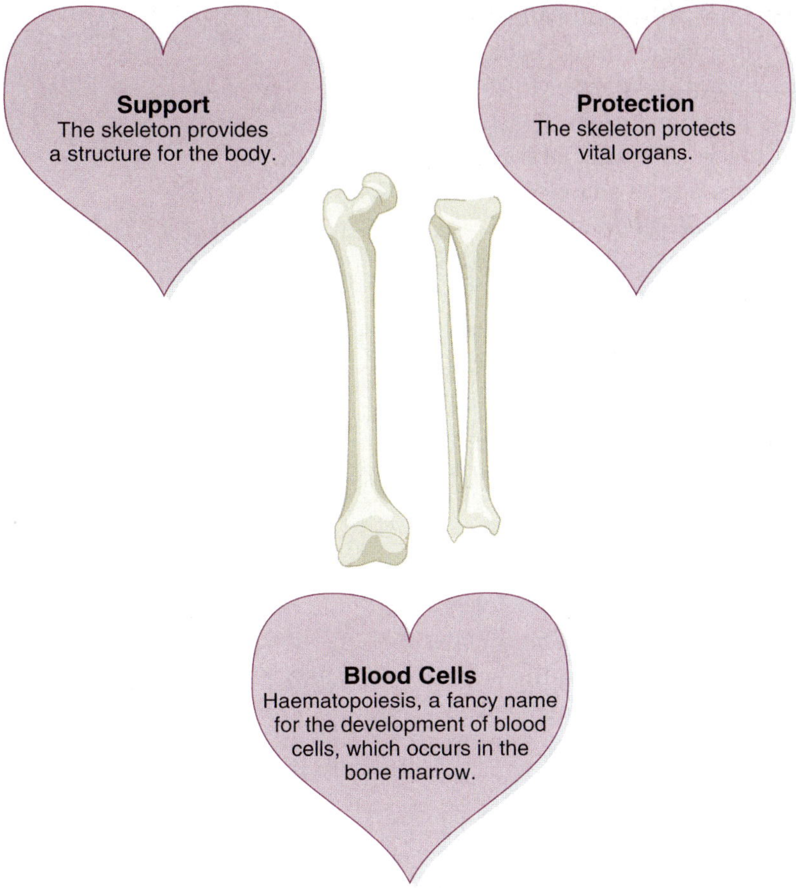

FIGURE 2.1 The Skeletal System.

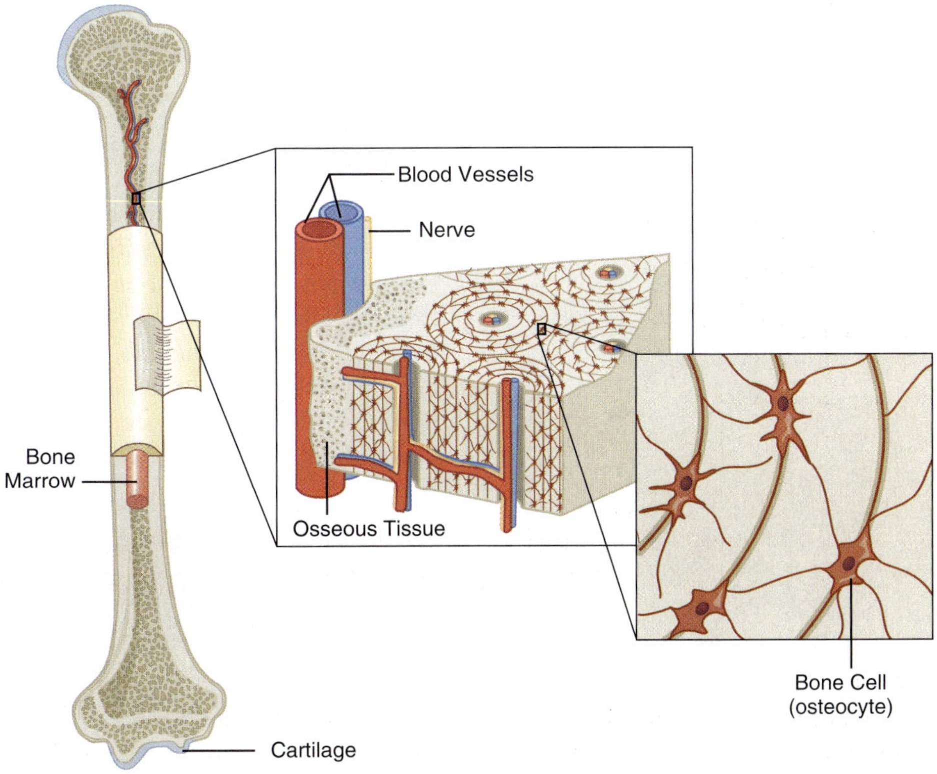

FIGURE 2.2 The Tissues of a Bone.

So, this text will establish some distinctions to lessen that. The word **bone** will refer to an entire organ. The term **bone tissue** (tissue of a bone) will refer to any or all of the different tissues of a bone organ. **Osseous tissue** will refer only to the specific tissue of a bone that is comprised of bone cells. **Bone cells**, also known as osteocytes, are the cells of osseous tissue.

Osseous tissue is a composite of about two-thirds inorganic matter and one-third organic matter. The inorganic portion is about 85% calcium phosphate, 10% calcium carbonate, and varying amounts of other minerals. These minerals calcify (harden) the bone. However, bones are not completely stiff and do have flexibility. The organic matter is bone cells (osseous cells) mixed with collagen and various protein-carbohydrate complexes.

Osseous tissue matures into either *compact osseous tissue* or *spongy osseous tissue* (**FIGURE 2.3**). **Compact osseous tissue** is dense layers of bone cells and collagen. There are no spaces between the layers of compact osseous tissue; however, there are passageways to allow for vascular and nerve supply. **Spongy osseous tissue** has the same composition as compact osseous tissue, but its appearance is different. Spongy osseous tissue develops slivers of tissue that leaves spaces somewhat like a honeycomb or lattice-work. Also, spongy bone is always encased by a thickness of compact bone.

Bone marrow is a soft tissue that occupies the spaces of spongy osseous tissue and the passageways of compact osseous tissue (**FIGURE 2.4**). Bone marrow begins as red bone marrow. **Red bone marrow** plays a significant role in producing red and white blood cells. In children, almost all of the bone marrow is red. In adults, red bone marrow is restricted to the axial skeleton and the proximal ends of the bones of the appendicular skeleton. The rest of it converts into yellow bone marrow. **Yellow bone marrow** is fatty tissue and no longer is involved in producing blood cells, but can revert back into red bone marrow if some severe condition creates a much greater than normal need for production.

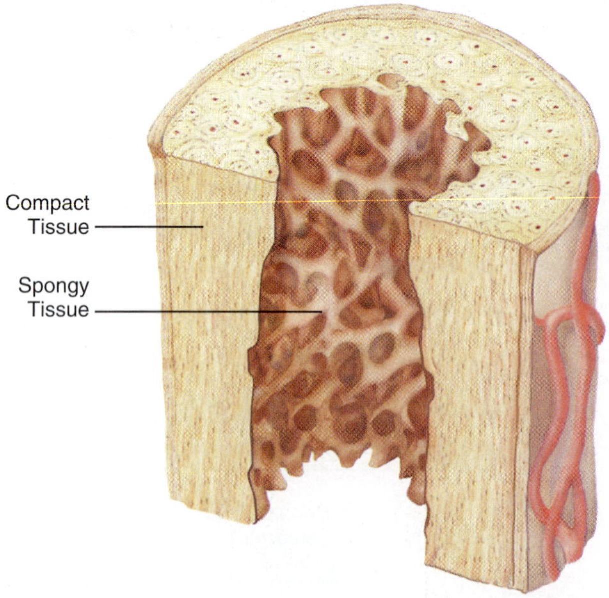

FIGURE 2.3 Compact and Spongy Osseous Tissue.

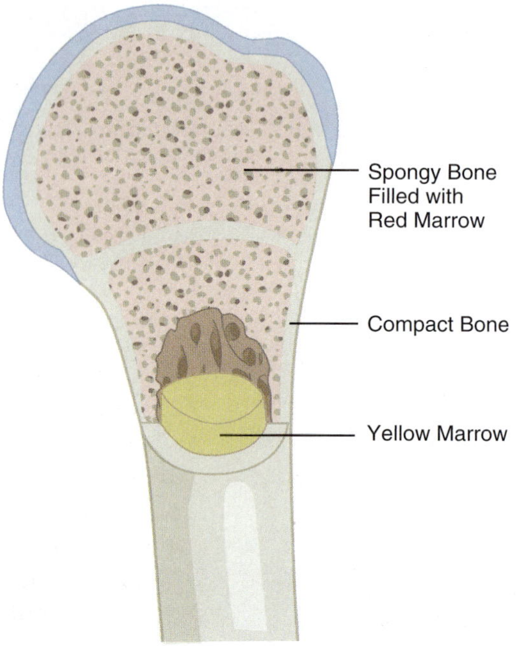

FIGURE 2.4 Bone Marrow.

Bones, like any other organ, need a blood supply and nerve supply. Vascular tissue (blood vessels) includes arteries and veins. Arteries deliver everything the cells of a bone need. Veins remove any waste produced as well as take away and deliver any cell products produced. For instance, bones play a role in the formation of blood cells which are needed throughout the body and not just in bones. A system of communication is also needed. Nerves pass into bones for this purpose. Examples include, through the nerves of the nervous system, bones can communicate about sensations and the nervous system can communicate to the bones about producing more or less blood cells.

PAUSE TO CHECK FOR UNDERSTANDING

1. What are the three main functions of the skeletal system?
2. Distinguish between bone, bone tissue, osseous tissue, and bone cell.
3. Which part of a bone is involved in blood formation?

Types of Bones

There are four main types of bones (**FIGURE 2.5**). They are *long, short, flat*, and *irregular*. Bones are three-dimensional objects which have height, width, and depth. Most of the bone types are classified according to the relationships of these dimensions.

Long bones include the longest ones in the body, but this is not the reason they are titled this. Rather, it is because their length in relation to their width is much greater. For instance, the femur (thigh bone) is much longer than a phalange (digit), but both are long bones because of the difference in their respective lengths and widths. Furthermore, the depth and width are relatively similar; like a cylinder. **Short bones**, such as the bones in the wrist, are cube-like. Their length, width, and depth are not necessarily the same, but similar (or close) in length. **Flat bones**, such as the scapula (shoulder blade), are called this because of its lack of depth in relation to the height and width. The rest of the bones are grouped as **irregular bones** because there is not a uniform relationship between its dimensions. An example is a vertebra which has a very unusual shape.

In addition to the traditional list of bone types are *sesamoid* and *sutural* bones (**FIGURE 2.6**). **Sesamoid bones** develop within tendons because of stress. One example that develops in all people and is the largest sesamoid bone is the patella (kneecap). Most of the others, if they develop, are in the hands and feet. A **sutural bone** is an extra bone in the cranium.

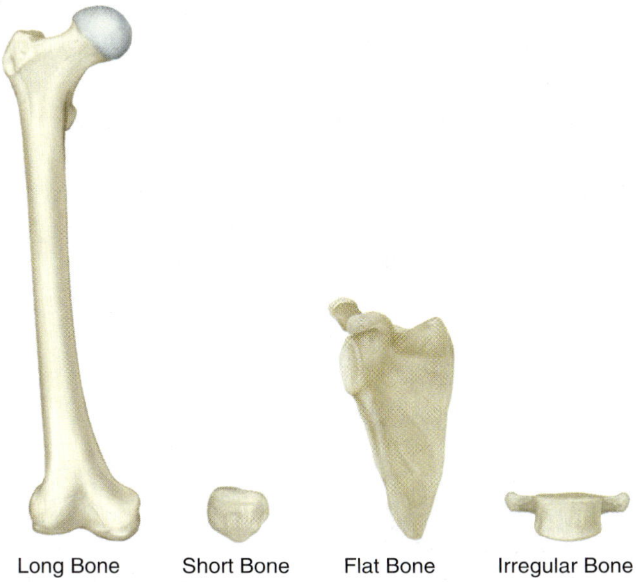

FIGURE 2.5 The Types of Bones.

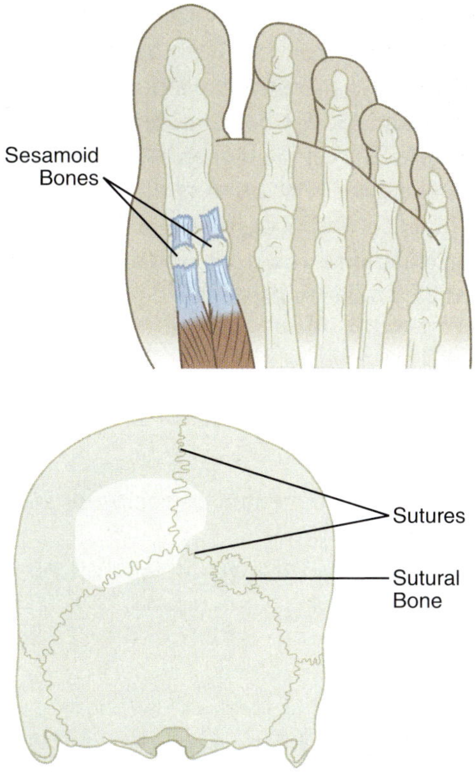

FIGURE 2.6 Example of a Sesamoid Bone (top) and a Sutural Bone (bottom).

Long Bone Anatomy

Long bones have several parts that are important to realize (**FIGURE 2.7**). The **diaphysis** (shaft) is the cylindrical middle region and is composed of compact bone. It has a hollowed out center called the **medullary cavity**, also known as the **marrow cavity** because it is filled

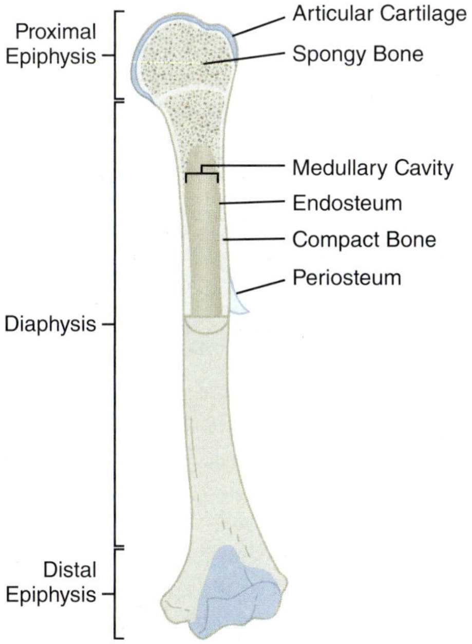

FIGURE 2.7 Lone Bone Anatomy.

with bone marrow. This cavity is lined with a connective tissue called the **endosteum**. The **epiphyses** (*singular – epiphysis*) are the expanded ends of the bone and they contain spongy bone inside their outer layer of compact bone. The very end of each epiphysis is covered by hyaline cartilage that protects it. This tissue is also referred to as **articular cartilage** because it is part of a joint (articulation). The rest of the outer layer of bone is covered by a collagen tissue called **periosteum**.

PAUSE TO CHECK FOR UNDERSTANDING

1. List and describe the four types of bones.
2. What is the difference between epiphysis and diaphysis?
3. Which connective tissue wraps around a long bone?
4. Which connective tissue covers the ends of long bones?

Anatomical Features of Bones

Whether long, short, flat, or irregular, bones have features, sometimes called landmarks. The features are important because they are sites that other tissue either connects to bone or passes through it. The anatomical features of bones have been divided into four main categories: *articulations*, *projections* (sometimes called *extensions*), *depressions*, and *passageways*.

Projections are convex "bumps" that extend from the main surface of a bone. **Depressions** are concavities that dip inwardly from the main surface of a bone. **Articulations** are features of bones related to the structure of a joint. They can be either concave or convex. **Passageways** are tunnel-like features that extend into the inner parts of bone or all the way through to the other side. **TABLES 2.1–2.4** list and briefly define the several terms associated with each category.

TABLE 2.1 Projection Terms

Term	Definition
Crest	narrow edge
Epicondyle	expanded region superior to a condyle
Line	slightly raised, elongated ridge
Process	bony prominence
Protuberance	bony outgrowth
Spine	sharp, slender narrow process
Trochanter	massive process (only found on femur)
Tubercle	small, rounded process
Tuberosity	rougher than tubercle process

TABLE 2.2 Depression Terms

Term	Definition
Alveolus	pit or socket
Fossa	shallow, broad, or elongated basin
Sulcus	groove for a tendon, nerve, or blood vessel

TABLE 2.3 Articulation Terms

Term	Definition
Condyle	round knob
Facet	smooth, slightly concave or convex surface
Head	proximal, expanded end of some bones

TABLE 2.4 Passageway Terms

Term	Definition
Canal	tubular, tunnel-like passage
Fissure	slit
Foramen	opening into a canal
Sinus	air filled space

PAUSE TO CHECK FOR UNDERSTANDING

1. List and define the four categories of bone features.
2. Which category does the feature called head belong to?
3. Which category does the feature called foramen belong to?
4. Which category does the feature called epicondyle belong to?
5. Which category does the feature called fossa belong to?

The Skeleton

The skeleton is the organization of all the bones (**FIGURE 2.8**). The average adult skeleton contains 206 bones. Like the body regions, the skeleton is divided into **axial** and **appendicular** regions. The **axial** region includes bones of the skull, the vertebral column, and

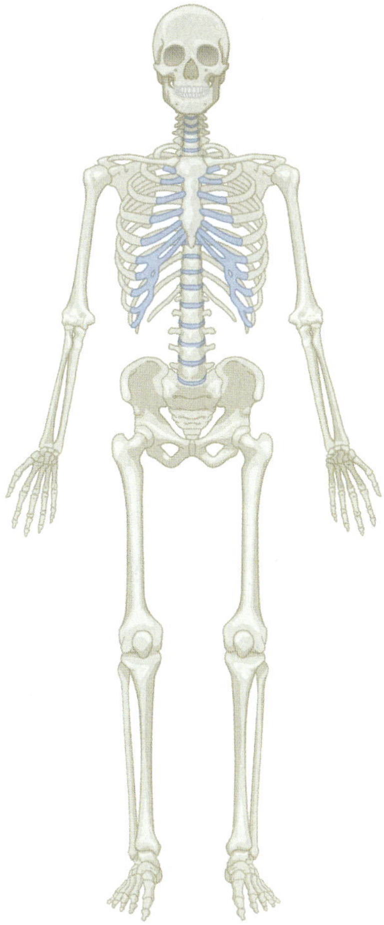

FIGURE 2.8 The Skeleton.

the thoracic cage. The **appendicular** region contains bones in the upper and lower extremities. Chapters 5–7 will be more specifically examine the bones.

Joints

The bones of the skeleton are arranged about *joints*, also known as *articulations*. A **joint** is the intersection between two bones. There is one exception to this rule. In the ankle, three bones articulate to form one

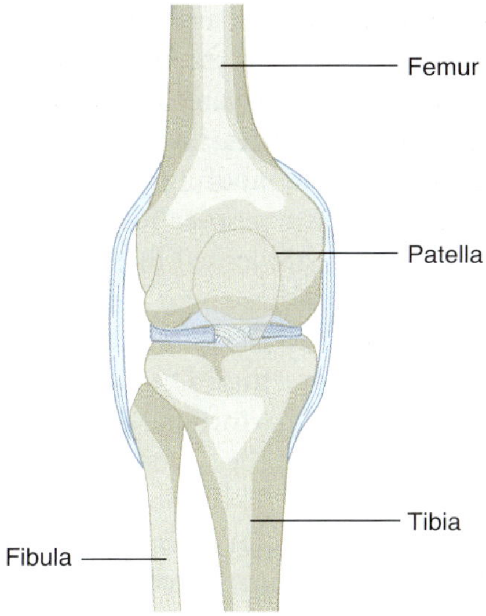

FIGURE 2.9 The Joints of the Knee.

joint. Chapter 9, The Ankle, will describe this in more detail. Interestingly, many body regions that have been called joints are actually a compilation of more than one joint.

An example is the knee which is actually a region formed by three joints (FIGURE 2.9). They are the articulations between 1) the femur and tibia, 2) the patella and femur, and 3) the tibia and fibula.

Classifications of Joints

Joints can be classified as either *functionally* by their ability to move or *structurally* by how the tissues that stabilize them. **Functional classes** include 1) **synarthroses** which are non-movable, 2) **amphiarthroses** which are slightly movable, or 3) **diarthroses** which are freely movable.

The **structural joint classifications** are *bony, fibrous, cartilaginous*, or *synovial* (FIGURE 2.10). **Bony joints** (synarthrodial) are where two bones have actually fused together and appear more like one bone. For example, an infant has two frontal bones, but this

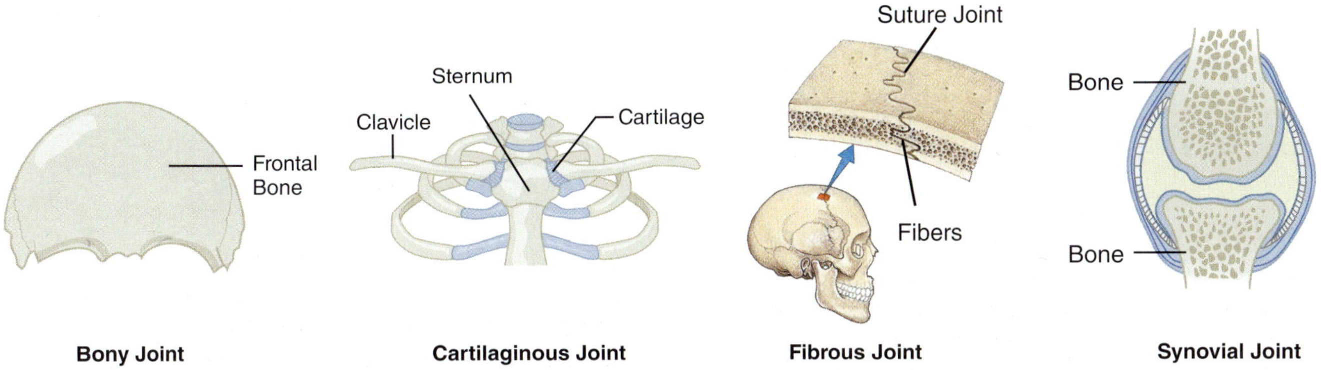

FIGURE 2.10 Categories of Joints.

joint closes during maturity and only one bone is left. In a **fibrous joint**, such as the suture joints between cranial bones, the bones are connected by thick, tough collagen fibers. These types of joints restrict movement to almost nothing (amphiarthrodial). **Cartilaginous joints** are so called because the two bones of these joints are joined by some type of cartilage. The cartilage of these joint types is not as thick and strong as the collagen fibers in fibrous joints. So, movement is more than fibrous joints, but still quite restricted (amphiarthrodial). The articulation between the clavicle and the sternum is an example of a cartilaginous joint. **Synovial joints**, like the elbow or knee, allow for great movement (diarthrodial) and because of this require many accessory tissues to stabilize them.

Synovial Joint Anatomy

The anatomy of a synovial joint is more complex than the other types of joints (**FIGURE 2.11**). At the articulation between the bones of a synovial joint is a small space called the **joint cavity**. This cavity is enclosed by a **joint capsule**. The inner liner of this membrane is the **synovial membrane**. The synovial membrane secretes **synovial fluid**, a slippery lubricant that nourishes structures within the joint cavity, rinses waste products from them, and reduces friction which can wear out the joint.

Synovial joints also have **ligaments** which are tough, bungy cord-like connective tissue. Ligaments span from one bone to another and are the primary stabilizers as one bone articulates (moves) across the other. The ends of all bones are covered by hyaline cartilage to protect and since these ends are where two bones articulate, this cartilage is also called **articular cartilage**.

Some synovial joints have extra pieces of relatively tough, dense cartilage called **fibrocartilage** to further protect and absorb forces placed upon it. It also improves the fit and stability of the adjoining bone and provides a smooth surface for a bone to rotate across another. Fibrocartilage is referred to by different names depending on the joint such as the *meniscus* in the knee.

Classification of Synovial Joints

There are six classes of joints. They are the *ball and socket, hinge, pivot, saddle, condylar* (also called *ellipsoid*), *plane* (also called *gliding*) joints. Illustrations and examples of each joint classification are in **FIGURES 2.12–2.14**. With each class of joints, one bone end is convex and the other is concave.

Ball and socket joints are joints with one bone having a spherical head (ball) that articulates with a relatively deep depression (socket) on the other bone. **Hinge joints** are similar in appearance to condylar joints, but have more rounded ends rather than oval ends that articulate with each other.

Pivot joints are so called for the convex end of one bone rotating about its long axis within the concave groove of the other bone. The radioulnar joints of the elbows and the atlantoaxial joint (articulation between the first two vertebrae) of the neck are the only pivot joints in the body. **Saddle joints** have such sharp curvatures to their bone ends that they resemble a saddle.

Condyloid joints have corresponding ends that are more oval rather than round. With **plane joints**, the

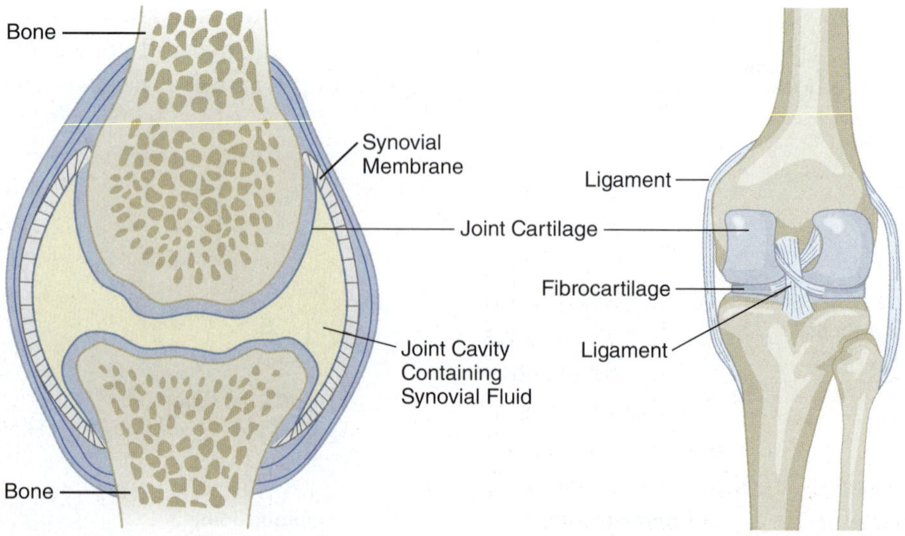

FIGURE 2.11 Synovial Joint Anatomy.

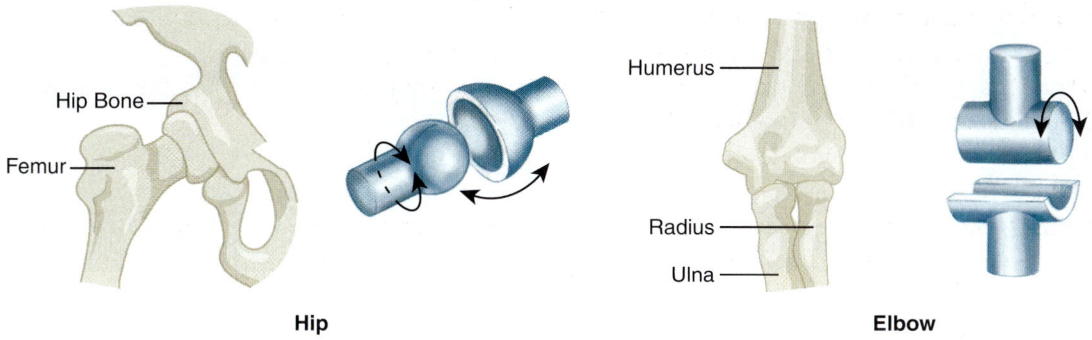

FIGURE 2.12 Examples of a Ball and Socket Joint (left) and a Hinge Joint (right).

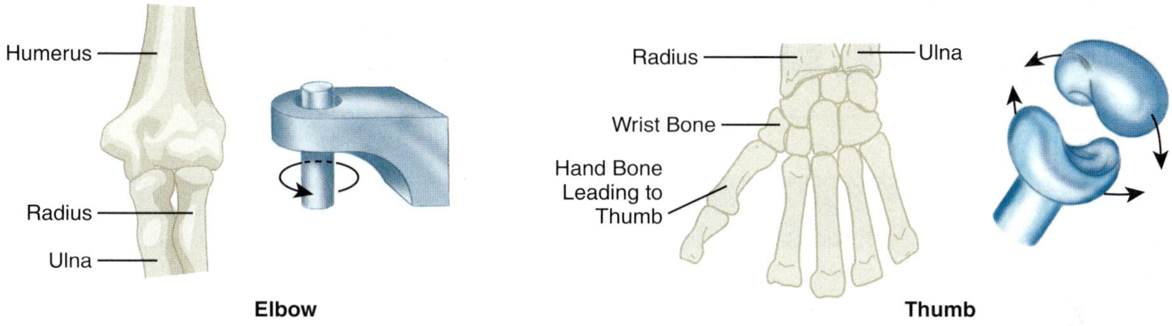

FIGURE 2.13 Examples of a Pivot Joint (left) and a Saddle Joint (right).

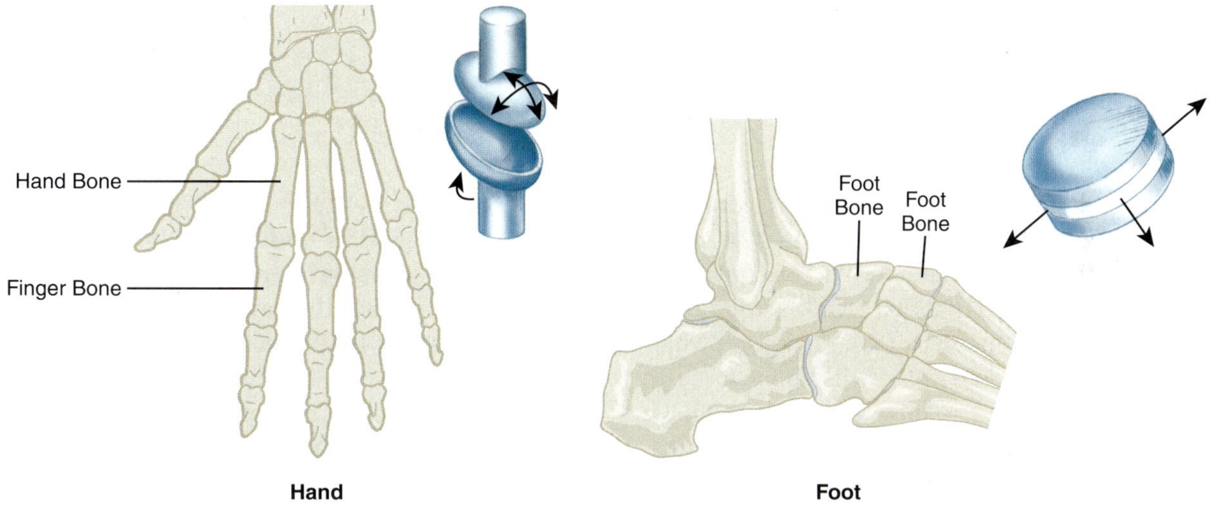

FIGURE 2.14 Examples of a Condyloid Joint (left) and a Plane Joint (right).

bone ends are much flatter and the movement is more of a gliding (or sliding) motion rather than a rotation.

Synovial joints are also described in terms of their ability to move about one or more of the axes of rotation. **Multi-axial**, also called tri-axial, joints can move through all three planes. The only multi-axial joints are classified as ball and socket. **Bi-axial** joints are limited to two planes and include condylar, saddle, and plane joints. **Mon-axial** joints only move in one body plane. Mon-axial includes hinge and pivot joints.

PAUSE TO CHECK FOR UNDERSTANDING

1. What are the four types of structural joints and how much movement does each one allow?
2. List the main structures of a synovial joint and what they do (i.e., a ligament …).
3. List and briefly describe the six categories of synovial joints.
4. Distinguish between mon-, bi-, and tri-axial.

CHAPTER 3

The Muscular System

CHAPTER OBJECTIVES

After completing this chapter, the student will be able to:

1. describe general anatomy of the muscular system;
2. classify muscles by shape;
3. understand the three types of muscle actions (contractions);
4. employ the joint motions;
5. realize the functional roles of muscles; and
6. recognize how muscles are named.

The overall purpose of the **muscular system** is to move the body and bodily fluids. As stated in chapter 1, the characteristic of contractility is unique to muscular tissue. And, a **muscular contraction** causes tension (force). Therefore, whatever a muscle is attached to, it will apply force to that body part. And, if the force is great enough, it will move that body part. The use of the term muscular contraction is quite common. However, the word *contraction* implies that something is shrinking or shortening. As you shall learn in this chapter, shortening is liken to only one of the three types of muscular contractions. With this in mind, there is a relatively modern term that does not rely on muscle length and more indicative of all the types of muscle actions. The simple term is **muscular action** since the muscle *acts* to produce force. This term will be used hereafter.

There are three types of muscles. They are *skeletal, cardiac*, and *smooth* (**FIGURE 3.1**). **Skeletal muscles** apply force to and move the bones of the skeleton and hence the body. **Cardiac muscle** is only found in the heart and it generates the force that causes blood, a bodily fluid, flow through the blood vessels. **Smooth muscles** are found in several organs and are the cause of other bodily fluids, such as digestive fluids, to flow.

Skeletal muscles are the only type that pertains to the main focus of this text and so the rest of this chapter and other chapters will be dedicated to it. Unless otherwise noted, any further mention of a muscle will refer to a skeletal muscle which is aligned with the main purpose of this text.

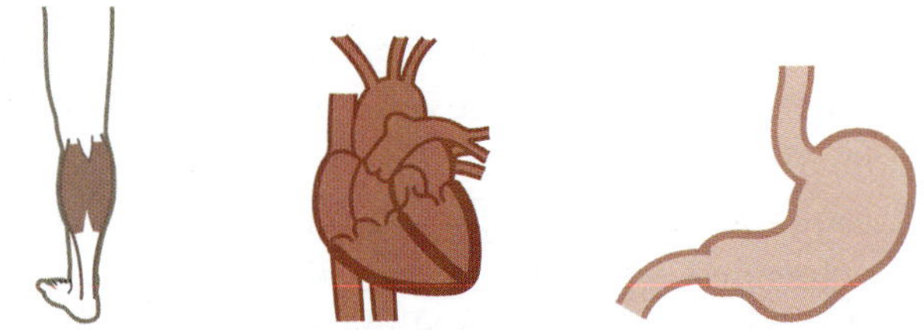

FIGURE 3.1 Types of Muscles.

Anatomy

Each individual muscle of the muscular system is an organ. And, like all organs, they are composed of tissues. These tissues include muscular, vascular, nervous, and connective (**FIGURE 3.2**). Muscular tissue is compartmentalized in rod-like areas called **fascicles**. Fascicles are bundles of muscle cells. Muscle cells are called **muscle fibers**.

Vascular tissue and nervous tissue are part of a muscle organ like they are for all types of organs. The vascular tissue (blood vessels) supplies the muscle fibers (cells) with oxygen and nutrients and drains any waste products. The nervous tissue not only supplies the route for sensory information, but it stimulates the muscles to act.

The connective tissues of a muscle include *fascia, epimysium, perimysium*, and *endomysium*, and *tendon* (Figure 3.2). **Fascia** is the outermost layer of connective tissue that serves to separate muscles and muscle groups from adjacent ones as well as from subcutaneous tissue. The **epimysium** is relatively indistinguishable to fascia lying just beneath it. It is the outer covering of a single muscle. The **perimysium** surrounds the fascicles and the **endomysium** wraps around individual muscle fibers (cells). **Tendons** are the "bridge" that connects a muscle to a bone.

Tendons may attach at a point, several points, or along a line. Also, when a muscle acts, one end of the muscle's attachment(s) is called the origin and the end of attachment(s) is referred to as the insertion. A muscle's **origin** is usually more proximal and is the relatively stationary attachment site(s). The **insertion** of a muscle is usually more distal and is the more dynamic or mobile attachment site(s). When a muscle acts, it is the insertion end of the muscle that is either pulled toward the origin or lowered away from the origin. **FIGURE 3.3** illustrates the types of attachments and the motion of the insertion toward the origin.

PAUSE TO CHECK FOR UNDERSTANDING

1. What two main things do muscles move?
2. Define muscular action.
3. List the connective tissues and tell their general location.
4. List the muscular tissues and tell their general location.
5. Differentiate between an origin and insertion.

Classifications of Muscles by Shape

Muscles are classified by shape. These classifications are *parallel, fusiform, convergent*, and *pennate* (**FIGURE 3.4**). The shape is dictated by the angle that fascicles are arranged from the tendons. The fascicles of **parallel** muscles, hence the name, are parallel to its tendons and their width is relatively uniform. The fascicles of

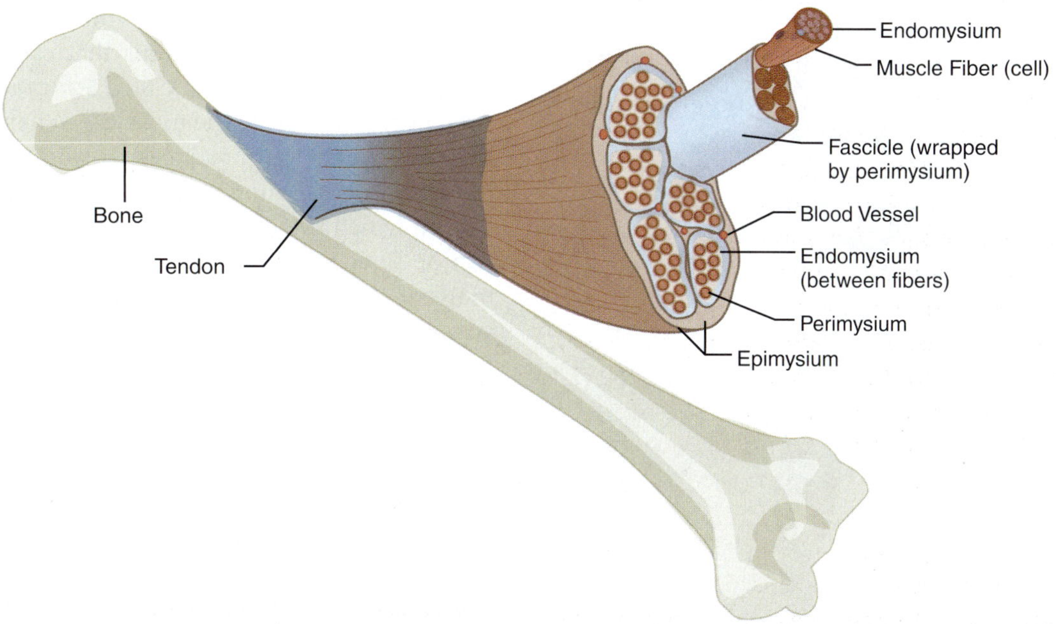

FIGURE 3.2 Tissues of a Muscle.

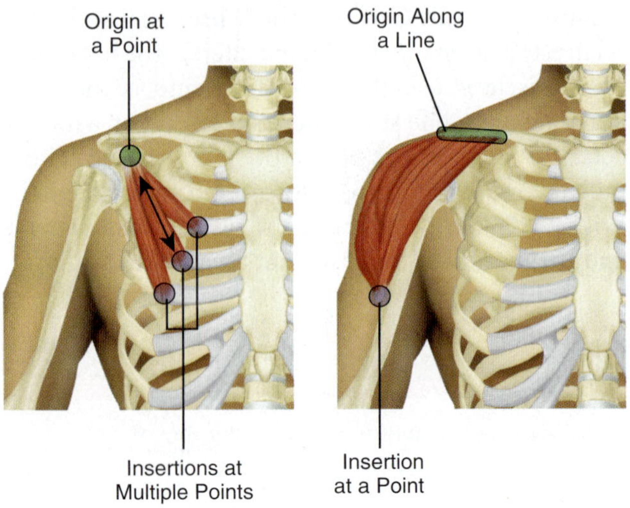

FIGURE 3.3 Examples of Origin and Insertion Attachment Sites.

fusiform muscles diverge out from relatively tight arrangements on both ends. The divergence of fascicles creates a wider mid-section called the belly. The fascicles of **convergent** (also called **triangular**) muscles merge at the end of a tendon. They are broad at one end and narrow at the other creating a triangular shape.

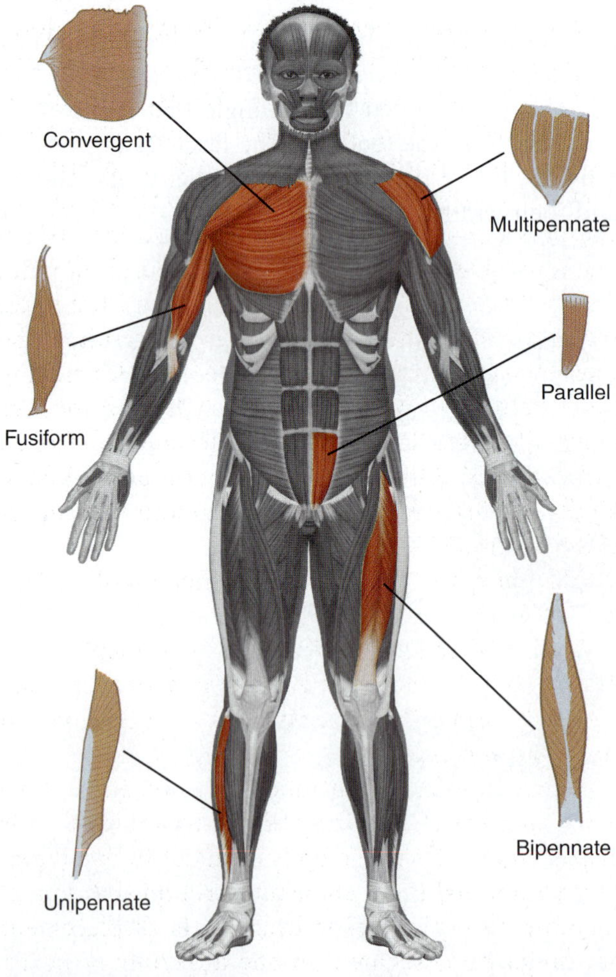

FIGURE 3.4 Muscle Shape Classifications.

Instead of connecting at one end of a tendon, pennate muscles are feather-like. Their fascicles connect at adjacent points along a section of the tendon that extends some distance into the belly of the muscle. There are three types of pennate muscles. The fascicles of **unipennate** muscles insert into the tendon from one side, **bipennate** from two sides, and **multipennate** from multiple sides.

These muscle shapes effect force transmission and range of motion. Recall that fascicles are bundles of muscle fibers. And so, the angle of the fascicles is congruent with the angle of the fibers. When the fibers act, the force generated is transferred on to the tendon to move the bone. The greater the angle of the fibers, within their fascicles, relative to the tendon, the less of the force produced is conveyed to the tendon. For instance, when moving a heavy object, you would line yourself up with it to push it rather than standing off to the side. The muscle fibers of parallel muscles have almost no angle to their tendons and thus transmit most of their force to them. However, fusiform, triangular, and pennate muscles impart less of their force to the tendons because of their greater angle of pull. Additionally, the greater the angle of fascicles and its fibers, the shorter the distance the tendon is moved.

Actions of Muscles

Muscles do not act independent of some stimulus. Under normal conditions, the nervous system generates and carries the stimulus that causes a muscle to act. This connection is called **innervation** as in, "The nerves innervate the muscles." When stimulated, a muscle reacts and produces tension (force). When the force is great enough, the muscle will pull on the tendon which moves the bone it is attached to. In the section titled *Motor Units* of Chapter 4 you will learn more of this interrelated function between the nervous system and muscles.

There are three types of muscular actions. They are concentric, eccentric, and isometric. A **concentric muscular action** causes the length of a muscle to decrease and is used to lift objects (**FIGURE 3.5**). An **eccentric muscular action** is the opposite of a concentric action causing the length of a muscle to increase (**FIGURE 3.6**). Eccentric actions are used to lower objects. The application of force by a muscle without movement and therefore, no change in muscle length is called an **isometric muscular action** (**FIGURE 3.7**).

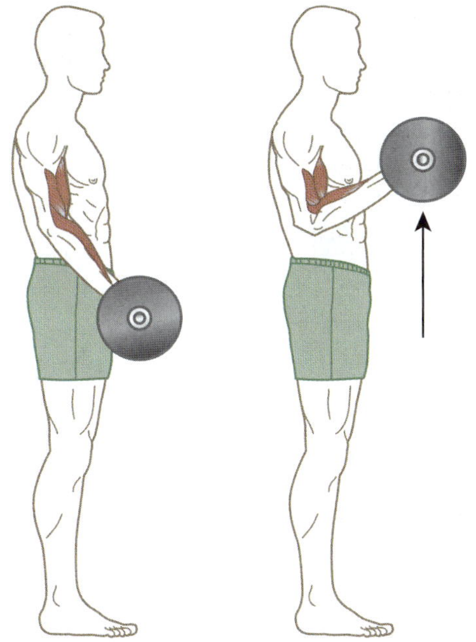

FIGURE 3.5 A Concentric Action Lifting an Object.

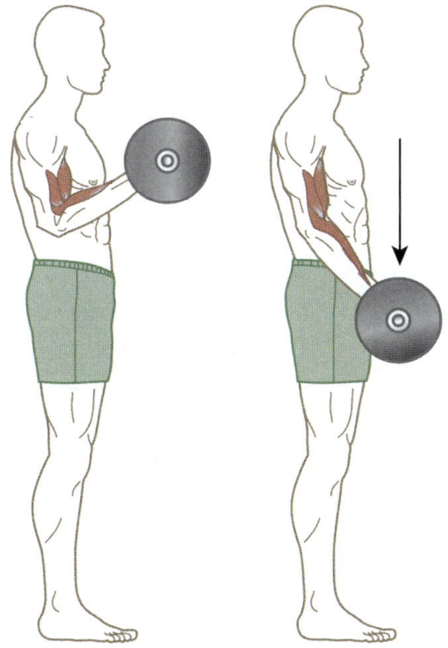

FIGURE 3.6 An Eccentric Action Lowering an Object.

PAUSE TO CHECK FOR UNDERSTANDING

1. What are the four muscle classifications by shape?
2. Which type of muscular action causes a muscle to lengthen?
3. Which type of muscular action holds an object motionless?
4. Which type of muscular action lowers an object?

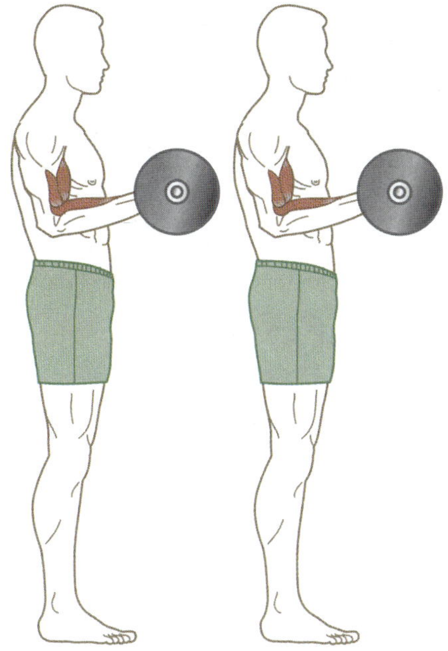

FIGURE 3.7 An Isometric Action Holding an Object.

Motion

Muscular actions, except for isometric actions, cause motion of body parts (segments) at the joints. **Motion** is a change in the position (displacement) of an object over time. Therefore, human motion is a change in position of a human body over time. Further, the entire body can change position or just one or more segments can change position. When a body segment moves, it does so around its joint. For instance, the lower leg moves around the knee. When the entire body moves, it is the result of the combined efforts of some number of muscles moving some number of body segments at their respective joints. For example , when a person stands up from a chair, there is, at the least, motion at the ankle, knees, hips, and waist. Additionally, some motion occurs at many other joints.

To clarify the possibilities of human motion, this text makes some distinctions. First, **joint motion** is the displacement of a single body segment. **Body motion** is a simultaneous displacement of more than one body segment. Furthermore, a joint motion can be simple or complex.

Furthermore, a joint motion can be either simple or complex (**FIGURE 3.8**). **Simple joint motion** is displacement of a single body segment in one direction within one body plane and around one axis of rotation. **Complex joint motion** is displacement of a single body segment in one direction; however, it is moved through more than one body plane and

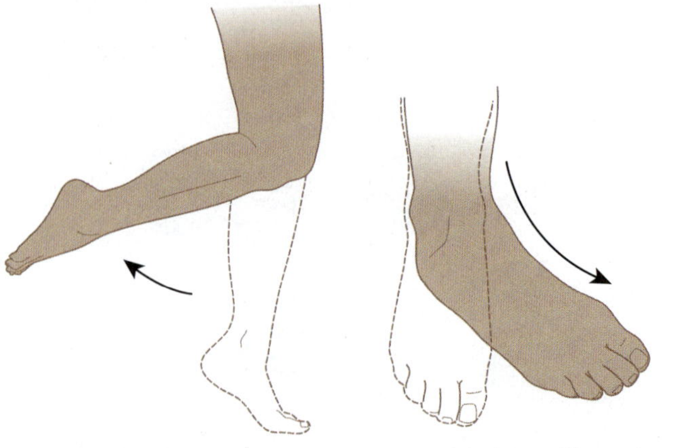

FIGURE 3.8 Simple Joint Motion (left) and Complex Joint Motion (right).

FIGURE 3.9 Simple Body Movement (left) and Complex Body Movement (right).

therefore more than one axis of rotation. Another way of describing complex joint motion is that it is more than one simple joint motion occurring simultaneously.

Like joint motion, body motion can either be simple or complex (**FIGURE 3.9**). **Simple body motion** is the simultaneous displacement of more than one segment without the entire body being displaced. **Complex body motion** is the simultaneous displacement of more than one segment whereas the entire body is displaced.

Ultimately, all motion is the result of joint motion(s). There is a common list of simple joint motions used to indicate the displacement of segments about their joints (**TABLE 3.1**). Remember, simple joint motions only occur in one body plane around one axis of rotation. Furthermore, each simple joint motion only occurs in one direction; so, each one has a counter (opposite) motion so a body segment can be returned to its former location. For instance, extension is a motion that displaces a body part in one direction and flexion is a motion in the exact opposite direction.

TABLE 3.1 Simple Joint Motions

Motion	Definition	Plane	Axis
Flexion	decreases the angle of a joint; bending	Sagittal	Frontal
Extension	increases the angle of a joint; straightening	Sagittal	Frontal
Hyperextension	extension beyond anatomical position	Sagittal	Frontal
Lateral Flexion	tilting of a body segment that resides on the midline (head or trunk) to the *right* or *left* side	Frontal	Sagittal
Abduction	to "take away" a body segment from the midline	Frontal	Sagittal
Adduction	to "add" a body segment towards the midline	Frontal	Sagittal
Horizontal Abduction	same as abduction, but in different plane	Transverse	Vertical
Horizontal Adduction	same as adduction, but in different plane	Transverse	Vertical
Elevation	a shoulder motion towards the superior	Sagittal	Frontal
Depression	a shoulder motion towards the inferior	Sagittal	Frontal

Protraction	a shoulder motion towards the anterior	Transverse	Vertical
Retraction	a shoulder motion towards the posterior	Transverse	Vertical
Rotation	turning of a body segment that resides on the midline (the head and trunk) to the *right* or *left* side	Transverse	Vertical
Medial (Internal) Rotation	turning around the longitudinal axis towards the midline or medially	Transverse	Vertical
Lateral (External) Rotation	turning around longitudinal axis away from the midline or laterally	Transverse	Vertical
Pronation	medial rotation of the elbow	Transverse	Vertical
Supination	lateral rotation of the elbow	Transverse	Vertical
Ulnar Deviation	lateral flexion of the wrist towards the ulna which is on the medial side	Frontal	Sagittal
Radial Deviation	lateral flexion of the wrist towards the radius which is on the lateral side	Frontal	Sagittal
Opposition	thumb motion across the palm towards the little finger	Frontal	Sagittal
Reposition	thumb motion back across palm from little finger to anatomical position	Frontal	Sagittal
Dorsiflexion	ankle motion causing foot to rise	Sagittal	Frontal
Plantar Flexion	ankle motion causing foot to lower	Sagittal	Frontal
Inversion	ankle motion causing medial rotation of foot	Frontal	Sagittal
Eversion	ankle motion causing lateral rotation of foot	Frontal	Sagittal

PAUSE TO CHECK FOR UNDERSTANDING

1. What are the differences between joint motion and body motion?
2. Provide an example of each simple joint motion.
3. For each joint motion, tell which plane and axis of rotation it corresponds with.

Functional Roles of Muscles

Depending on the motion or movement, muscles function in one of four ways. The terms that describe the role a muscle plays during a given motion are *agonist, synergist, antagonist*, and *stabilizer*.

An **agonist** (sometimes called primary mover) is the muscle or group of muscles primarily responsible for the given motion that is occurring. **Synergists** (sometimes called secondary movers) are the muscle(s) that assist the agonists in a motion. These secondary muscles work from a variety of angles to help and refine the work of the primary muscle(s). Furthermore, they assist more so during certain portions of the motion and less during others. Still other muscles called **stabilizers** increase the stability of certain bones in order for the agonists and/or synergists to work off of a more firm foundation. Finally, **antagonists** regulate motion from the opposite side of the agonists. The antagonists serve as the brake when it is time to slow and eventually stop the motion. For a given motion, all of the muscles involved in play-

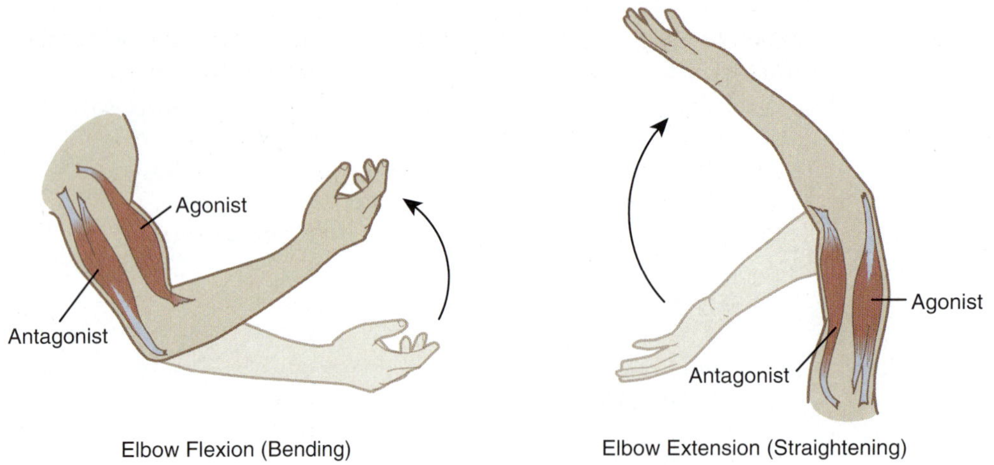

FIGURE 3.10 The Varying Roles of Muscles During Elbow Motion.

ing the different roles function in concert to produce incredibly fluid, smooth movement.

It is very important to understand that all muscles will serve as either an agonist, an antagonist, a synergist, or a stabilizer depending on the situation. That is to say, during a specific motion, a muscle functions as an agonist. But, if the opposite motion is performed that same muscle now would play the role of an antagonist. Also, if a different motion is performed, this same muscle can function as a synergist or a stabilizer. Whatever the motion, a specific muscle will only play one role, but it will play different roles during different motions.

A simple example of the varying roles of muscles can be illustrated with elbow motion (**FIGURE 3.10**). During bending (flexion) of the elbow, the biceps brachii muscle is the main agonist. Other muscles play the role of synergists. Still other muscles stabilize the humerus so the forearm can rotate around a solid elbow. On the opposite side of the arm, the triceps brachii acts as an antagonist. When the elbow motion is changed to straightening (extension), the roles of the muscles change as well. For the most part, the muscles that stabilize the humerus remain the same, but the triceps brachii now acts as the agonist while the biceps brachii and synergists now play the role of antagonists.

How Muscles Are Named

The names of muscles may seem like a difficult to impossible task. However, realizing how they are named can make learning them easier. The muscles are named by one or some combination of their location, shape, action, attachment site, region that the belly resides within, number of divisions, or relative depth (**TABLE 3.2**). For instance, the biceps brachii is named for both the number of divisions (or heads) and its

TABLE 3.2 Examples of Meanings of Muscle Names

Characteristic	Example Muscle	Reference
Shape	**deltoid**	shaped like a triangle
Size	gluteus **maximus** gluteus **medius** gluteus **minimus**	largest relatively medium sized smallest
Number of Divisions	**quadriceps** femoris **triceps** brachii	four muscles in group three divisions (heads)
Direction of Fibers	**rectus** femoris **transverse** abdominis external abdominal **oblique**	"straight" vertically "across", straight horizontally slanted

Attachment Site	**coraco**brachialis **infraspin**atus	attaches on coracoid process attaches on infraspinous fossa
Regional Location	biceps **brachii** **tibialis posterior**	within brachial region within posterior tibia
Action	**flexor** digitorum longus	action causes toe flexion
Depth	flexor digitorum **superficialis**	closer to skin surface relative to other muscles
	flexor digitorum **profundus**	farther from skin surface relative to other muscles

location. It has two heads and lies within the brachial region. Another example is the deltoid muscle. Deltoid means triangular and that is the general shape of the deltoid. A final example is the flexor hallucis longus. Hallucis refers to the first toe and this muscle flexes it.

Intrinsic and Extrinsic Muscles

Some muscles are described by the terms *intrinsic* and *extrinsic*. These terms are restricted to the muscles that move the fingers and toes and indicate general location. An **intrinsic** muscle is entirely (origin, belly, and insertion) located within the region that it acts upon. For instance, two examples of intrinsic muscles are the abductor pollicis brevis muscle and the extensor hallucis brevis muscle (**FIGURE 3.11**). Both of these muscles completely reside within the hand and foot, respectively.

Extrinsic muscles are ones whose origin and belly are outside the region of the joint it acts upon (**FIGURE 3.12**). The abductor pollicis longus muscle and the extensor hallucis longus muscle are examples of extrinsic muscles.

PAUSE TO CHECK FOR UNDERSTANDING

1. Define agonist, antagonist, synergist, and stabilizer.
2. Review how muscles are named. The more familiar you are with this, the easier it will be to locate and understand them.
3. What is it called when a muscle's origin, insertion, and belly are all contained within the area it moves?

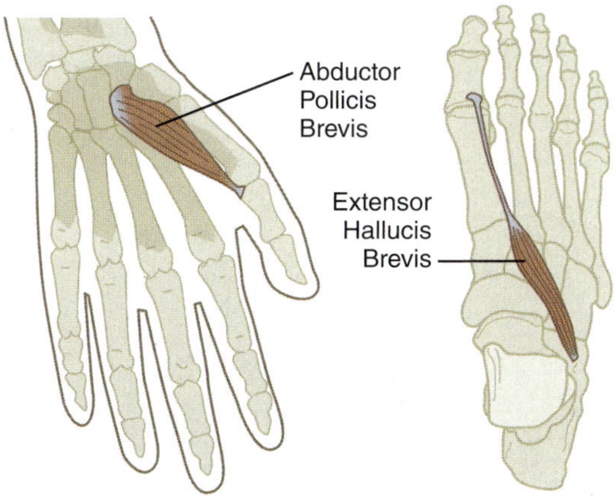

FIGURE 3.11 Sample Intrinsic Muscles.

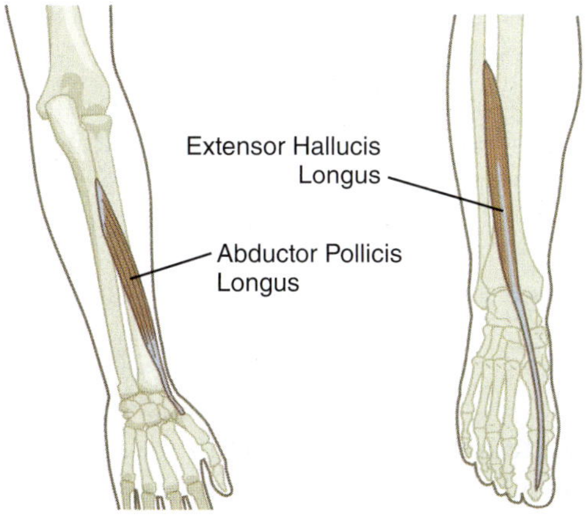

FIGURE 3.12 Sample Extrinsic Muscles.

CHAPTER 4

The Nervous System

CHAPTER OBJECTIVES

After completing this chapter, the student will be able to:

1. understand the divisions of the nervous system;
2. comprehend the anatomy and basic function of a neuron (nerve cell);
3. summarize the anatomy and basic function of the central nervous system;
4. describe the anatomy and basic function of the peripheral nervous system; and
5. explain the role of a motor unit.

The nervous system is the most complex and mysterious in the body. As much as scientists have discovered about it, there is still so much yet to discover. The overall purpose of the nervous system is to control, regulate, and coordinate all the other systems of the body (**FIGURE 4.1**). For instance, body temperature needs to be maintained at a relatively narrow range with an average of 98.6 degrees. Certain body systems can and will raise body temperature and others lower it. The nervous system regulates temperature by activating the appropriate systems to raise or lower it.

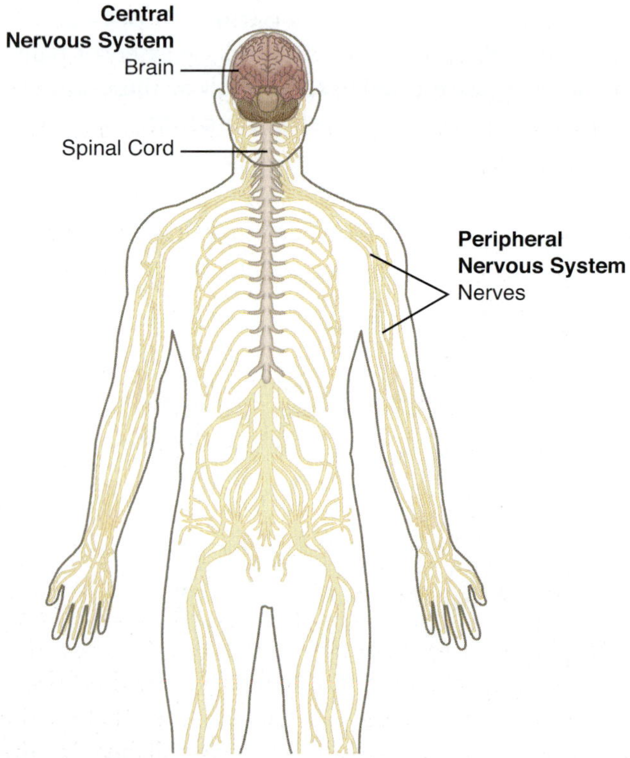

FIGURE 4.1 The Nervous System.

Divisions of the Nervous System

The nervous system is composed of cells, tissues, and organs like any other organ system. Before describing these however, it is best to first learn the divisions of the nervous system. The nervous system is traditionally separated into two main divisions (**FIGURE 4.2**). These are the central nervous system and the peripheral nervous system.

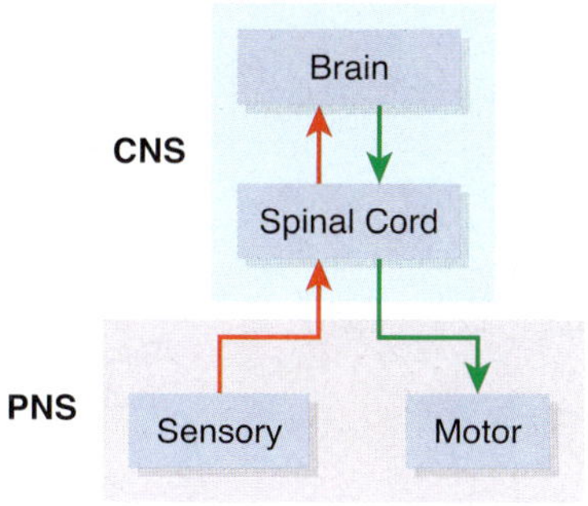

FIGURE 4.2 Divisions and Subdivisions of the Nervous System.

The **central nervous system** (CNS) contains the brain and spinal cord. The overall function of the **CNS** is to:

1. receive information from the body;
2. interpret the information;
3. make decisions; and
4. send messages to the body to cause an effect.

The main function of the **peripheral nervous system** (PNS) is to provide a route of communication between the body and the CNS. The PNS is further divided into sensory and motor divisions (**FIGURE 4.2**). The **sensory division** is responsible for monitoring the status of the body and delivering information (sensory messages), hence its name, to the CNS. The **motor division** of the PNS carries messages from the CNS to the motors. **Motors** are organs that can cause an effect or change and includes all the various muscles and the glands. The PNS can be divided even further, but this text will restrict its discussion to these two main divisions.

PAUSE TO CHECK FOR UNDERSTANDING

1. Restate the main functions of the CNS & PNS.
2. Which PNS subdivision sends messages from the body to the CNS?
3. Which PNS subdivision sends messages from CNS to the body?

Cellular Anatomy

The cells of the nervous system (nerve cells) are called **neurons**. As with all cells, the neurons are the "heart" of the functions of the nervous system. That is, the functions of an organism are accomplished by the combined efforts of the cells and so the function of the nervous system, both in the CNS and PNS, is performed by the neurons.

Neurons are specialized with the characteristics of excitability and conductivity. **Excitability** refers to its ability to respond to stimuli. **Conductivity** means that it can transmit (carry) electrical energy. So, when stimulated, neurons conduct electrical energy (messages or information). Furthermore, they transmit (carry) these messages in only one direction.

Since neurons can only carry messages in one direction, more than one is needed to send messages in all directions. There are three types of neurons. They are *sensory*, *motor*, and *interneurons*. **Sensory neurons** conduct information related to the state of the body to the CNS. They are also called **afferent** (or **ascending**) **neurons** because "afferent" literally means "carry towards" which in this case means towards the CNS. **Motor neurons** conduct messages from the CNS to motors (muscles or glands) to cause an effect. They are also sometimes called **efferent** (or **descending**) **neurons** because "efferent" literally means "carry away." **Interneurons** complete the communication route by connecting the sensory neurons and the motor neurons. Interneurons are only found in the CNS and are also called **associative or connecting neurons**.

Not all neurons have the same anatomy, but the following is indicative of the main features relevant to this text (**FIGURE 4.3**). The **soma** is the body of a neuron which also contains the nucleus. **Dendrites** are the numerous, relatively short extensions of the soma. The **axon**, also called a **nerve fiber**, is the relatively long extension from the soma. Messages are conducted in

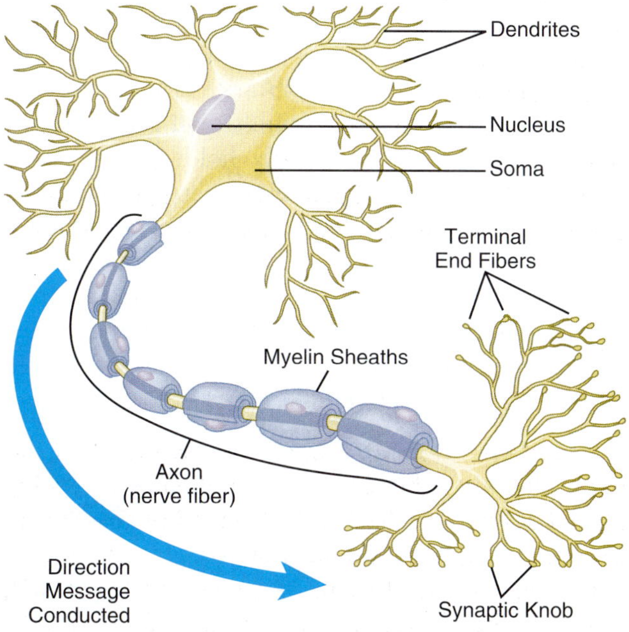

FIGURE 4.3 Basic Anatomy of a Typical Neuron.

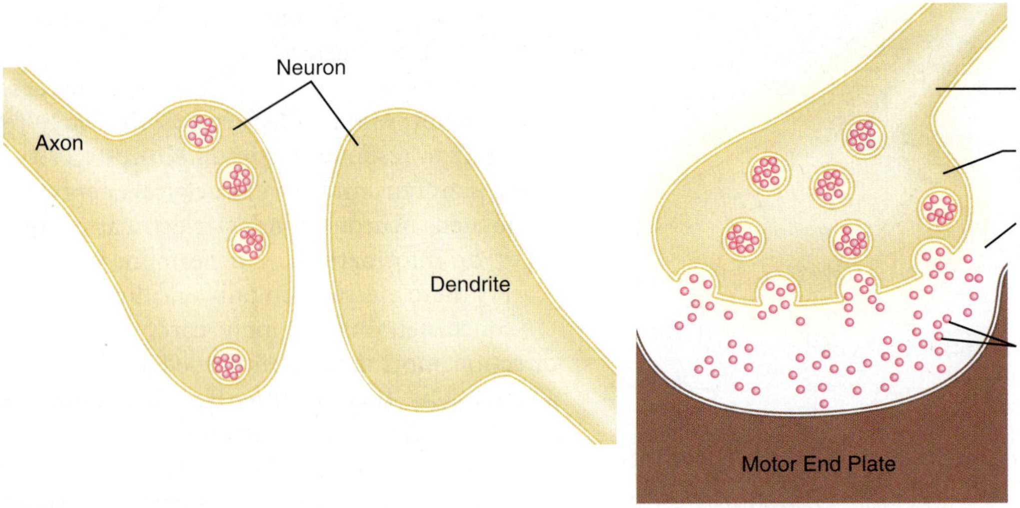

FIGURE 4.4 Synapse of Two Neurons (left) and of a Neuron and Muscle Cell (right).

one direction from dendrites through the soma and down the axon.

At the end of the nerve fiber (axon) is the *terminal arborization*. The terminal arborization is a group of short branches from the end of the nerve fiber. Each branch is called a **terminal end fiber**. Furthermore, the very end of each terminal end fiber widens into what is called the **synaptic knob** (also known as **synaptic bulb**). The synaptic knob is part of a synapse.

A **synapse** is the junction between a neuron and another cell. And, neurons only synapse with either another neuron, a sensory receptor, or a motor (muscle or gland) (**FIGURE 4.4**). In the case of muscle, the synapse between a neuron and the muscle cell (fiber) is called a **neuromuscular junction**. However, for all types of synapses, there is no direct contact between the cells. Instead, there is a gap called the **synaptic cleft**.

Within the synaptic knob are vesicles. **Vesicles** are bubble-like structures found in all kinds of cells and they serve to temporarily hold and transport substances in bulk. Within the synaptic vesicles are neurotransmitters. **Neurotransmitters** are chemical messengers. When a message reaches the end of a neuron, vesicles are stimulated to move to the tip of the synaptic knob and release the neurotransmitters. The neurotransmitters cross the synaptic cleft and stimulate the next cell (**FIGURE 4.4**).

PAUSE TO CHECK FOR UNDERSTANDING

1. Draw and label the parts of a neuron.
2. List the three types of neurons. What does each kind do?
3. What are neurotransmitters?

CNS Anatomy

The main organs of the CNS are the spinal cord and the brain (**FIGURE 4.5**). There are also structures called tracts which exist in both the spinal cord and the brain.

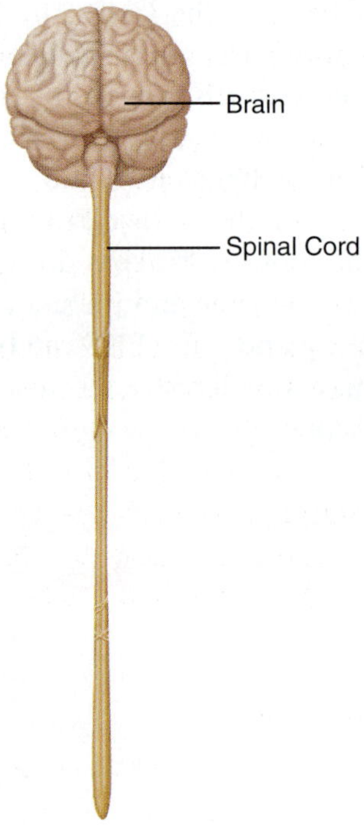

FIGURE 4.5 The Brain and Spinal Cord.

CNS Tracts

The **tracts** of the CNS are bundles of nerve fibers (axons) (**FIGURE 4.6**). Tracts serve as routes of communication between all the different parts of the CNS. In the spinal cord, a tract only contains either sensory or motor nerve fibers. **Ascending tracts** are bundles of sensory nerve fibers. They transmit sensory

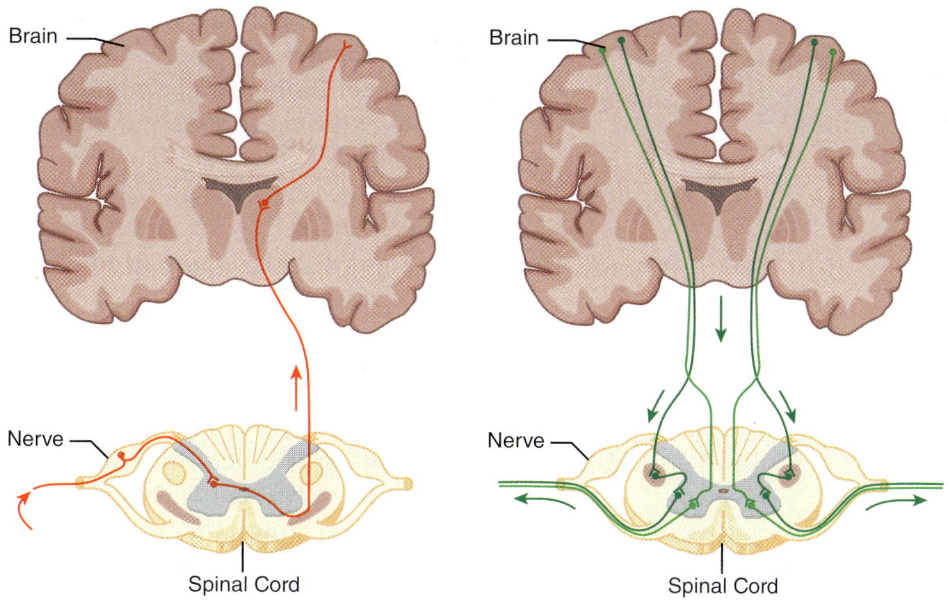

FIGURE 4.6 Ascending (left) and Descending (right) Spinal Cord Tracts.

information from PNS sensory neurons to the brain. **Descending tracts** are bundles of motor nerve fibers that transmit motor messages from the brain to motor neurons of the PNS.

There are also tracts (bundles of nerve fibers) in the brain (**FIGURE 4.7**). The tracts in the brain connect all its parts so that information can be shared between different decision-making centers. For instance, a sensory organ in the inner ear called the vestibular apparatus sends messages to the brain related to balance. However, other organs, such as the eyes, also send information pertinent to maintaining balance. Yet, the initial information from these two sensory organs comes to the brain from different locations. The tracts allow both sets of information as well as others to be shared in order to enhance balance.

There are three classifications of tracts in the brain (**FIGURE 4.7**). **Projection tracts** connect the spinal cord with the various main parts of the brain. **Commissural tracts** connect each hemisphere (side) of the brain. **Association tracts** connect different areas within each hemisphere of the brain.

The Spinal Cord

The **spinal cord** is a cylinder of nerve tissue that extends from the base of the brain through the vertebral foramina of the spinal column (**FIGURE 4.8**). As already learned, it contains ascending and descending tracts (bundles of nerve fibers). Along its length, the spinal nerves of the PNS exit. The main function of the spinal is to make one of two relatively simple decisions which are far more limited than the decision capabilities of the brain.

One of the simple decisions is whether or not sensory information from the body is an "emergency" or

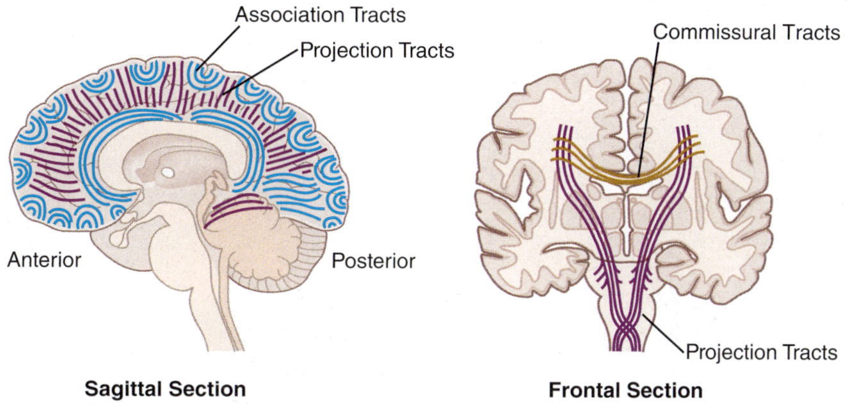

FIGURE 4.7 CNS Tracts in the Brain.

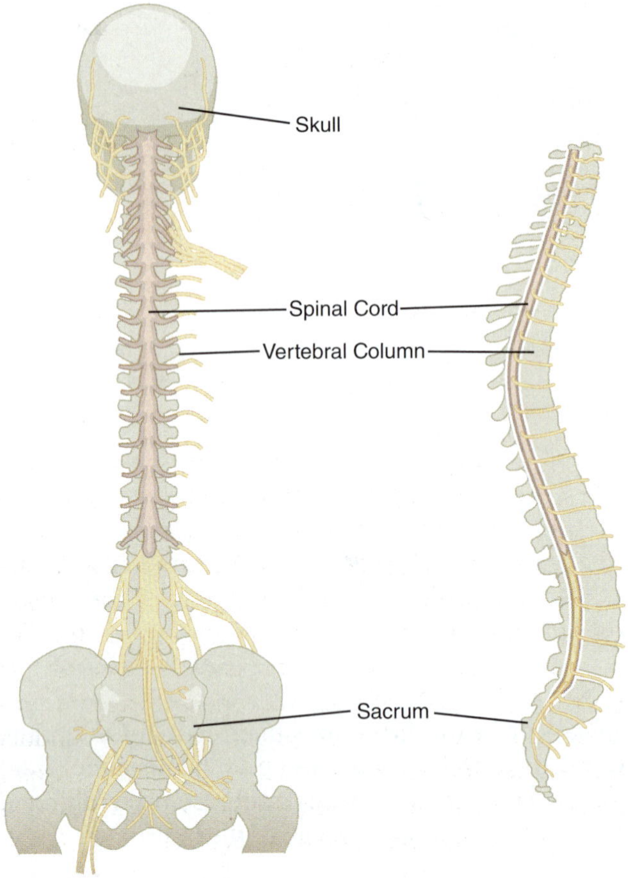

FIGURE 4.8 The Spinal Cord: Posterior View (left) and Lateral View (right).

not. For emergencies, it instantly sends commands back through the PNS to specific organs to alleviate the "emergency." The reaction to an "emergency" is called a *reflex*. A **reflex** is a rapid, involuntary reaction to stimuli when needed. For non-emergencies, the spinal cord simply relays the information to the brain on one of its ascending tracts.

The other simple decision is related to continuing stereotypical (non-changing) movements (or motions). Non-emergency movement is always initiated by the brain. And, any change to that motion is regulated by the brain. However, as long as a movement does not change (stereotypical), like taking a long walk at the same speed in the same direction, the spinal cord continues to stimulate the muscles engaged in that without brain input.

The Brain

The main parts of the brain are the cerebrum, cerebellum, and brain stem (**FIGURE 4.9**). The **cerebrum** is the largest part of the brain (83%). It is divided longitudinally down the middle into two hemispheres (sides) and each side is divided into areas called lobes (**FIGURE 4.9**). Generally speaking, each lobe has some basic functions. The **frontal lobe** controls motor function, memory, and behavior. The **parietal lobe** receives and interprets sensory input from the tongue, skin, and muscles. The **temporal lobe** is responsible for hearing and smell. The **occipital lobe** controls vision.

The **cerebellum** is about 10% of the mass of the brain and is located inferior to the posterior end of the cerebrum. The cerebellum receives information about joint movements and body position, decides which parts of the brain need to act on this input, and relays that to them.

The **brainstem** (7% of brain) resides inferior to and beneath the inner core of the cerebrum (**FIGURE 4.10**). It begins at the superior end of the spinal cord and is divided into four regions. From the most inferior, they are the *medulla oblongata*, *midbrain*, the *pons*, and the *thalamus*. The main functions of the **medulla oblongata** are to regulate

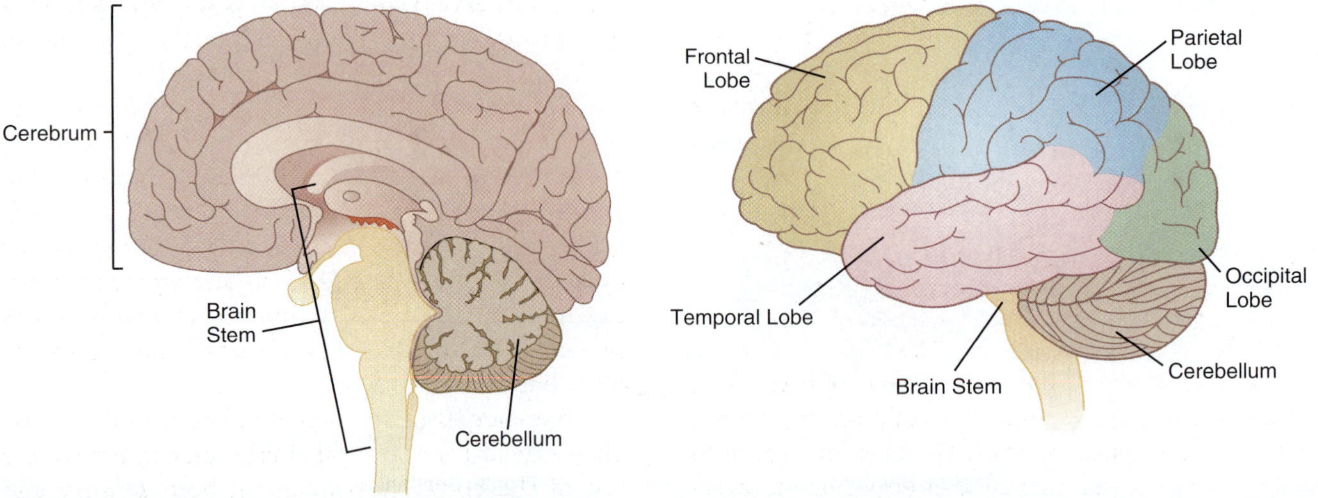

FIGURE 4.9 The Brain: The Main Parts (left) and the Lobes of the Cerebrum (right).

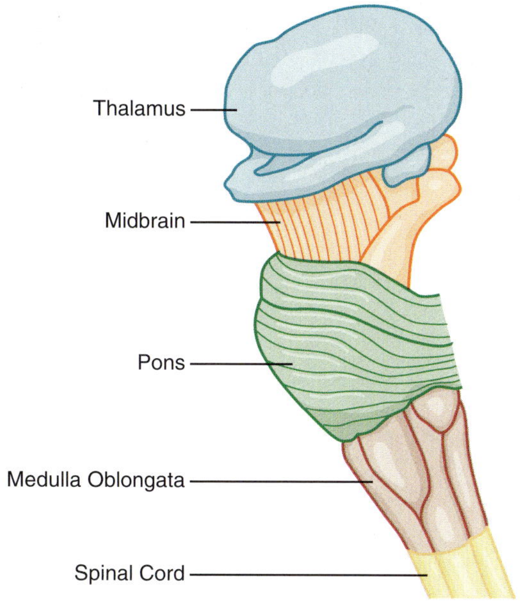

FIGURE 4.10 The Brainstem.

the vital body functions of heart rate, breathing, and blood pressure. The **midbrain** and the **pons** mostly serve to send information to and from various parts of the brain. The **thalamus** is sometimes referred to as the "sensory gateway" because it receives and relays sensory input to the proper location of the brain.

PAUSE TO CHECK FOR UNDERSTANDING

1. Can the spinal cord make decisions?
2. What are the three main parts of the brain?
3. List the lobes of the brain and state their general location.

PNS Anatomy

The main function of the PNS is to monitor the status of the body and provide a two-way route of communication between the body and the CNS. To accomplish this, the PNS uses its two main organs which are the sensory receptors and nerves.

Sensory Receptors

Sensory receptors are specialized to react to a stimulus. There are various types of stimuli (energy) that different types of sensory receptors react to (**FIGURE 4.11**). These forms of energy include pressure, light, sound, taste, smell, and heat. And, each

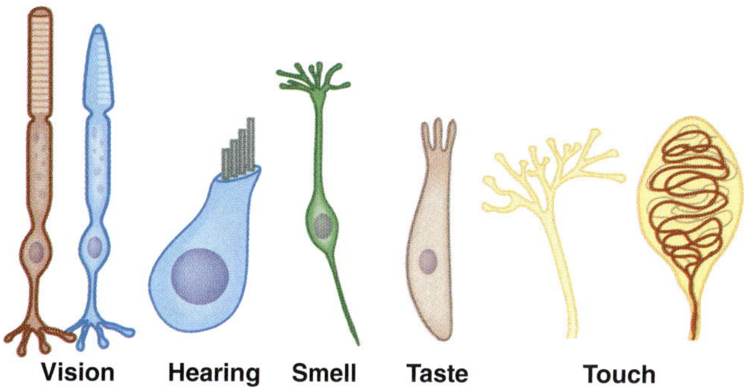

FIGURE 4.11 Types of Sensory Receptors.

type of sensory receptors will only respond to a specific type of energy. For instance, sensory receptors for light will only react to light energy.

Sensory receptors are also sometimes called **transducers** because they transduce (convert) the energy they respond to into electrical energy. If a stimulus is great enough, the sensory receptor converts that energy into electrical energy. Sensory receptors synapse with sensory neurons which are able to conduct this electrical energy to the CNS. Hence, a "sensory message" about the environment has been delivered to the CNS for interpretation and decision.

Nerves

Nerves are conduits ("pipes") filled with nerve fibers (axons) that extend from the spinal cord and mark the beginning of the PNS (**FIGURE 4.12**). Almost all nerves are said to be mixed. A **mixed nerve** refers to containing the axons of both sensory and motor neurons. This design conserves mass by having one structure supplying one area. For instance, there are sensory receptors within muscles connected to sensory neurons. The same muscle synapses with motor neurons. Rather than having two nerves going to the same place, one nerve provides a passageway for both types of neurons. Nerves are restricted to the PNS and so interneurons do not reside within them.

Nerves are classified as either *cranial nerves* or *spinal nerves*. There are twelve (12) pairs (total of 24) of **cranial nerves** that extend directly from the brain and innervate organs mainly in the head and neck regions. Although these nerves are vital for every day function and life itself, they do not innervate the muscles responsible for joint motions. Innervation refers to the connection or supply of nerves and neurons to other body parts.

There are 31 pairs (total of 62) of **spinal nerves**. They extend from the spinal cord and innervate the rest of the body. Nerves contain both sensory and motor axons (nerve fibers). Therefore, nerves provide

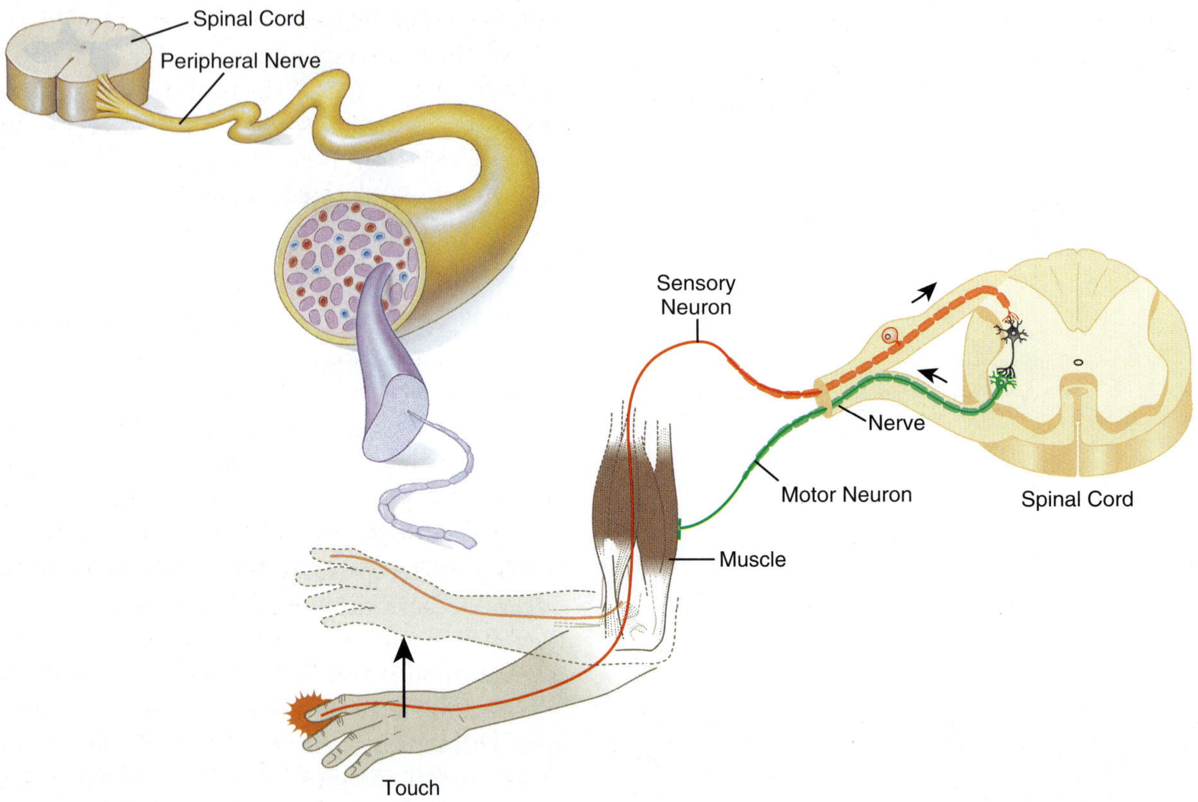

FIGURE 4.12 A Peripheral Nerve.

the route to send messages to motors and provide the route to receive sensory information. Because of this, spinal nerves are sometimes called mixed because of their dual role in sending and receiving messages.

Each spinal nerve extends from the spinal cord from two locations (**FIGURE 4.13**). The **dorsal root** extends from the posterior spinal cord and contains sensory neurons bound for an ascending tract. The **ventral root** attaches to the anterior spinal cord and has motor neurons from a descending tract within it. Still within the vertebrae, these two roots merge, pass through its corresponding intervertebral foramen, and then splits into two branches. One branch is called the **posterior ramus**. This branch innervates structures close in proximity to the spinal column. The other branch is called the **anterior ramus** and this is the main nerve that innervates structures distal from the spinal column. The anterior ramus is also known as a **nerve root**; not to be confused with the anterior or posterior nerve root.

Each pair of spinal nerves typically is abbreviated with a capital letter and number that corresponds to their exit location from the vertebrae (**FIGURE 4.14**). For instance, the C1 nerve root exits from the top of the first cervical (C1) vertebra. Eight (8) pairs of nerve roots (C1–C8) extend from the cervical region, yet there are only seven (7) cervical vertebrae. The first seven spinal nerves exit from the superior side of each cervical vertebrae. The eighth spinal nerve exits from the inferior side of the seventh cervical vertebra. As

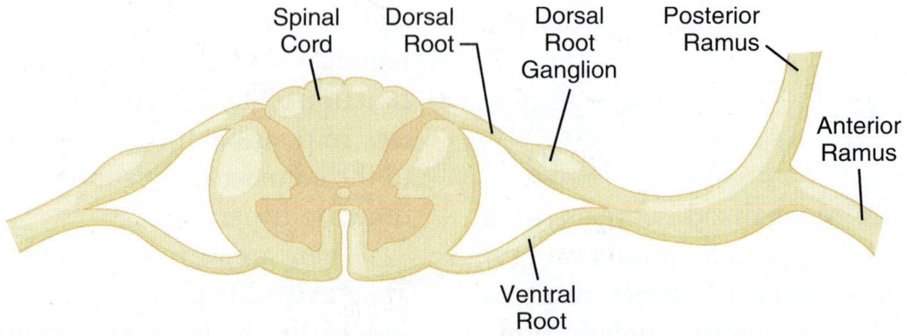

FIGURE 4.13 A Spinal Nerve and Its Connection to the Spinal Cord.

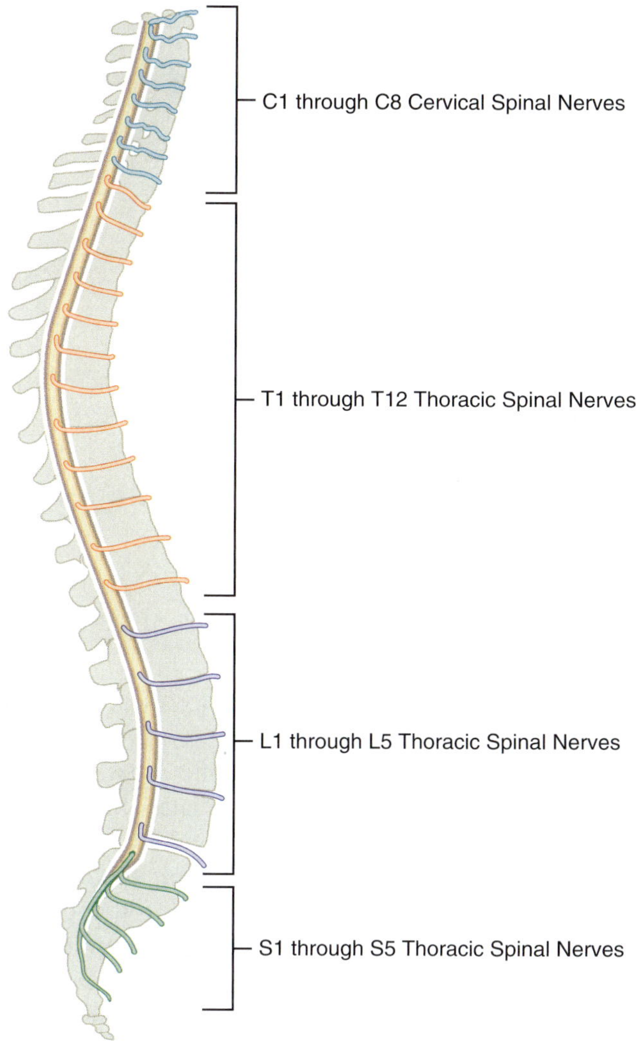

FIGURE 4.14 Nerve Roots.

well, the rest of the spinal nerves exit from the inferior side of their corresponding vertebrae. So, there are twelve (12) pairs (T1–T12) from the thoracic region, five (5) pairs (L1–L5) from the lumbar region, five (5) pairs (S1–S5) from the sacral region, and one (1) pair (Co1) from the coccygeal region.

Nerve Plexuses and Their Major Nerves

Shortly after leaving the vertebral column, nerve roots merge with and branch from each other a number of times to form a web-like structure called a **plexus**. There are five nerve plexuses. They are the *cervical plexus*, the *brachial plexus*, the *lumbar plexus*, the *sacral plexus*, and the *coccygeal plexus*. Spinal nerve roots from the thoracic region do not form plexuses. From these plexuses or nerve roots are the major nerves which extend to certain regions of the body for innervation.

The **cervical plexus** is formed by C1–C5 nerve roots. The nerves of this plexus innervate structures in the head and neck regions. The **coccygeal plexus** is formed by S4–Co1 nerve roots. The nerves of this plexus mostly innervate structures in the region of the genitalia. The functions of the nerves of these two plexuses are not relevant to the purpose of this text; and therefore, will not be discussed further.

The **brachial plexus** is formed by C5–T1 nerve roots. Thirteen (13) pairs of nerves relevant to this text emerge from these nerve roots and/or the brachial plexus (**FIGURE 4.15**). These 13 nerve pairs mostly innervate structures in the pectoral, shoulder, and upper extremity regions. The specific muscles these nerves innervate will be indicated in later chapters.

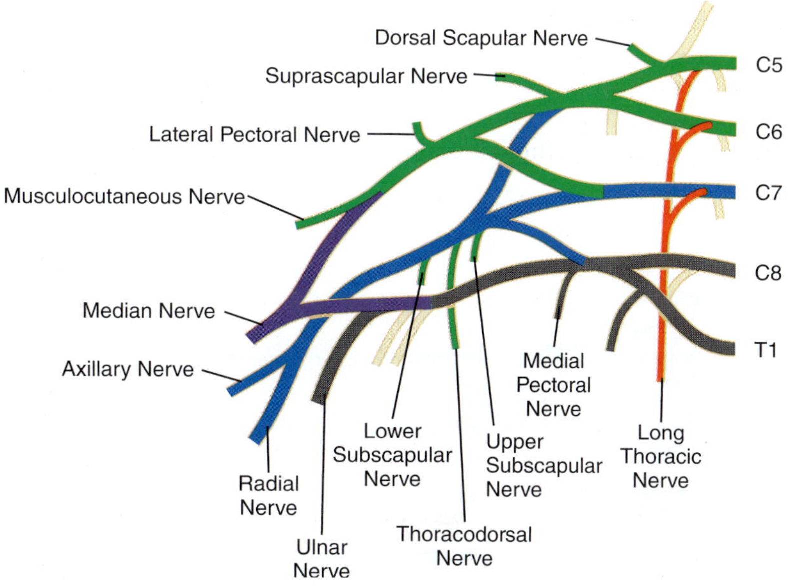

FIGURE 4.15 The Brachial Plexus and Its Major Nerves.

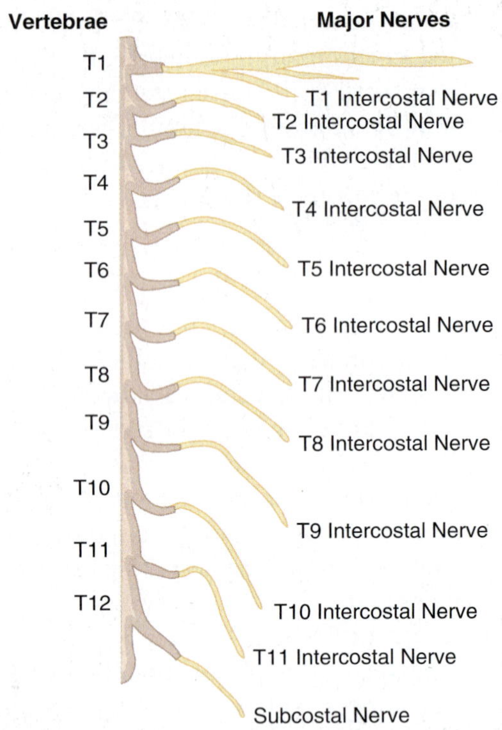

FIGURE 4.16 The Major Nerves from the Thoracic Nerve Roots.

The nerve roots of the thoracic region do not form a plexus. Instead, the twelve (12) nerve roots form nerves. The first eleven (11) form eleven **intercostal nerves** (**FIGURE 4.16**). However, the T1 nerve root forms its intercostal nerve differently than the rest. One branch forms part of the brachial plexus and the other becomes the T1 intercostal nerve. The T12 nerve root forms the **subcostal nerve**. These twelve (12) pairs of nerves innervate structures in the thoracic and abdominal regions. Some of these nerves innervate the muscles most directly responsible for movement and will be indicated in later chapters.

The **lumbar plexus** is formed by L1–L4 nerve roots. Four (4) pairs of nerves relevant to this text emerge from these nerve roots and/or the plexus (**FIGURE 4.17**). These four nerve pairs mostly innervate structures in the abdominal, hip, and some of the lower extremity regions. The specific muscles these nerves innervate will be indicated in later chapters.

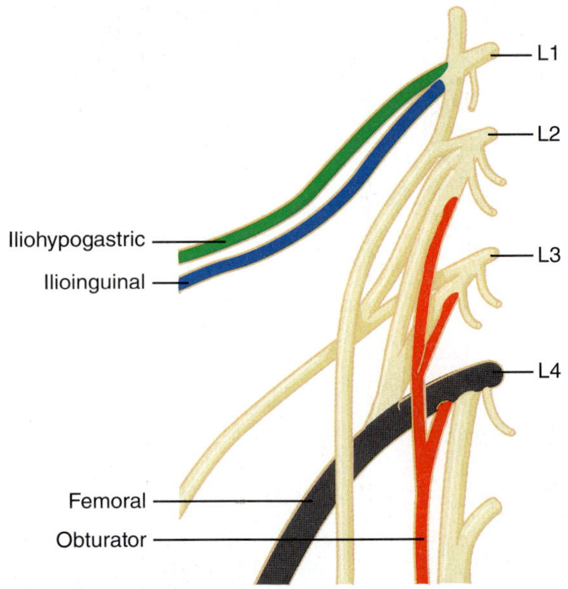

FIGURE 4.17 The Lumbar Plexus and Its Major Nerves.

The **sacral plexus** is formed by L4–S3 nerve roots. Five (5) pairs of nerves relevant to this text emerge from these nerve roots and/or the plexus (**FIGURE 4.18**). These five nerve pairs mostly innervate the rest of the lower extremity regions not supplied by the lumbar plexus. The specific muscles these nerves innervate will be indicated in later chapters.

Motor Units

It is essential for the student of kinesiology to understand the motor unit. Under normal conditions, a muscle will only contract when it is stimulated by the nervous system; more specifically, a motor neuron. A **motor unit** is a motor neuron and all the muscle fibers (cells) that it innervates (**FIGURE 4.19**). No one

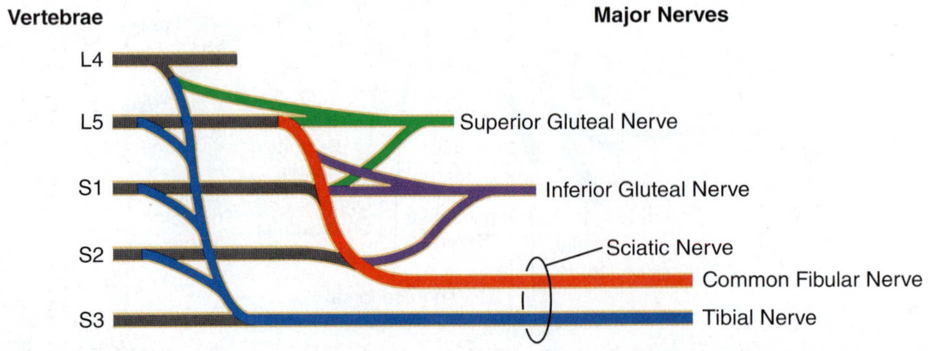

FIGURE 4.18 The Sacral Plexus and Its Major Nerves.

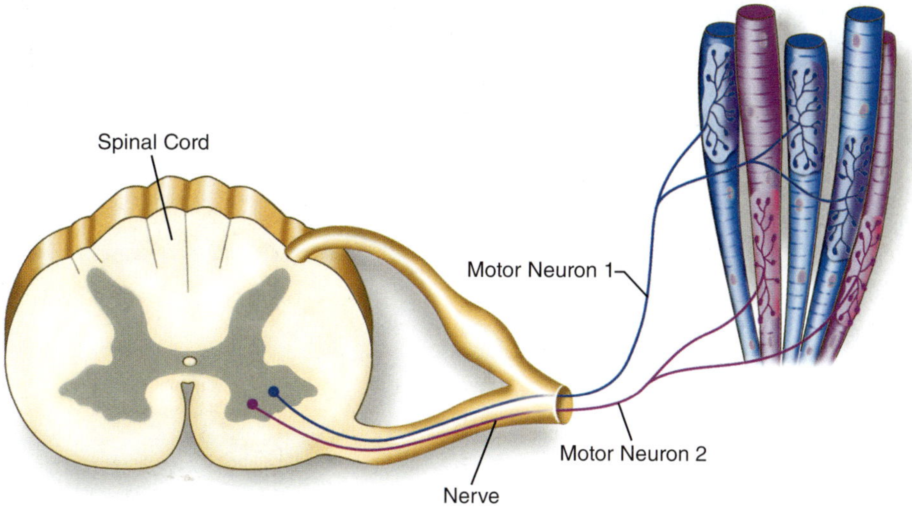

FIGURE 4.19 Two Motor Units Innervating Muscle Fibers of the Same Muscle.

motor unit innervates an entire muscle. Instead, it takes a number of motor neurons and therefore motor units to innervate one muscle.

Motor units are categorized as large or small. A **large motor unit** is one motor neuron innervating a large number, up to 1,000s, of muscle fibers. Additionally, this means that fewer motor units are needed to innervate the whole muscle. Large motor units are used for muscles where strength and power, such as a thigh muscle, is needed. This is called *gross motor control*. **Small motor units** innervate far fewer muscle fibers. In fact, some small motor neurons innervate as few as three (3) muscle fibers. This in turn means that more motor units are needed to stimulate the muscle. Small motor units are used for *fine motor control* where precision is more important than strength such as innervating a finger muscle while playing a piano.

PAUSE TO CHECK FOR UNDERSTANDING

1. What structure is the start or beginning of the sensory pathway?
2. List the four plexuses and which main region they innervate.
3. Does one motor neuron innervate one muscle fiber, one muscle, more than one muscle fiber, or more than one muscle?

© BraunS/E+/Getty Images

SECTION 2

The Bones and Their Landmarks

CHAPTER 5

The Bones of the Axial Skeleton

CHAPTER OBJECTIVES

After completing this chapter, the student will be able to recall:

1. the names and locations of the bones in the skull;
2. the significant features of the skull bones;
3. the names and locations of the bones in the vertebral column;
4. the significant features of the vertebral column bones;
5. the names and locations of the bones in the thoracic cage; and
6. the significant features of the thoracic cage bones.

It is imperative for the kinesiology student to master the bones and their features significant to body movement presented in this chapter and the next two. Knowledge of this material prepares the student for understanding the joints, the muscle attachments, and the motions produced which appears in the final sections and chapters. The bones and features covered are restricted to what is significant to the study of anatomical kinesiology. For instance, other than appearing in the table listing the bones of the skeleton, the auditory ossicles will not be mentioned again because they play no direct role in kinesiology.

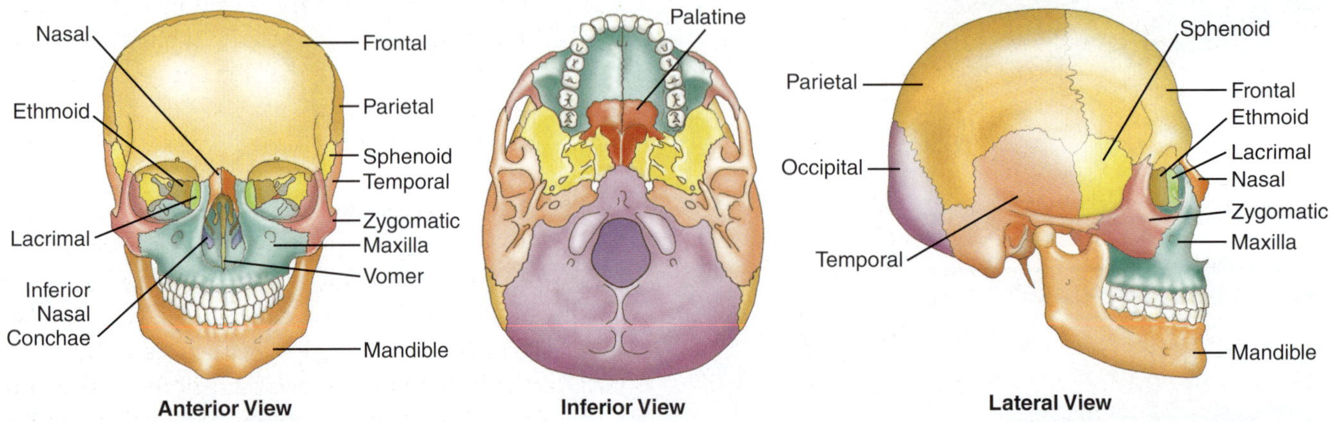

FIGURE 5.1 The Bones of the Skull.

The Bones of the Skull

There are 22 bones in the skull (**FIGURE 5.1**). These bones are subcategorized into facial and cranial bones. There are 14 **facial bones**. The facial bones include one **mandible**, two **maxillae**, two **palatine**, two **zygomatic**, two **lacrimal**, two **nasal**, one **vomer**, and two **inferior nasal conchae**.

There are eight **cranial bones**. They are one **frontal**, two **parietal**, one **occipital**, two **temporal**, one **sphenoid**, and one **ethmoid**. Only the temporal bone and occipital bone have features significant to the study of anatomical kinesiology. On the temporal bone, the **mastoid process** is a prominent bump just behind the ear (**FIGURE 5.2**). The significant features of the occipital bone and their description are shown in **FIGURE 5.3**.

PAUSE TO CHECK FOR UNDERSTANDING

Review all the bones of the skull and features. Here are a few samples to get you started:

1. Which skull bone corresponds with the forehead?
2. Which skull bone corresponds with the upper jaw and which one with the lower jaw?
3. What is the bump (projection) on the inferior end of the midline of the skull bone in the posterior head?

Another suggestion is to produce some drawings that represent well and make sense to you (they don't have to be pretty) and label them.

FIGURE 5.2 The Temporal Bone and its Features.

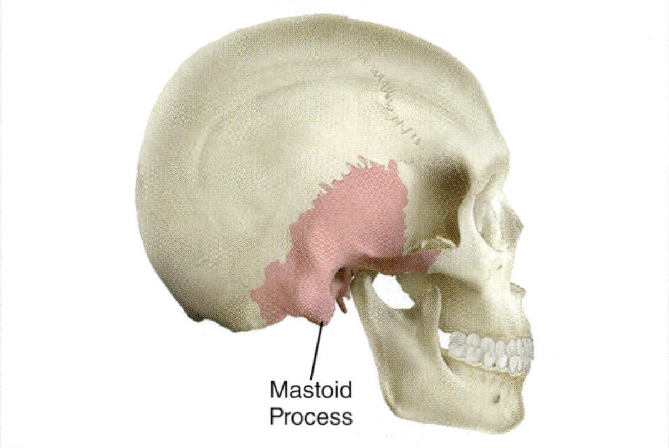

FEATURES	DESCRIPTION
Mastoid Process	relatively large "bump" posterior to ear

FIGURE 5.3 The Occipital Bone and its Features.

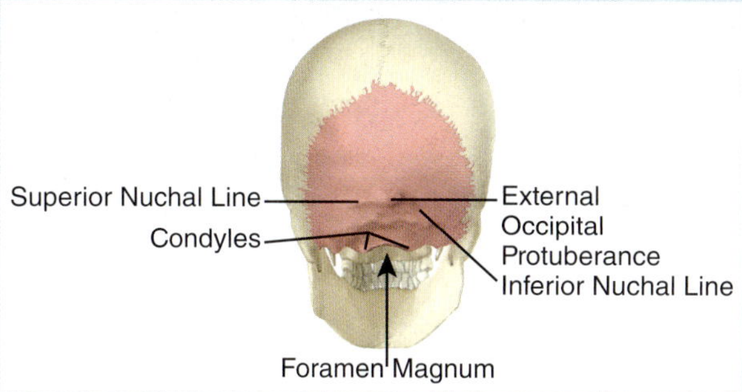

FEATURES	DESCRIPTION
Foramen Magnum	passageway on inferior side for spinal cord
Occipital Condyles	smooth knobs that articulate with C1 vertebra
External Occipital Protuberance	bump on inferior end of midline of posterior bone
Superior Nuchal Line	horizontal edge inferior to protuberance
Inferior Nuchal Line	non-palpable horizontal edge inferior to superior nuchal line

The Bones of the Vertebral Column

The vertebral column is divided into five regions. The regions are called the **cervical** which means neck, **thoracic** which pertains to the thorax, **lumbar** (low back), **sacral** for sacrum, and **coccygeal** (coccyx) (**FIGURE 5.4**). The vertebrae are commonly abbreviated by a "C" for cervical, a "T" for thoracic, an "L" for lumbar, an "S" for sacral, or a "Co" for coccygeal and are numbered from superior to inferior. For example, the most superior vertebra in the cervical spine is abbreviated "C1."

There are seven **cervical vertebrae** (C1–C7) which are relatively small. This entire section of the column creates a lordotic (inward) curve from the bottom of the skull. The first vertebra (C1) is also called **atlas** because it supports the head. This vertebra articulates with the occipital bone of the cranium. One distinctive feature of C1 which is unlike other vertebrae is that it has no body. The second

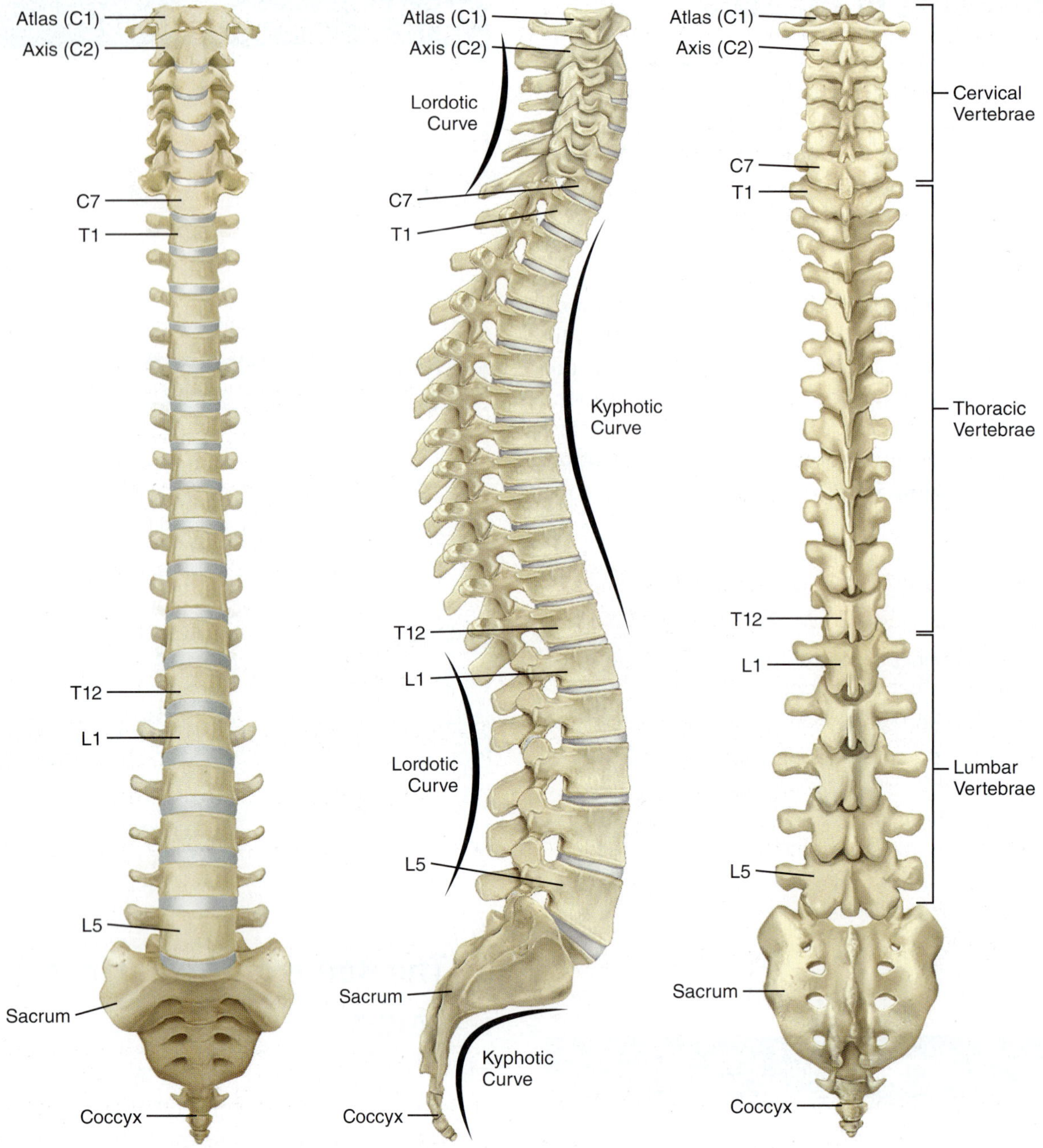

FIGURE 5.4 The Regions of Vertebral Column.

vertebra (C2) also has a name which is **axis**. There are 12 **thoracic vertebrae** (T1–T12). This section of the spine creates a kyphotic (outward) curve. These vertebrae also have the function of articulating with all the ribs of the thoracic cage. There are five **lumbar vertebrae** (L1–L5) which form a lordotic curve. These vertebrae are the largest because of bearing all of the weight of the upper body as well as serving as a solid anchor point for a number of large muscles. In between each vertebrae is a shock absorbing **intervertebral disc**. The outer ring of each disc is a tough fibrocartilage while the inner tissue is a more gelatinous form of cartilage.

The Vertebrae

Except for atlas (C1) and axis (C2), most of the vertebrae share some common features. The significant features of a typical vertebra and their description are in **FIGURE 5.5**.

The Sacrum & Coccyx

The sacrum and coccyx are the last two bones of the vertebral column (**FIGURE 5.6**). The **coccyx** articulates with the posterior end of the sacrum. The coccyx originates as four separate, very small vertebrae that fuse into one bone. There are no significant features of the

FIGURE 5.5 A Typical Vertebrae and its Features.

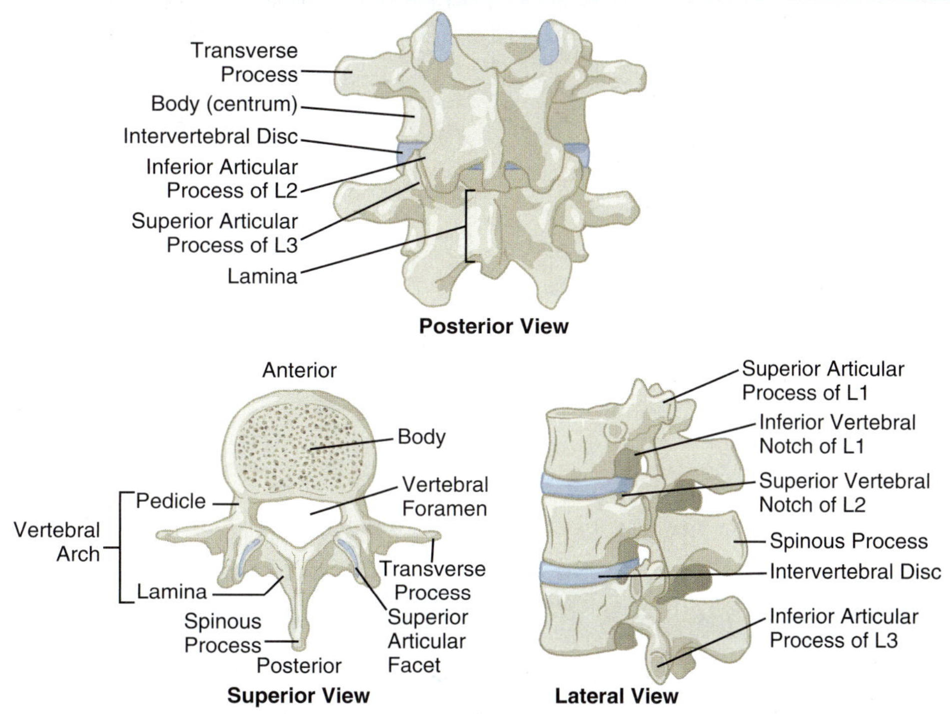

FEATURES	DESCRIPTION
Body	weight bearing, main section where vertebrae stack one upon another
Vertebral Arch	curved tissue extending posterior from body that forms vertebral foramen
Vertebral Foramen	passageway for spinal cord
Spinous Process	single projection from the posterior apex of vertebral arch
Transverse Processes	two projections from each side of the vertebral arch that extend laterally
Superior Articular Processes	two projections from each side of the arch that extend superiorly towards next vertebra
Inferior Articular Processes	two projections from each side of the arch that extend inferiorly towards next vertebra
Superior Articular Facets	articulating surface of each superior articular process
Inferior Articular Facets	articulating surface of each inferior articular process
Superior Vertebral Notches	concavity on the superior side of each vertebral arch
Inferior Vertebral Notches	concavity on the inferior side of each vertebral arch
Intervertebral Foramen	passageway formed by inferior and superior vertebral notches of two articulating vertebrae for exit of nerve from column

coccyx. The **sacrum** is triangular shaped and forms a kyphotic curve. The sacrum originates as five separate vertebrae that fuse into one bone. Some of the features of the sacrum listed below are remnants of this fact. Before fusing together, the median sacral crest was the spinous processes of the five vertebrae. Also, the lateral sacral crests were the transverse processes. This text will give the sacrum "dual membership" as part of the vertebral column and the pelvic bones. This is because it articulates with the 5th lumbar vertebra on its superior surface and the two hip bones laterally. The significant features of the sacrum and their description are in **FIGURE 5.6**.

FIGURE 5.6 The Coccyx and Sacrum with its Features.

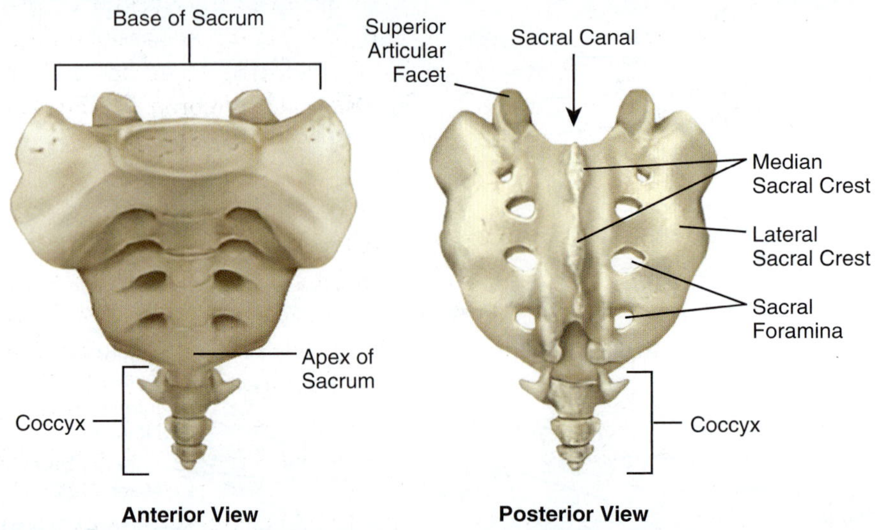

FEATURES	DESCRIPTION
Superior Articular Processes	superior projections leading to articulation with inferior articular facets of L5 vertebra
Superior Articular Facets	articulating surface of each superior articular process
Base	roundish superior slight depression that articulates with the body of the 5th lumbar vertebra
Sacral Canal	passageway for nerves
Sacral Hiatus	inferior opening of sacral canal
Sacral Foramina	anterior and posterior passageways for spinal nerves & blood vessels
Median Sacral Crest	rough ridge of bumps along the mid posterior side
Lateral Sacral Crest	less prominent ridges on either side of median sacral crest

PAUSE TO CHECK FOR UNDERSTANDING

Review all the bones of the vertebral column and features. Here are a few samples to get you started:

1. What is the name of the second most superior vertebra?
2. What is the pointed projection extending straight from the back of vertebrae?
3. What is the name of the triangular shaped bone inferior to L5?

Another suggestion is to produce some drawings that represent well and make sense to you and label them.

The Bones of the Thoracic (Rib) Cage

The thoracic cage (rib cage) is formed by a sternum and 12 pairs (24 total) of ribs (**FIGURE 5.7**). One of the purposes of this cage is to protect organs such as the heart and lungs. However, the cage is not solid; rather it is composed of several separate bones that allow for expansion and contraction. The lungs are attached to the cage and this motility of the lungs and ribs is imperative for ventilations (breathing).

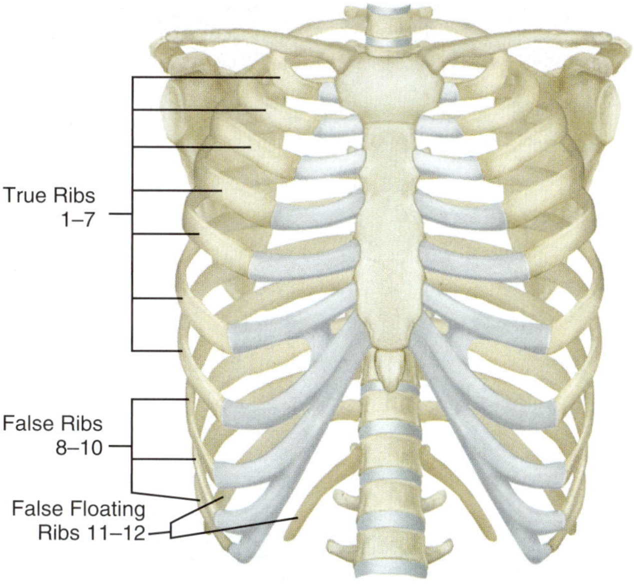

FIGURE 5.7 The Rib (Thoracic) Cage.

The Ribs

The ribs are arranged in 12 pairs (24 total bones). They are titled into groups called *true ribs*, *false ribs*, and *false floating ribs* based on their anterior "connection" (Figure 5.7). All ribs articulate posteriorly with the thoracic vertebrae. Hyaline cartilage (called costal cartilage) extends from the anterior end of all ribs. The cartilage of **true ribs** 1–7 directly connect to the sternum. The cartilage of **false ribs** 8–10 merge with the cartilage of various true ribs rather than articulating with the sternum. The cartilage of **false floating ribs** 11 and 12 does not merge with any other cartilage and without the soft tissues of the abdomen would appear to be floating. Not all ribs look alike or articulate the same way with vertebrae. Differences can be seen as well as descriptions of the significant features are in **FIGURE 5.8**. Also, ribs that share the same features are identified in parentheses next to the feature.

FIGURE 5.8 The Ribs, Their Variations, and Features.

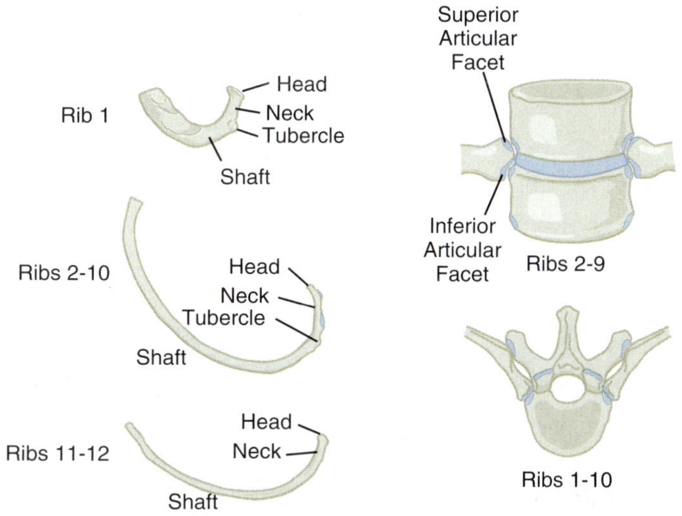

FEATURES	DESCRIPTION
Head (R1–R12)	end of rib that articulates with the vertebrae
Neck (R1–R12)	narrowed portion distal to head
Tubercle (R1–R10)	projection that articulates with transverse process of vertebra
Shaft (R1–R12)	curved, flattened part of the rib from the tubercle to the anterior end
Superior Articular Facets(R2–R9)	superior part of head that articulates with the vertebra superior to it
Inferior Articular Facets (R2–R9)	inferior part of head that articulates with the vertebra inferior to it

The Sternum

The sternum is a flat bone in the center of the anterior thorax. It rests superficially to the heart and encloses the anterior side of the thoracic cage by articulating with the ribs via their costal cartilage. It also articulates with two clavicles that extend to the shoulder. The significant features of the sternum and their description are in **FIGURE 5.9**.

PAUSE TO CHECK FOR UNDERSTANDING

Review all the bones of the thoracic cage and features. Here are a few samples to get you started:

1. Why are true, false, and false floating ribs called such as they are?
2. At the superior end of the sternum, what bones articulate with the two notches?
3. What is the inferior end of the sternum called?

Another suggestion is to produce some drawings that represent well and make sense to you and label them.

FIGURE 5.9 The Sternum and its Features.

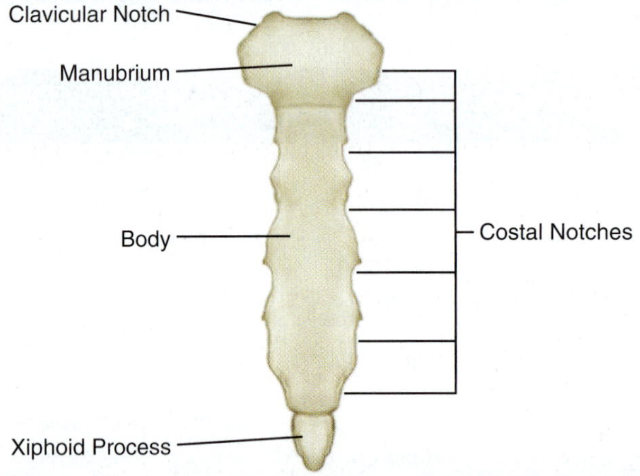

FEATURES	DESCRIPTION
Manubrium	broad superior portion shaped like the knot of a necktie
Suprasternal Notch	slight depression on the superior edge of the manubrium
Clavicular Notches	lateral depressions on the manubrium where the clavicles articulate
Body of [Sternum]	middle section of the sternum
Costal Notches	depressions along lateral edges of manubrium and body that articulate with the cartilage of true ribs 1–7
Xiphoid (xyphoid) process	arrowhead shaped inferior projection

© BraunS/E+/Getty Images

CHAPTER 6

The Bones of the Upper Extremities

CHAPTER OBJECTIVES

After completing this chapter, the student will be able to recall:

1. the names and locations of the bones in the shoulder;
2. the significant features of the shoulder bones;
3. the names and locations of the bones in the forearm;
4. the significant features of the forearm bones;
5. the names and locations of the bones in the wrist and hand; and
6. the significant features of the wrist and hand bones.

This chapter continues with the presentation and description of the bones in the upper extremities and their features significant to kinesiology. Recall that knowledge of this material prepares the student for understanding the joints, the muscle attachments, and the motions produced which appears in the final sections and chapters.

The Bones of the Shoulder

There are three bones in each shoulder region (**FIGURE 6.1**). They are the clavicle, scapula, and humerus bones. The clavicle runs from the sternum to the shoulder. The scapula rests on the posterior side of the ribs. The humerus passes from the shoulder region to the elbow.

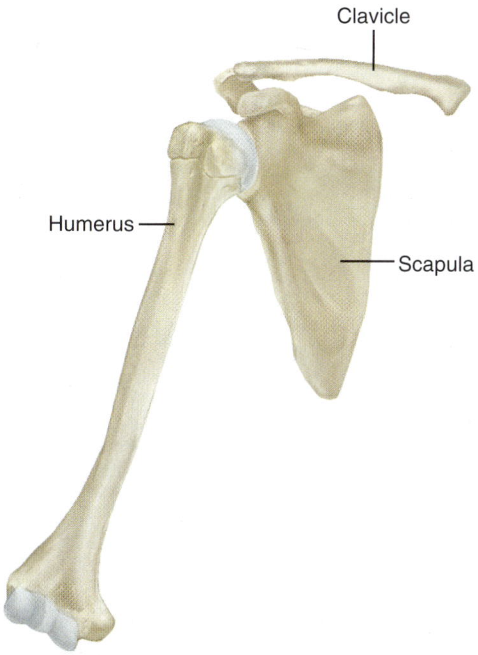

FIGURE 6.1 The Bones of the Shoulder Girdle.

The Clavicle

The **clavicle** is an elongated S-shaped bone (**FIGURE 6.2**). Its proximal end, also known as the sternal end,

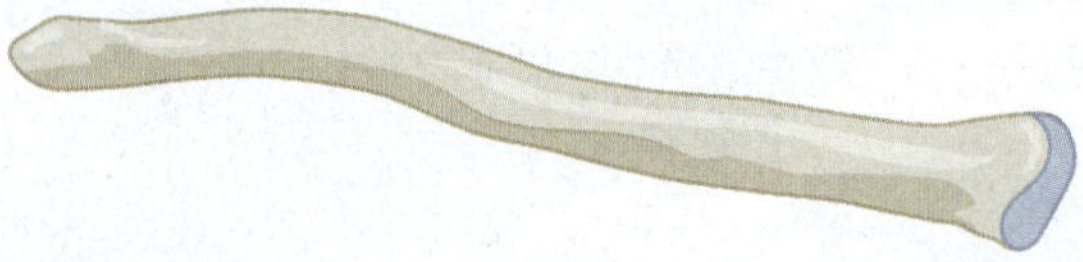

FIGURE 6.2 The Clavicle.

articulates with the sternum. Its distal end, also known as the acromial end, terminates in the superior shoulder region and articulates with the acromial process (also known as the acromion) of the scapula. There are no key features significant to anatomical kinesiology. Rather, muscles that attach on it are on sections of the diaphysis (shaft). Because of this (and other reasons) the clavicle is typically divided into three sections called the proximal one-third, middle one-third, and distal one-third.

The Scapula

The **scapula** is named for its shovel-like appearance. It is a triangular shaped flat bone that rests superficially to the posterior side of ribs 2–7. It not only articulates with the distal end of the clavicle, but also with the humerus. The scapula and its significant features are shown and described in **FIGURE 6.3**.

The Humerus

The **humerus** plays a role in the shoulder and is the only bone that resides in the arm. It is a long bone that articulates with the scapula at its proximal end and the radius and ulna at its distal end. The humerus and its significant features are shown and described in **FIGURE 6.4**.

FIGURE 6.3 The Scapula and its Features.

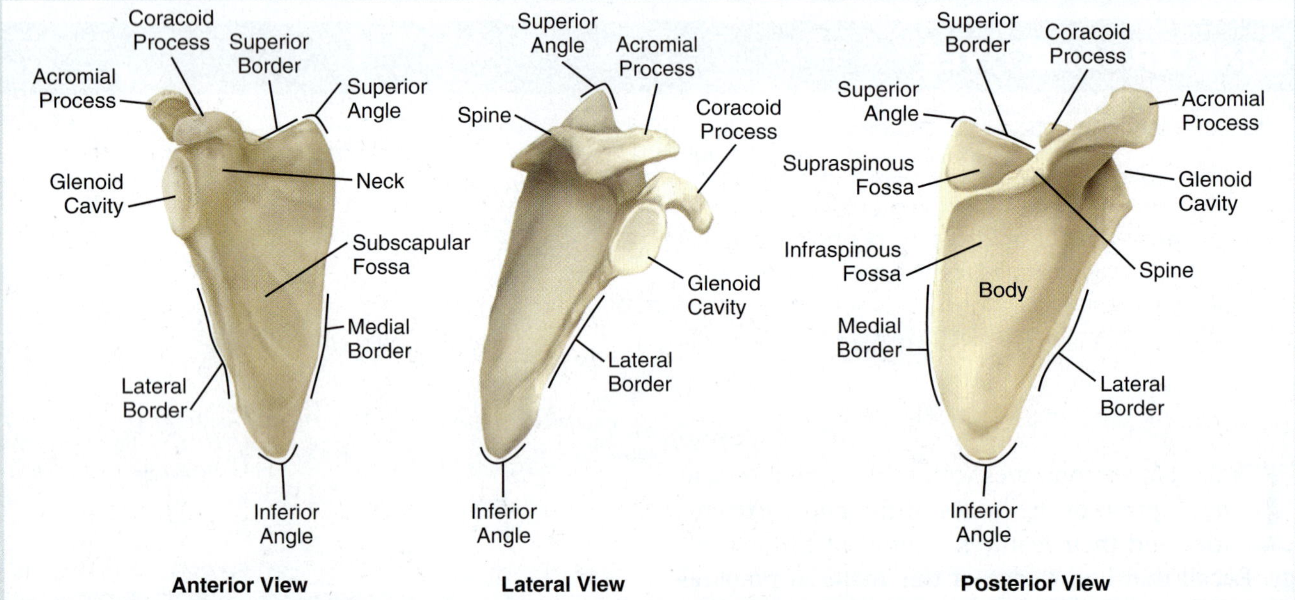

FEATURES	DESCRIPTION
Superior Border	superior, horizontal edge
Medial (Vertebral) Border	medial, vertical edge closet to the vertebral column
Lateral (Axillary) Border	lateral, oblique edge adjacent to the axilla (armpit)
Superior Angle	corner where superior and medial borders merge
Inferior Angle	corner where the medial and lateral borders merge
Lateral Angle	corner where the superior and lateral borders merge
Glenoid Cavity (Fossa)	concavity from lateral angle that articulates with humerus
Subscapular Fossa	slightly concave anterior side

Coracoid Process	extension from subscapular fossa just medial to lateral angle
Spine [of the Scapula]	transverse, oblique ridge crossing posterior side
Acromial Process (Acromion)	extension from the lateral end of the scapular spine articulating with clavicle
Supraspinous Fossa	deep depression superior to the scapular spine
Infraspinous Fossa	relatively shallow concavity inferior to the spine

FIGURE 6.4 The Humerus and its Features.

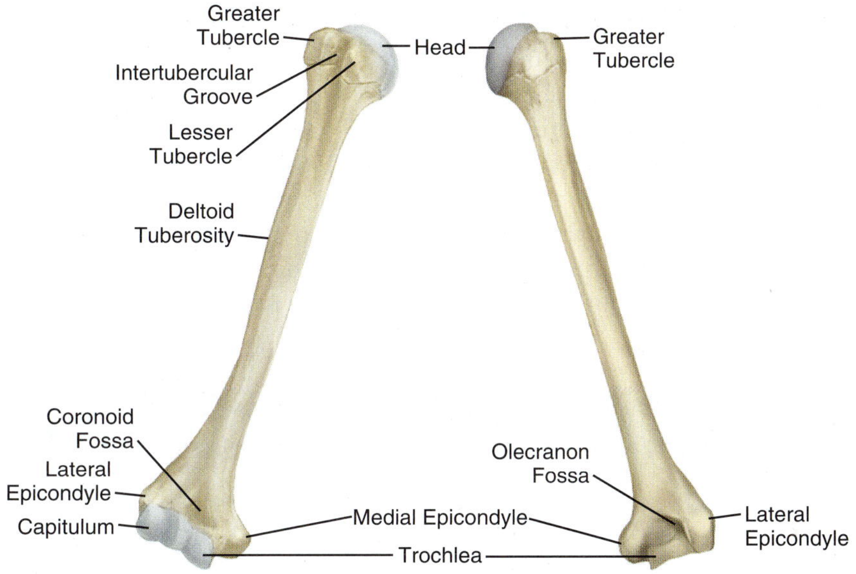

FEATURES	DESCRIPTION
Head	hemispherical proximal end that articulates with the glenoid cavity of the scapula
Greater Tubercle	larger, more lateral "bump" close to head
Lesser Tubercle	smaller, more medial "bump" close to head
Intertubercular Groove (Sulcus)	depression in between the greater & lesser tubercles
Deltoid Tuberosity	relatively small bump on lateral diaphysis where deltoid muscle attaches
Capitulum	lateral condyle at the distal end that articulates with the radius
Trochlea	medial condyle at the distal end that articulates with the ulna
Epicondyles	relatively small bumps projecting from superior side of the condyles
Olecranon Fossa	posterior, distal depression between the epicondyles and superior to the trochlea
Radial Fossa	anterior, distal depression just superior to capitulum and lateral to coronoid fossa
Coronoid Fossa	anterior, distal depression just superior to trochlea and medial to radial fossa

PAUSE TO CHECK FOR UNDERSTANDING

Review all the bones and features of the forearm. Here are a few samples to get you started:

1. Which bone is more lateral than the other?
2. What is the posterior bump of the elbow called?
3. What feature corresponds to the bump at the wrist?

Another suggestion is to produce some drawings and label them. They don't have to be pretty, but should represent well and make sense to you.

The Bones of the Forearm

There are two bones in the forearm (**FIGURE 6.5**). In anatomical position, the *ulna* resides on the medial side of the forearm while the *radius* is on the lateral side from the elbow to the wrist. When the forearm is rotated so the palm is facing posteriorly (pronation), the proximal radius (head) rotates around the ulna and the bones take on a long, slender "x" appearance. At the distal ends of the bones during this motion, the ulna also rotates around the radius, but to a much lesser degree than the proximal radius.

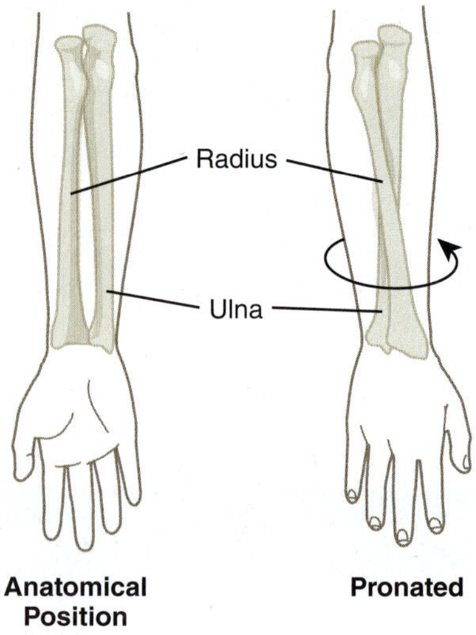

FIGURE 6.5 The Ulna and Radius.

The Ulna

The **ulna** is a long, relatively slender bone narrowing from its proximal end to the smaller distal epiphysis at the wrist (**FIGURE 6.6**). From a lateral view, it is easy to see the C-shape of the proximal epiphysis. This dramatic anatomical feature enhances the stability of the elbow.

FIGURE 6.6 The Ulna and its Features.

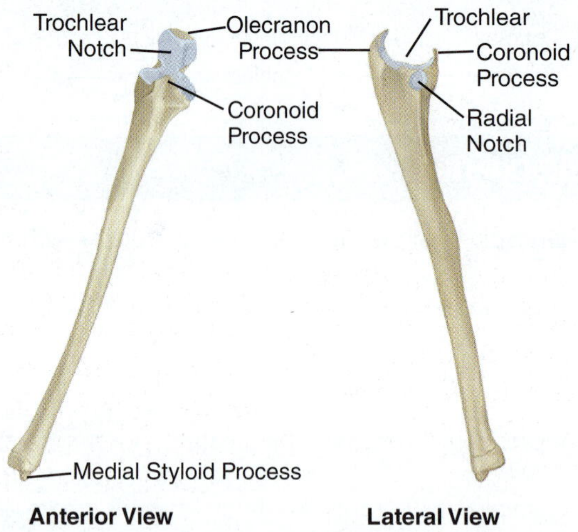

FEATURES	DESCRIPTION
Trochlear Notch	proximal c-shaped concavity that articulates with the trochlea of the humerus
Olecranon Process	proximal projection on posterior side; prominent "point" of posterior elbow
Coronoid Process	proximal projection on anterior side
Radial Notch	groove where radial head rotates
Ulnar (Medial) Styloid Process	distal projection on medial side (medial "wrist bump")

FIGURE 6.7 The Radius and its Features.

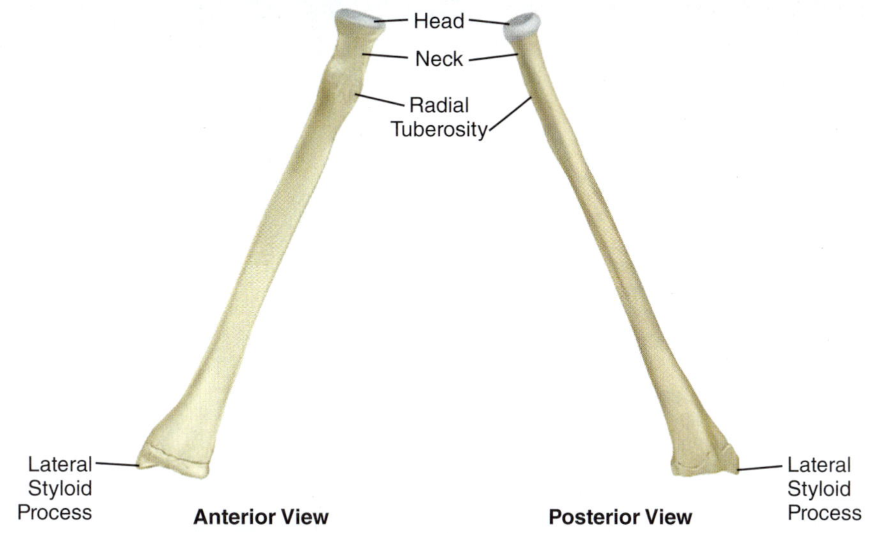

FEATURES	DESCRIPTION
Head	proximal end that articulates with the capitulum of the humerus and also the radial notch of the ulna
Radial Tuberosity	extension inferior to radial head
Neck	region between the head and the radial tuberosity
Radial (Lateral) Styloid Process	distal projection on lateral side (lateral "wrist bump")

The Radius

Similar to the ulna, the **radius** is also a long, slender bone (**FIGURE 6.7**). In contrast to the ulna, the distal epiphysis of the radius is larger than the proximal end. At its proximal end, it articulates with both the ulna and humerus. Distally, it comprises most of the wrist intersecting with the proximal carpal bones.

PAUSE TO CHECK FOR UNDERSTANDING

Review all the bones and features of the shoulder. Here are a few samples to get you started:

1. Which bone runs across the anterior chest?
2. What feature of the humerus articulates with the scapula?
3. What is the most inferior tip of the scapula called?

Another suggestion is to produce some drawings and label them. They don't have to be pretty, but should represent well and make sense to you.

The Bones of the Wrist and Hand

The wrist is the region between the forearm and hand. The hand is the region beyond the wrist and includes the fingers. In each wrist are a group of bones called the carpal bones. The bones in the palm region of the hand are called metacarpals. The bones in the fingers are called phalanges.

The Carpals

There are eight **carpal bones** in each wrist (**FIGURE 6.8**). The carpals are arranged in two rows of four. The proximal row from medial to lateral are the pisiform, triquetrum (tri-QUEE-trum), lunate, and scaphoid. The distal row from medial to lateral includes the hamate, capitate, trapezoid, and trapezium. There are no key features of the carpals significant to anatomical kinesiology.

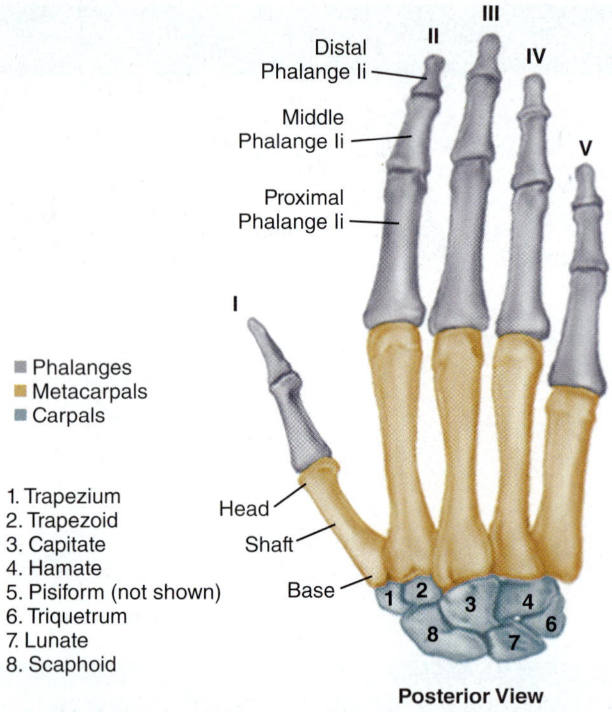

FIGURE 6.8 The Bones of the Wrist and Hand.

The Metacarpals

There are five metacarpal bones in each hand (palm) (Figure 6.8). The metacarpals articulate with the carpals at their proximal ends and the phalanges at their distal ends. The metacarpals are distinguished by a numbering system. The numbers are commonly expressed either by words, numbers, or roman numbers (i.e., first metatarsal, 1st metatarsal, or metatarsal I). The most lateral that leads to the thumb is the **first metacarpal** and then sequentially from there are the **second metacarpal**, **third metacarpal**, **fourth metacarpal**, and **fifth metacarpal**. There are just two features significant to the focus of this text which are listed below.

Base	proximal end of a metacarpal
Head	distal end of a metacarpal

The Finger Phalanges

There are fourteen (14) **phalanges** in all the fingers of each hand (Figure 6.8). Two of them are in the first finger (thumb). The other four fingers contain three phalanges per digit. All of the phalanges that are adjacent to a metacarpal are called a **proximal phalange**. The most distal is called a **distal phalange**. In the 2nd–5th fingers, the bone in between the proximal and distal phalange is called the **middle (or intermediate) phalange**. All the names of the phalanges can be written the same three ways mentioned with the metacarpals (i.e., first proximal phalange, 1st proximal phalange, or proximal phalange I). Therefore, the list of these bones may be summarized as the first proximal phalange, the **first distal phalange**, the second–fifth proximal phalanges, the second–fifth middle phalanges, and the second–fifth distal phalanges. Just like the metacarpals, **base** and **head** indicate the proximal and distal ends, respectively, of the phalanges.

PAUSE TO CHECK FOR UNDERSTANDING

Review all the bones and features of the wrist and hand. Here are a few samples to get you started:

1. How many rows of carpal bones are there?
2. What is the collective name of the bones in the hand (palm, not fingers)?
3. What is the proximal end of phalange called?

Another suggestion is to produce some drawings and label them. They don't have to be pretty, but should represent well and make sense to you.

CHAPTER 7

The Bones of the Lower Extremities

CHAPTER OBJECTIVES

After completing this chapter, the student will be able to recall:

1. the names and locations of the bones in the pelvis;
2. the significant features of the pelvis bones;
3. the names and locations of the bones in the thigh;
4. the significant features of the thigh bones;
5. the names and locations of the bones in the lower leg;
6. the significant features of the lower leg bones;
7. the names and locations of the bones in the foot; and
8. the significant features of the foot bones.

Like the previous chapter, this chapter continues with the presentation and description of the bones in the lower extremities and their features significant to kinesiology. Recall that knowledge of this material prepares the student for understanding the joints, the muscle attachments, and the motions produced which appears in the final sections and chapters.

The Bones of the Pelvis

The pelvis, also known as the pelvic girdle, is a compilation of a few bones that form a somewhat odd bowl shape (**FIGURE 7.1**). Two coxal (hip) bones and a sacrum comprise the pelvic bones. The two coxal bones curve inward to enclose the anterior and lateral portions of the "bowl." The sacrum resides in between the two coxal bones enclosing the posterior side. This text gives the sacrum "dual membership"

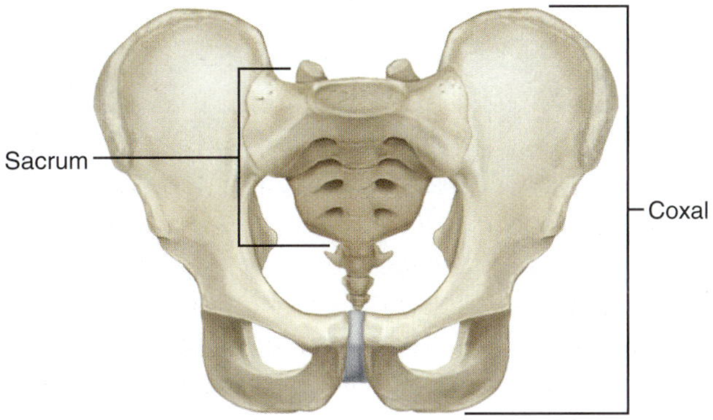

FIGURE 7.1 The Bones of the Pelvis.

here in the pelvis and as part of the vertebral column. However, since the sacrum was already discussed in Chapter 5, it will not be repeated in this chapter. Furthermore, this text will not include the coccyx as part of the pelvis even though it articulates with the sacrum.

The Coxal Bone

The **coxal** (hip) **bone** is formally named the os coxa (singular; os coxae - plural). The coxal bone is an irregular bone that embryonically begins as three bones. By birth, these bones have fused together into one coxal bone. The three main parts (formally bones) of the coxal bone are the *ilium*, *ischium*, and *pubis*. All three extend out from the acetabulum ("socket"). The **ilium** is the superior part, the ischium is the posterior part, and the **pubis** is the anterior part. The coxal bone, its parts, and significant features are shown and described in **FIGURE 7.2**.

PAUSE TO CHECK FOR UNDERSTANDING

Review all the bones and features of the hip. Here are a few samples to get you started:

1. What are the three main parts of the coxal bone?
2. What is the superior edge of the hip called?
3. Where are the ischial tuberosities?

Another suggestion is to produce some drawings and label them. They don't have to be pretty, but should represent well and make sense to you.

FIGURE 7.2 The Coxal Bone & Its Features.

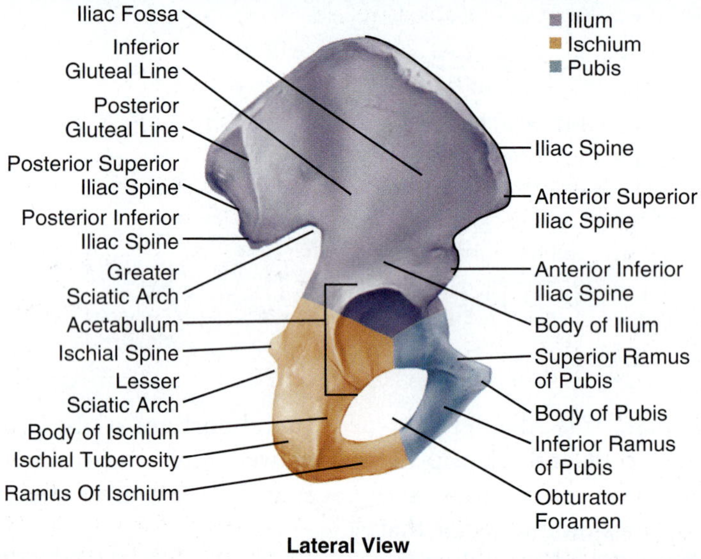

PUBIS

FEATURES	DESCRIPTION
Ramus [of the Pubis]	tissue that extends from the acetabulum that forms the anterior portion of the obturator foramen
Superior Ramus	superior section of the ramus
Inferior Ramus [of the Pubis]	inferior section of the ramus
Body [of the Pubis]	angle of the ramus; its medial surface articulates with the corresponding surface of the contralateral pubis

ILIUM	
FEATURES	**DESCRIPTION**
Iliac Crest (Iliac Spine)	superior ridge of the ilium
Anterior Superior Iliac Spine	rounded anterior end of the iliac spine
Anterior Inferior Iliac Spine	projection on anterior ilium inferior to ASIS
Posterior Superior Iliac Spine	rough, relatively sharp posterior end of the iliac spine
Posterior Inferior Iliac Spine	projection on posterior ilium inferior to PSIS
Iliac Fossa	shallow depression on medial ilium inferior to iliac spine (crest)
Greater Sciatic Notch	concavity in bone inferior to the PIIS where the sciatic nerve passes
ISCHIUM	
FEATURES	**DESCRIPTION**
[Ischial] Body	tissue that extends from the acetabulum that forms part of the posterior portion of the obturator foramen
[Ischial] Ramus	continuation of tissue from the ischial body that forms the rest of the posterior portion of the obturator foramen
[Ischial] Spine	relatively sharp posterior projection from the ischial body
Lesser Sciatic Notch	concavity in the bone tissue just inferior to the ischial spine
[Ischial] Tuberosity	rough projection inferior to the lesser sciatic notch that supports the body in a seated position

The Bones of the Thigh

There are two bones in the thigh. They are the *femur* and the *patella*. The femur articulates with the coxal bone at its proximal end and the tibia at its distal end. The patella (knee cap) sits anteriorly to the distal femur.

The Femur

The **femur** is a long bone and is the strongest and longest of the body. Its length is about 25% of total body height. The femur and its significant features are shown and described in **FIGURE 7.3**.

The Patella

The **patella**, commonly known as the kneecap, is a sesamoid bone. A sesamoid bone is a bone that develops within a tendon. The patella is anterior to the femur at its distal end. It is a triangular shaped bone with a wider proximal end called the **base** and narrower distal end called the **apex**. The femur and its significant features are shown and described in **FIGURE 7.4**.

PAUSE TO CHECK FOR UNDERSTANDING

Review all the bones and features of the thigh. Here are a few samples to get you started:

1. Where is the linea aspera?
2. Is the apex of the patella proximal or distal?
3. Are epicondyles categorized as articulations?

Another suggestion is to produce some drawings and label them. They don't have to be pretty, but should represent well and make sense to you.

FIGURE 7.3 The Femur.

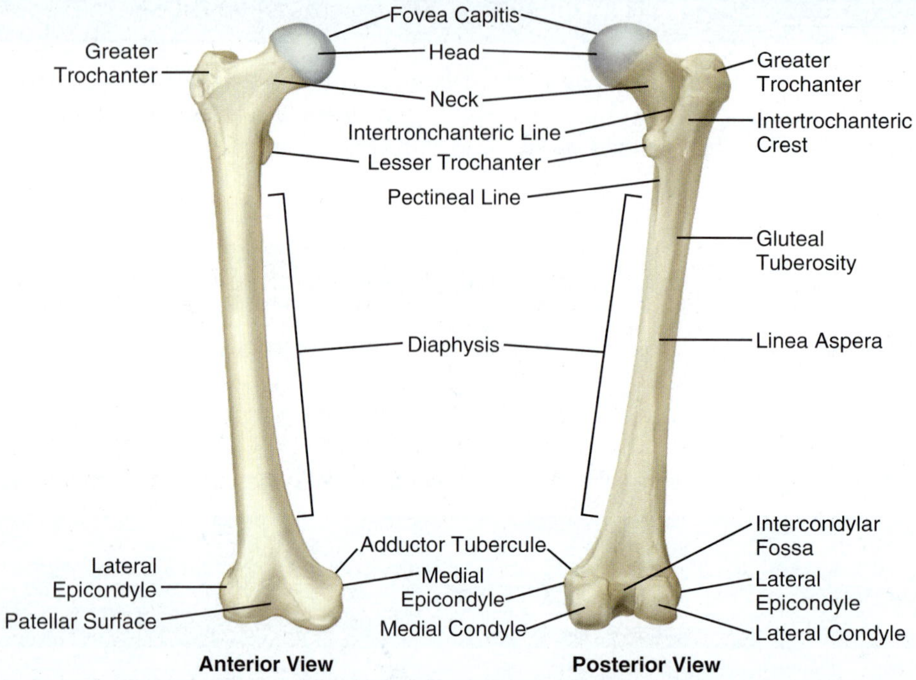

FEATURES	DESCRIPTION
Head	hemispherical, proximal end that articulates with the acetabulum of coxal bone
Fovea Capitis	pit in the head that receives a ligament from the acetabulum
Neck	relatively constricted region of tissue between the head and diaphysis
Greater Trochanter	superior, prominent bump adjacent to the neck
Lesser Trochanter	inferior, less prominent bump than greater trochanter adjacent to the neck
Intertrochanteric Line	subtle, anterior ridge between trochanters
Linea Aspera	longitudinal ridge on mid posterior diaphysis
Spiral (Pectineal) Line	medial fork at proximal end of linea aspera
Gluteal Tuberosity	lateral fork at proximal end of linea aspera
Medial Supracondylar Line	medial fork at distal end of linea aspera
Lateral Supracondylar Line	lateral fork at distal end of linea aspera
Adductor Tubercle	small projection just proximal to the medial epicondyle
Condyles	medial & lateral, rounded distal ends that articulate with the proximal tibia
Epicondyles	medial & lateral, relatively small projections from superior side of condyles
Intercondylar Fossa	inferior and posterior groove between the condyles
Patellar Surface	anterior region between condyles that articulates with patella

FIGURE 7.4 The Patella.

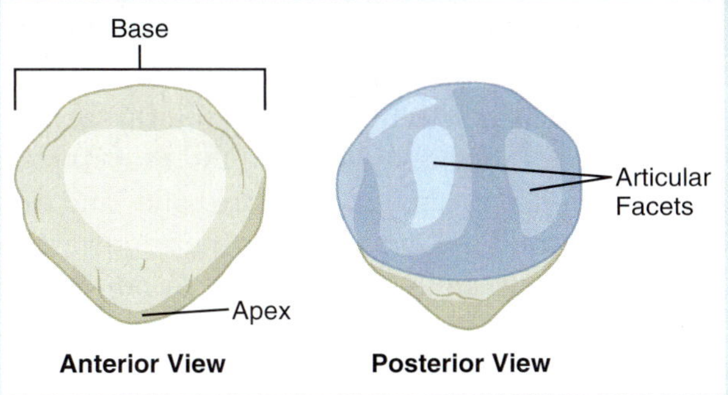

FEATURES	DESCRIPTION
Base	broad proximal end; also known as the superior pole
Apex	distal, blunted point end; also known as the inferior pole
Articular Facets	regions on posterior side that articulate with the patellar surface of the femur

The Bones of the Lower Leg

The lower leg, also known as the crural region, is from the knee to the ankle. There are only two bones in this region. They are the tibia and the fibula (**FIGURE 7.5**). The **tibia**, sometimes called the shin, is the larger, weight bearing. Lateral from the tibia is the fibula. It is much more slender and does not bear any significant weight of the body.

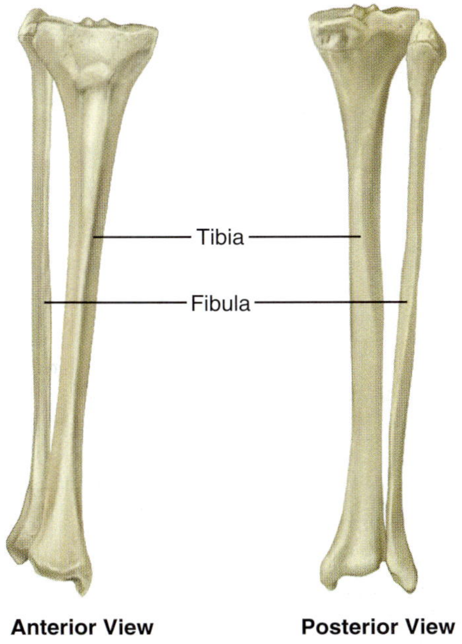

FIGURE 7.5 The Tibia and Fibula.

The Tibia

The **tibia** is a long bone articulating with the femur at its proximal end to form part of the knee. At its distal end, it forms part of the ankle as it articulates with the talus bone of the foot. The tibia and its significant features are shown and described in **FIGURE 7.6**.

FIGURE 7.6 The Tibia.

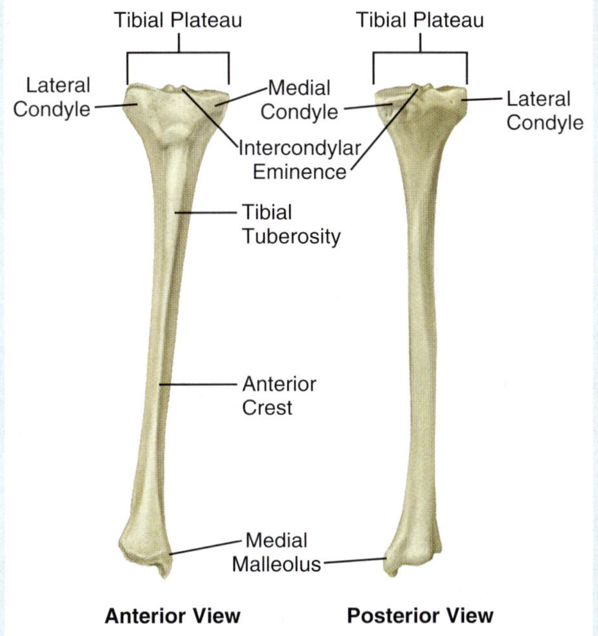

FEATURES	DESCRIPTION
Tibial Plateau	superior plain of the proximal epiphysis
Condyles	medial & lateral shallow concavities that articulate with the femoral condyles
Intercondylar Eminence	projection of tissue between the tibial condyles
Tibial Tuberosity	rough "knot" on the mid anterior bone just inferior to the proximal epiphysis
Tibial (Anterior) Crest	longitudinal ridge on the anterior shaft
Medial Malleolus	"knobby" extension of the distal epiphysis ("medial ankle bump")

FIGURE 7.7 The Fibula.

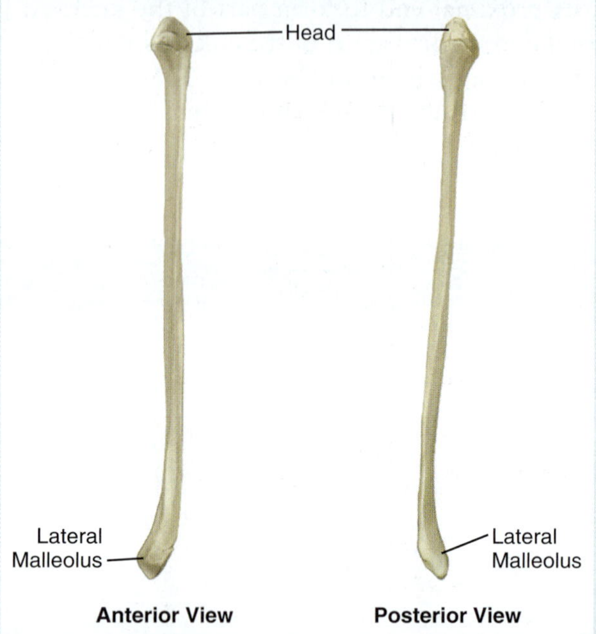

FEATURE	DESCRIPTION
[Fibular] Head	proximal epiphysis that articulates with the tibia
Lateral Malleolus	"knobby" extension of the distal epiphysis ("lateral ankle bump")

The Fibula

The **fibula** is a long bone lateral to the tibia in the crural region. It articulates with the tibia at its proximal and distal ends. At its distal end, it also articulates with the talus bone forming a significant part of the ankle. The fibula and its significant features are shown and described in **FIGURE 7.7**.

PAUSE TO CHECK FOR UNDERSTANDING

Review all the bones and features of the lower leg. Here are a few samples to get you started:

1. What is the bump below the anterior knee called?
2. What is the proximal end of the fibula called?
3. What are the ankle bumps called?

Another suggestion is to produce some drawings and label them. They don't have to be pretty, but should represent well and make sense to you.

The Bones of the Foot

The foot can be divided into three regions; the hind foot, the mid foot, and the forefoot (**FIGURE 7.8**). The hind foot is the proximal end of the foot and contains a group of bones collectively called the **tarsals**. The group of bones in the midfoot are the **metatarsals**. The prefix "meta" means "after" or "beyond" and these bones are beyond the tarsals. The **phalanges** are the name of the bones in the toes (digits). As noted in the previous chapter, the same arrangement of toe bones is found in the fingers.

The Tarsals

There are seven tarsal bones in each foot. The **tarsals** include the talus, the calcaneus, the navicular, the cuboid, and the cuneiforms (**FIGURE 7.9**). The **talus** lies inferior to the tibia and fibula. The **calcaneus** (commonly called heel) is inferior to the talus. The **navicular** bone lies just anterior (distal) to the talus. The **cuboid** is lateral to the navicular and anterior to the calcaneus. The cuneiforms are anterior to the navicular. Their names describe their location in relation to the foot and are the **medial cuneiform**, **middle** (or intermediate) **cuneiform**, and **lateral cuneiform**. The tarsals and there significant features are shown and described in **FIGURE 7.7**.

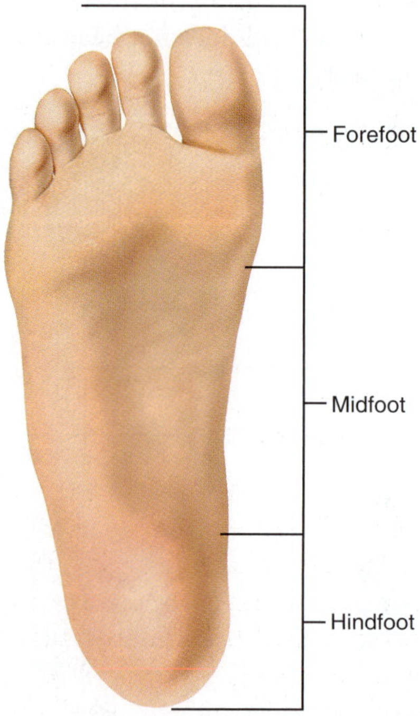

FIGURE 7.8 The Regions of the Foot.

FIGURE 7.9 The Foot Bones.

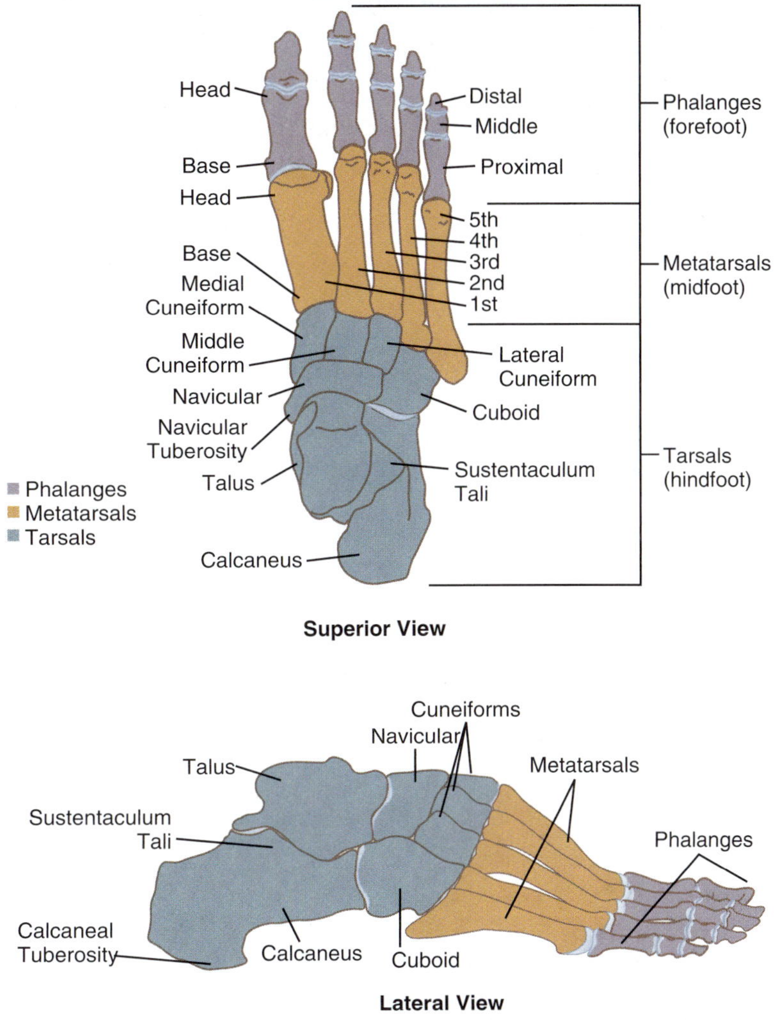

FEATURE	DESCRIPTION
Calcaneal Tuberosity	inferior projection of the posterior calcaneus
Sustentaculum Tali	projection from medial side of calcaneus that supports talus bone
Navicular Tuberosity	projection on medial side of bone

The Metatarsals

There are five **metatarsals** (**FIGURE 7.9**). They articulate with the tarsals at their proximal ends and the phalanges at their distal ends. The metatarsals are distinguished by a numbering system just like the metacarpals in the hand. The numbers are commonly expressed either by words, numbers, or roman numbers (i.e., first metatarsal, 1st metatarsal, or metatarsal I). The most medial is the **first metatarsal** and sequentially, from medial to lateral, are the **second metatarsal**, **third metatarsal**, **fourth metatarsal**, and **fifth metatarsal**. There are only two features significant to the focus of this text which are listed below.

Base	proximal end of a metatarsal
Head	distal end of a metatarsal

The Toe Phalanges

There are fourteen (14) **phalanges** in the toes of each foot (**FIGURE 7.9**). The first toe (also called the great toe) contains two phalanges. The remaining four toes have three phalanges in each digit. All of the phalanges that are adjacent to a metatarsal are called a **proximal phalange**. The most distal is called a **distal phalange**. In the 2nd–5th toes, the bone in between the proximal and distal phalange is called the **middle** (or intermediate) **phalange**. All the names of the phalanges can be written the same three ways noted with the metatarsals (i.e., first proximal phalange, 1st proximal phalange, or proximal phalange I). Therefore, the list of these bones may be summarized as the first proximal phalange, the **first distal phalange**, the second–fifth proximal phalanges, the second–fifth middle (or intermediate) phalanges, and the second–fifth distal phalanges. Just like the metatarsals, **base** and **head** indicate the proximal and distal ends, respectively, of the phalanges.

PAUSE TO CHECK FOR UNDERSTANDING

Review all the bones and features of the foot. Here are a few samples to get you started:

1. Which bone is also known as the heel?
2. What is the collective name of the long bones running from the distal tarsals to the phalanges?
3. How many phalanges are in the first digit?

Another suggestion is to produce some drawings and label them. They don't have to be pretty, but should represent well and make sense to you.

© BraunS/E+/Getty Images

SECTION 3

The Lower Extremities

CHAPTER 8

The Foot

CHAPTER OBJECTIVES

After completing this chapter, the student will be able to:

1. summarize the joints of the foot region;
2. comprehend the motions of the toes;
3. identify the names and locations of the foot muscles;
4. recall the attachment sites and nerves of the foot muscles; and
5. apply the motions produced by the actions of the foot muscles.

The foot is the region of the body distal to the ankle (**FIGURE 8.1**). The superior side of the foot is called the dorsum and the inferior side is referred to as the plantar side. There are certainly joints between all the bones of the foot, but only the toes move significantly. The arrangement of the toe bones and the actions of the toe muscles are able to produce four motions.

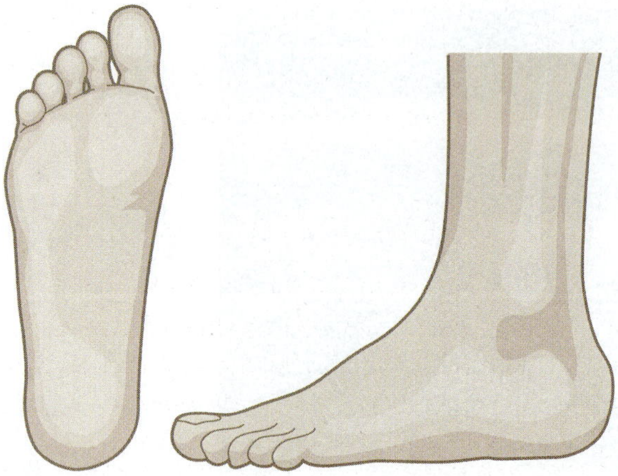

FIGURE 8.1 The Foot.

The Joints of the Foot

There are four groups of joints in the foot. They are the *tarsal* (or *intertarsal*) *joints*, the *tarsometatarsal joints*, the *metatarsophalangeal joints*, and the *interphalangeal joints* (**FIGURE 8.2**). The **intertarsal joints** are the collection of articulations between all the various tarsal bones.

The articulations between the five metatarsals and some of the tarsal bones are called the **tarsometatarsal (TM) joints** (**FIGURE 8.2**). The **first TM joint (TM joint I)** is between the first metatarsal (metatarsal I) and the medial cuneiform. The **second TM joint** is of second metatarsal and intermediate (or middle) cuneiform and the third TM joint is of the third metatarsal and the lateral cuneiform. The fourth and fifth metatarsals articulate with the cuboid bone forming the **fourth** and **fifth TM joints**, respectively.

The articulations between the metatarsals and the proximal phalanges are called the **metatarsophalangeal (MP) joints** (**FIGURE 8.2**). The MP joints are sequentially numbered from the most medial (first MP joint) to the most lateral joint (fifth MP joint).

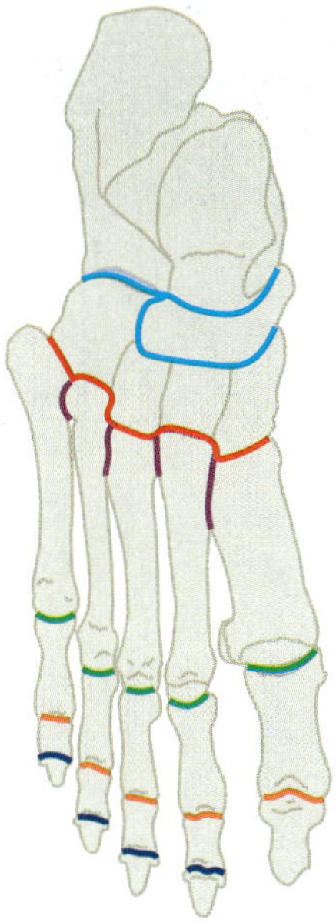

FIGURE 8.2 The Groups of Joints of the Foot.

The joints between the phalanges are called **interphalangeal (IP) joints** (**FIGURE 8.2**). All of the IP joints are also sequentially numbered from the medial side to the lateral side of the foot/toes. The hallucis (great toe) only contains two phalanges and therefore only has one IP joint called the **first IP joint**. The rest of the toes (II–V) contain three phalanges and so have two IP joints in each of them. The articulations between the proximal and intermediate (middle) phalanges are called the **proximal interphalangeal (PIP) joints**. The **distal interphalangeal (DIP) joints** are the articulations between the intermediate (middle) phalanges and the distal phalanges.

The Motions of the Foot

There are no significant motions of the tarsal joints and TM joints. The MP joints are capable of four motions. They are extension, flexion, abduction, and adduction (**FIGURE 8.3**). The motions of the IP joints are restricted

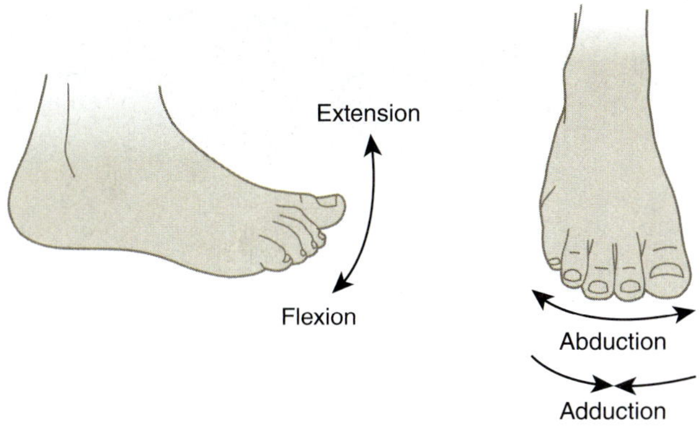

FIGURE 8.3 Toe Motions.

to extension and flexion. **Toe extension** and **flexion** straighten and bend the toes in the sagittal plane about the frontal axis. **Toe abduction** and **adduction** splay the toes in the frontal plane about the sagittal axis.

TABLE 8.1 summarizes the average ranges of motions of the toe joints. The MP, PIP, and DIP joints move together during **toe extension** and **flexion**. Average MP joint I extension is 75–85 degrees and flexion is 35–45 degrees. The degrees of motions for each distal joint are progressively less. Also, the degrees of motions of each toe from second–fifth are progressively less. There are no established standards for average toe abduction and adduction ranges of motions.

PAUSE TO CHECK FOR UNDERSTANDING

1. What bones are involved in an MP joint?
2. Is the first digit the big or little toe?
3. What motions are the toes capable of?

The Muscles that Move the Toes

One way that the toe muscles are described is by their general location. Toe muscles are commonly categorized as either *intrinsic* or *extrinsic*. **Intrinsic muscles**

TABLE 8.1 Average Toe Joint Ranges of Motions

Joint	Motion	Degrees of Motion
Metatarsophalangeal I	Extension Flexion Combined Extension-Flexion	75–85° 35–45° 110–130°
Metatarsophalangeal II–V	Extension Flexion Combined Extension-Flexion	decreases with each subsequent joint from MP I to MP V

TABLE 8.2 The Toe Muscles

Muscle	Location	Nerve
Toe Flexors		
Flexor Hallucis Longus	Extrinsic	Tibial
Flexor Digitorum Longus	Extrinsic	
Flexor Hallucis Brevis	Intrinsic	
Flexor Digitorum Brevis	Intrinsic	
Flexor Digiti Minimi	Intrinsic	
Lumbricals	Intrinsic	
Quadratus Plantae	Intrinsic	
Toe Extensors		
Extensor Hallucis Longus	Extrinsic	Deep Fibular
Extensor Digitorum Longus	Extrinsic	
Extensor Hallucis Brevis	Intrinsic	
Extensor Digitorum Brevis	Intrinsic	
Toe Abductors		
Abductor Hallucis Dorsal Interossei Abductor Digiti Minimi	Intrinsic	Tibial
Toe Adductors		
Adductor Hallucis Plantar Interossei	Intrinsic	Tibial

are muscles whose origin, belly, and insertion are located within the region of the joint(s) it acts upon. **Extrinsic muscles** are distinguished because their origin and belly are outside the region of the joint(s) it acts upon. The origins and bellies of the extrinsic toe muscles reside in the lower leg (crural region) whereas the intrinsic toe muscles completely reside within the foot. **TABLE 8.2** summarizes the toe muscles by dividing them into intrinsic and extrinsic categories as well as indicating the nerves that innervate them.

The intrinsic toe muscles are also commonly divided into layers of the foot (**TABLE 8.3**). One set of the intrinsic muscles reside within only one layer in the dorsal foot. The plantar (also called ventral) aspect of the foot contains four layers of intrinsic muscles with the most superficial being layer one and so on.

The rest of this chapter is dedicated to learning the toe muscles; their attachment sites; the *primary*, main motions they produce; and the nerves that innervate them. Each of the following sections is a muscle group based on the simple motion produced by their *concentric* action (i.e., the toe flexors flex the toes when they concentrically act.). However, it should not be forgotten that all muscles also can and will act and *eccentrically* and *isometrically*.

TABLE 8. 3 The Layers of Intrinsic Toe Muscles

Dorsal Layer	
Extensor Hallucis Brevis Extensor Digitorum Brevis	
Ventral Layer 1	**Ventral Layer 3**
Flexor Digitorum Brevis Abductor Hallucis Abductor Digiti Minimi	Flexor Hallucis Brevis Flexor Digiti Minimi Adductor Hallucis
Ventral Layer 2	**Ventral Layer 4**
Lumbricals Quadratus Plantae	Dorsal Interossei Plantar Interossei

PAUSE TO CHECK FOR UNDERSTANDING

1. What is the difference between an intrinsic and extrinsic toe muscle?
2. Create a three column chart. (title the first column "Toe Muscles," the second "Concentric" and the third "Eccentric")
   a. In the first column, list all the toe muscles.
   b. From memory, write what motion occurs during a concentric action in the second column. Double check your correct answers for accuracy.
   c. In the third column, record what motion occurs when the muscles act eccentrically.

Since there are four toe motions, there are four groups. The groups are the *toe flexors*, *toe extensors*, *toe abductors*, and *toe adductors*. Furthermore, each group of toe muscles will be summarized at the end of each section including the specific joints they act upon, their other motions, and the nerves that innervate them. It should be noted that each of these summary tables includes a column for "Other Motions." *Other Motions* means that this muscle also plays some lesser, *secondary* role in some other motion(s).

The Toe Flexors

There are ten muscles that flex the toes. Six of them are the *flexor hallucis brevis*, *flexor hallucis longus*, *flexor digitorum brevis*, *flexor digitorum longus*, *flexor digiti minimi*, *quadratus plantae*. The other four are commonly called the *lumbricals*. The toe flexors are summarized at the end of this section in **TABLE 8.4**.

The **flexor hallucis** (ha-LOO-sis) **brevis** lies in the plantar foot (**FIGURE 8.4**). It originates from the plantar side of the cuboid and lateral cuneiform and inserts on the plantar side of the base of the first proximal phalange.

The **flexor hallucis** (ha-LOO-sis) **longus** resides in the posterior lower leg (**FIGURE 8.5**). It originates from the posterior side of the distal two-thirds of the fibular diaphysis. Its tendon wraps behind the medial

FIGURE 8.4 The Flexor Hallucis Brevis Muscle.

ORIGINS
Cuboid
Lateral Cuneiform
INSERTION
base of proximal phalange I
NERVE
Tibial

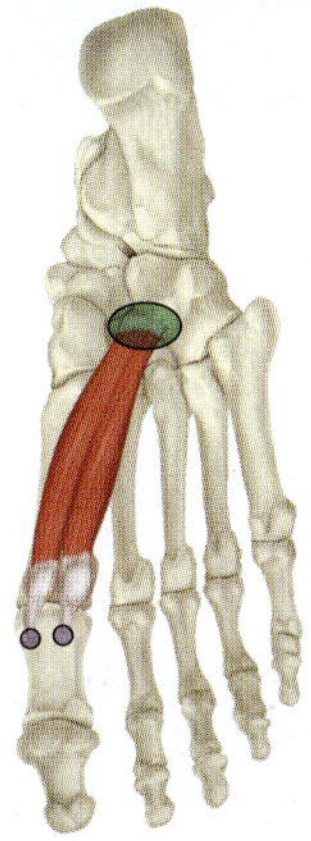

FIGURE 8.5 The Flexor Hallucis Longus Muscle.

ORIGIN
distal 2/3 of posterior fibula
INSERTION
plantar side of distal phalange I
NERVE
Tibial

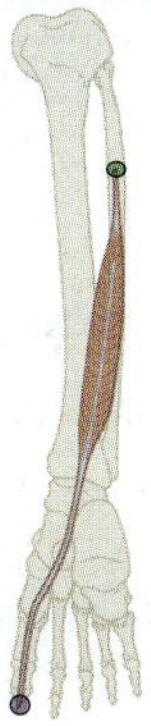

malleolus before passing under the first metatarsal and eventually inserting on the plantar side of the base of the first distal phalange.

The **flexor digitorum** (DIDJ-ih-TOE-rum) **brevis** lies longitudinally in the middle of the plantar side of the foot (**FIGURE 8.6**). It originates from the tuberosity of the calcaneus and branches into four bellies. The four tendons insert on the plantar side of the bases of the second through fifth intermediate (middle) phalanges.

The **flexor digitorum** (DIDJ-ih-TOE-rum) **longus** resides in the posterior crural region longitudinally over the tibia (**FIGURE 8.7**). It originates from the posterior side of the middle one-third of the tibial diaphysis. Its tendon wraps around the medial malleolus, splits into four branches, and inserts on the plantar side of the bases of the second through fifth distal phalanges.

The **flexor digiti** (DIDJ-ih-TIE) **minimi** (MEN-nah-my) lies on the lateral side of the plantar foot (**FIGURE 8.8**). It originates from the plantar side of the

FIGURE 8.6 The Flexor Digitorum Brevis Muscle.

ORIGIN	
Calcaneal Tuberosity	
INSERTIONS	
bases of middle phalanges II–V	
NERVE	
Tibial	

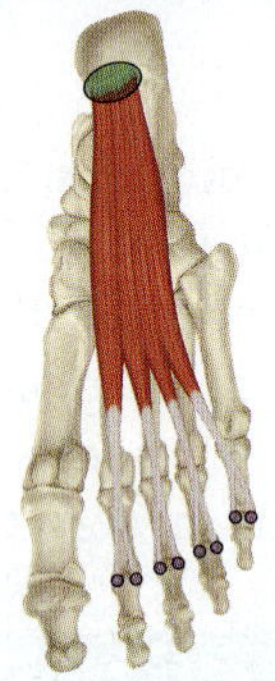

FIGURE 8.7 The Flexor Digitorum Longus Muscle.

ORIGIN	
middle 1/3 of posterior tibia	
INSERTIONS	
bases of distal phalanges II–V	
NERVE	
Tibial	

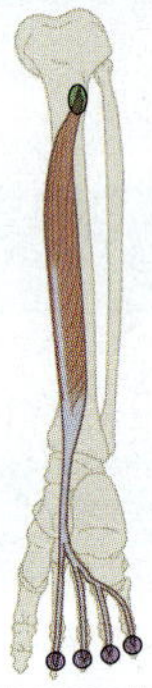

FIGURE 8.8 The Flexor Digiti Minimi Muscle.

ORIGIN	
base of metatarsal V	
INSERTION	
base of proximal phalange V	
NERVE	
Tibial	

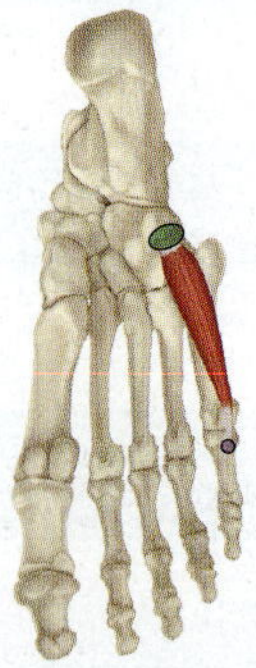

base of the fifth metatarsal and inserts on the plantar side of the base of the fifth proximal phalange.

The **quadratus plantae** (quad-RAY-tus PLAN-tee) lies in the plantar foot (**FIGURE 8.9**). It has two heads which originate from the medial and lateral sides of the tuberosity of the calcaneus and inserts on the tendon of the flexor digitorum longus tendon.

The **lumbricals** (LUM-brick-ulz) are actually a group of four separate muscles that reside in the plantar foot between the metatarsals (**FIGURE 8.10**). They originate from the four branches of the flexor digitorum longus tendon and insert on the second through fifth proximal phalanges.

FIGURE 8.9 The Quadratus Plantae Muscle.

ORIGINS

medial side of calcaneal tuberosity

lateral side of calcaneal tuberosity

INSERTION

flexor digitorum longus tendon

NERVE

Tibial

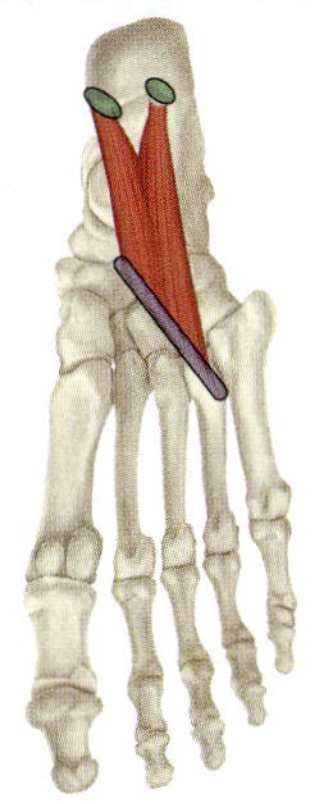

FIGURE 8.10 The Four Lumbrical Muscles.

ORIGIN

flexor digitorum longus tendon

INSERTIONS

proximal phalanges II–V

NERVE

Tibial

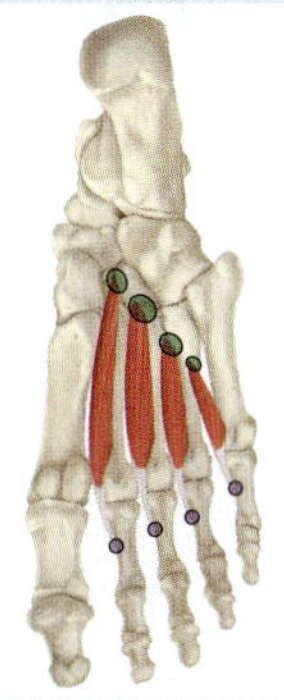

TABLE 8.4 The Muscles that Flex the Toes

Muscle	Joint(s) Acted Upon	Other Motion(s)	Nerve
Flexor Hallucis Brevis	1st MP	none	Tibial
Flexor Hallucis Longus	1st MP & IP	ankle inversion	
Flexor Digitorum Brevis	2nd–5th MP & PIP	none	
Flexor Digitorum Longus	2nd–5th MP, PIP, & DIP	ankle inversion	
Flexor Digiti Minimi	5th MP	none	
Quadratus Plantae	2nd–5th MP, PIP, & DIP		
Lumbricals	2nd–5th MP		

PAUSE TO CHECK FOR UNDERSTANDING

Table 8.4 should serve as a good review for much of the material, but you should also review all the origins and insertions.

The Toe Extensors

There are four muscles that extend the toes. They are the *extensor hallucis longus*, *extensor hallucis brevis*, *extensor digitorum longus*, and *extensor digitorum brevis* muscles. The toe extensors are summarized at the end of this section in **TABLE 8.5**.

The **extensor hallucis** (ha-LOO-sis) **brevis** resides in the dorsal foot (**FIGURE 8.11**). It originates from the anterolateral calcaneus. From there, it crosses obliquely over the dorsal foot and inserts on the dorsal side of the base of the first proximal phalange.

The **extensor hallucis** (ha-LOO-sis) **longus** passes longitudinally through the anterior crural region (**FIGURE 8.12**). It originates from the middle two-thirds of the anteromedial side of the fibular diaphysis, crosses the ankle, and inserts on the dorsal side of the first distal phalange.

The **extensor digitorum** (DIDJ-ih-TOE-rum) **brevis** resides in the dorsal foot just lateral to the extensor hallucis brevis (**FIGURE 8.13**). In fact, many classify these two muscles as one because they both originate from the same location on the anterolateral calcaneus. From its origin, the muscle splits into three bellies which pass over the second through fourth metatarsals and then merging with the tendons of the extensor digitorum longus.

The **extensor digitorum** (DIDJ-ih-TOE-rum) **longus** lies in the lateral side of the anterior lower leg passing longitudinally over the fibula (**FIGURE 8.14**). It has three origins: 1) the fibular head, 2) the lateral tibial condyle, and 3) the proximal two-thirds of the fibular diaphysis on its anterior side. It inserts on the dorsal side of the base of the second through fifth middle and distal phalanges.

FIGURE 8.11 The Extensor Hallucis Brevis Muscle.

ORIGIN	anterolateral calcaneus
INSERTION	proximal phalange I
NERVE	Deep Fibular

FIGURE 8.12 The Extensor Hallucis Longus Muscle.

ORIGIN	middle 2/3 of anteromedial fibula
INSERTION	distal phalange I
NERVE	Deep Fibular

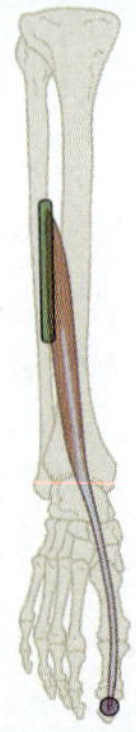

FIGURE 8.13 The Extensor Digitorum Brevis Muscle.

ORIGIN

anterolateral calcaneus

INSERTIONS

tendons of extensor digitorum longus

NERVE

Deep Fibular

FIGURE 8.14 The Extensor Digitorum Longus Muscle.

ORIGINS

fibular head

lateral tibial condyle

proximal 2/3 of anterior fibula

INSERTIONS

middle phalanges II–V

distal phalanges II–V

NERVE

Deep Fibular

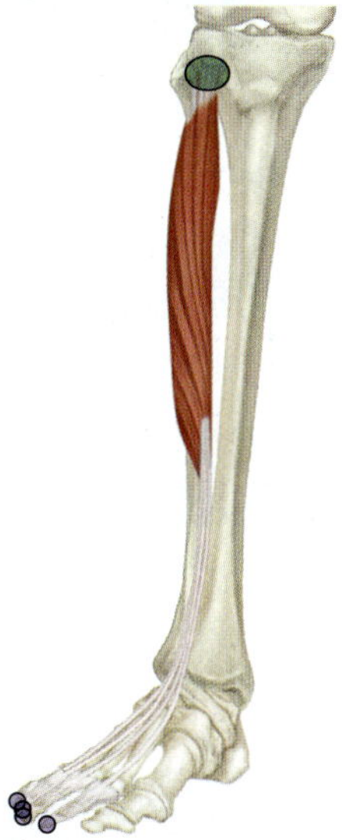

TABLE 8.5 The Muscles that Extend the Toes

Muscle	Joint(s) Acted Upon	Other Action(s)	Nerve
Extensor Hallucis Brevis	1st MP	none	Deep Fibular
Extensor Hallucis Longus	1st MP & IP	ankle dorsiflexion	
Extensor Digitorum Brevis	2nd–4th PIP	none	
Extensor Digitorum Longus	2nd–5th MP, PIP, & DIP	ankle dorsiflexion	

PAUSE TO CHECK FOR UNDERSTANDING

Table 8.5 should serve as a good review for much of the material, but you should also review all the origins and insertions.

The Toe Abductors

There are three main muscles that abduct the toes. They are the *abductor hallucis*, *dorsal interossei*, and *abductor digiti minimi* muscles. The toe abductors are summarized at the end of this section in **TABLE 8.6**.

The **abductor hallucis** (ha-LOO-sis) resides on the medial side of the plantar foot (**FIGURE 8.15**). It originates from the medial side of the calcaneal tuberosity. Its fibers pass through the medial arch of the foot and insert on the medial side of the base of the first proximal phalange.

The **dorsal interossei** (in-ter-ROSS-ee) is actually a group of four muscles lying between the metatarsals (**FIGURE 8.16**). Each of the four muscles has two heads that originate from the medial and lateral sides of the diaphyses of the first through fifth metatarsals. After the fibers converge in the middle, the first two, most medial tendons insert on the medial and lateral sides of the base of the second proximal phalange. The more lateral two muscles insert on the lateral sides of the bases of the third and fourth proximal phalanges.

The **abductor digiti** (DIDJ-ih-TIE) **minimi** (MEN-nah-my) resides in the plantar foot on the lateral side (**FIGURE 8.17**). It originates from the lateral half of the calcaneal tuberosity and inserts on the lateral side of the base of the fifth proximal phalange.

FIGURE 8.15 The Abductor Hallucis Muscle.

ORIGIN	
medial side of calcaneal tuberosity	
INSERTION	
medial base of proximal phalange I	
NERVE	
Tibial	

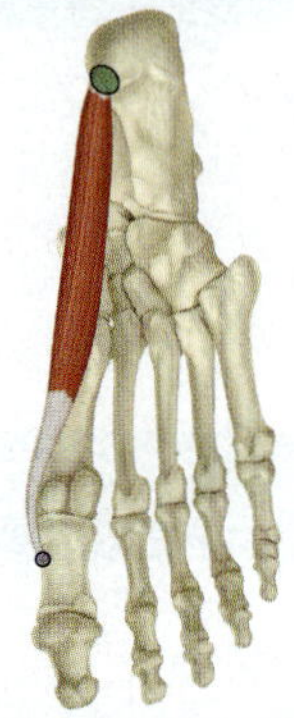

FIGURE 8.16 The Dorsal Interossei Muscles.

ORIGINS	
medial side of metatarsals I–V	
lateral side of metatarsals I–V	
INSERTIONS	
medial & lateral base of proximal phalange II	
lateral base of proximal phalanges III & IV	
NERVE	
Tibial	

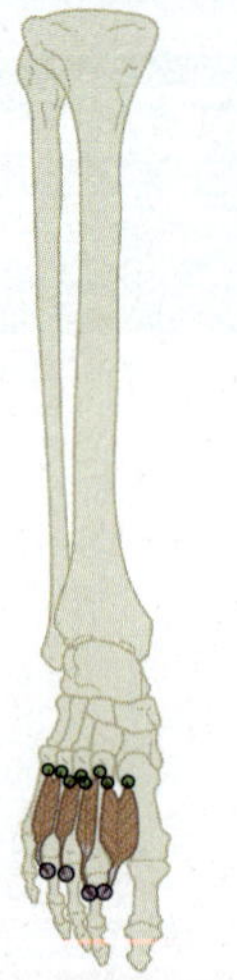

FIGURE 8.17 The Abductor Digiti Minimi Muscle.

ORIGIN

lateral half of calcaneal tuberosity

INSERTION

lateral base of proximal phalange V

NERVE

Tibial

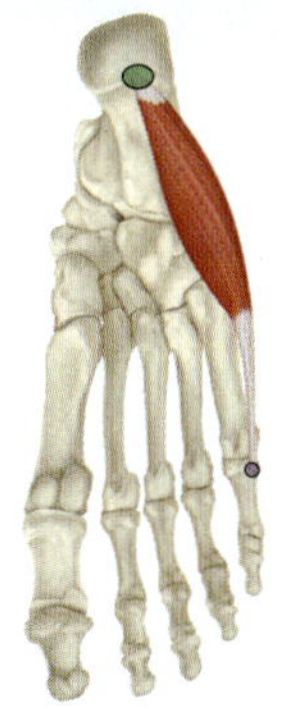

TABLE 8.6 The Muscles that Abduct the Toes

Muscle	Joint(s) Acted Upon	Other Actions	Nerve
Abductor Hallucis	1st MP		
Dorsal Interossei	2nd–4th MP	none	Tibial
Abductor Digiti Minimi	5th MP		

PAUSE TO CHECK FOR UNDERSTANDING

Table 8.6 should serve as a good review for much of the material, but you should also review all the origins and insertions.

The Toe Adductors

Only two muscles are primarily responsible for adducting the toes. They are the *adductor hallucis* and the *plantar interossei* muscles. The toe adductors are summarized at the end of this section in **TABLE 8.7**.

The **adductor hallucis** (ha-LOO-sis) is unique having two bellies that lie in the plantar aspect of the foot (**FIGURE 8.18**). The transverse belly originates from the heads (distal ends) of the third through fifth metatarsals as it passes across them. The oblique belly originates from the bases (proximal end) of the second through fourth metatarsals. The two bellies merge and insert on the lateral side of the base of the first proximal phalange.

The **plantar** (PLAN-tar) **interossei** (in-ter-ROSS-ee) are actually a group of three muscles lying between metatarsals II–V (**FIGURE 8.19**). They originate from medial side of the diaphyses of the third through fifth metatarsals. Each belly passes between their respective metatarsals and insert on the medial side of the base of the third through fifth proximal phalanges.

FIGURE 8.18 The Adductor Hallucis Muscle.

ORIGINS
heads of metatarsals III–V
bases of metatarsals II–V
INSERTION
lateral side of proximal phalange I
NERVE
Tibial

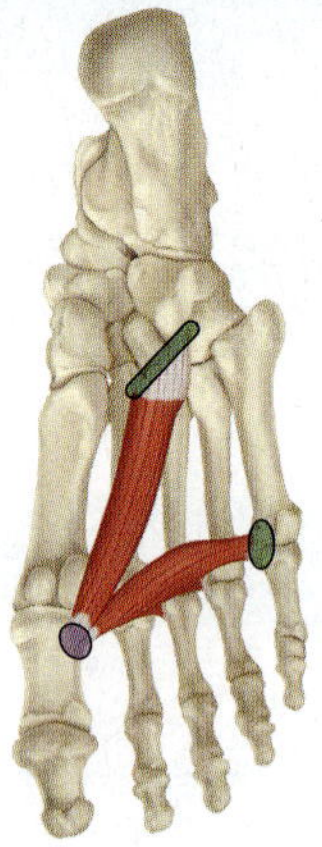

FIGURE 8.19 The Three Plantar Interossei Muscles.

ORIGINS
medial sides of metatarsals III–V
INSERTIONS
medial sides of proximal phalanges III–V
NERVE
Tibial

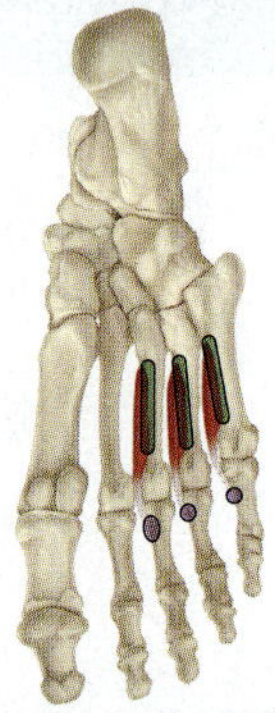

TABLE 8.7 The Muscles that Adduct the Toes

Muscle	Joint(s) Acted Upon	Nerve
Adductor Hallucis	1st MP	Tibial
Plantar Interossei	3rd–5th MP	

PAUSE TO CHECK FOR UNDERSTANDING

Table 8.7 should serve as a good review for much of the material, but you should also review all the origins and insertions.

© BraunS/E+/Getty Images

CHAPTER 9
The Ankle

CHAPTER OBJECTIVES

After completing this chapter, the student will be able to:

1. summarize the joints of the ankle region;
2. comprehend the ankle motions;
3. identify the names and locations of the ankle muscles;
4. recall the attachment sites and nerves of the ankle muscles; and
5. apply the motions produced by the actions of the ankle muscles.

The ankle is a region of the body where the lower leg and foot intersect. The ankle is actually a compilation of three joints formed by the tibia, fibula, talus, and calcaneus (**FIGURE 9.1**). The arrangement of the bones and the actions of the ankle muscles are able to produce four motions.

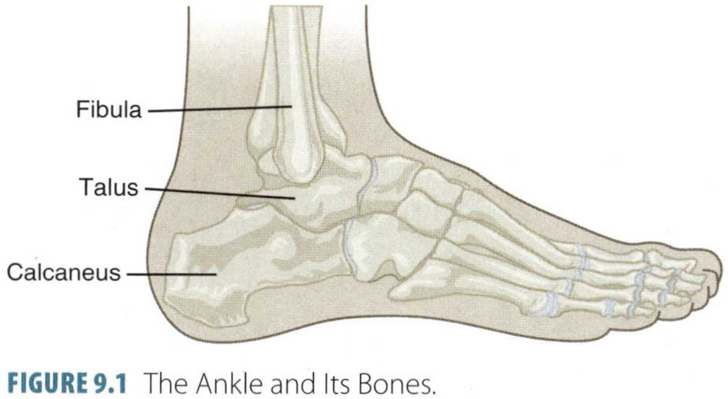

FIGURE 9.1 The Ankle and Its Bones.

The Joints of the Ankle

Three joints comprise the ankle. They are the *distal tibiofibular joint*, the *talocrural joint*, and the *subtalar joint*. The **distal tibiofibular joint**, as the name implies, is the articulation between the distal fibula and the distal tibia (**FIGURE 9.2**). The **distal anterior tibiofibular ligament** and the **distal posterior tibiofibular ligament** stabilize the tibia and fibula at the ankle. The tibia and fibula are also held together by a band of tissue called the **interosseous membrane** that runs across the length of the bones.

The **talocrural joint** (**FIGURE 9.3**) is the articulation between the talus and the bones in the crural region which are the tibia and fibula. Therefore, the talocrural joint is the articulation between the talus and the crural bones (tibia and the fibula). This is the only joint that is an exception to the rule that an articulation (joint) is the intersection of two bones. This joint has been described as a mortise and tenon.

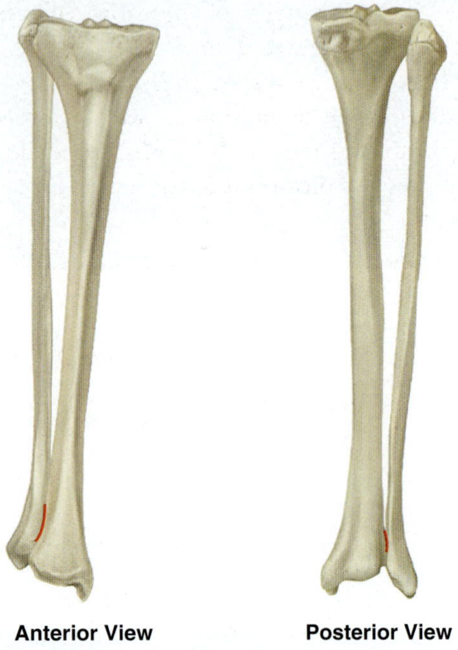

FIGURE 9.2 The Distal Tibiofibular Joint.

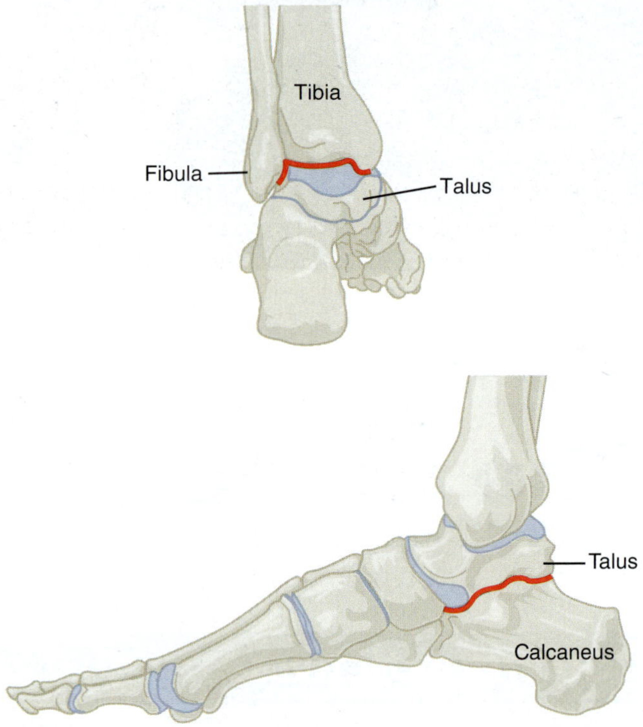

FIGURE 9.3 The Talocrural (above) and Subtalar Joints (below).

A mortise and tenon is a type of joint in wood working where the tenon (peg) fits into a mortise (slot). In this joint, the talus is the "tenon" or "peg" which fits into the "mortise" or "slot" created by the distal tibia and fibula.

The third joint is the **subtalar joint** (Figure 9.3). "Sub" means below and the calcaneus is below, or inferior, to the talus. So, this joint is the articulation between the talus and the calcaneus.

Four main ligaments stabilize the talocrural and subtalar joints (**FIGURE 9.4**). Three of them stabilize the lateral ankle and the remaining one stabilizes the medial ankle. The lateral ligaments extend from the lateral malleolus of the fibula to locations on either the talus or calcaneus. The **anterior talofibular ligament** connects the fibula and talus on the anterolateral side. The **posterior talofibular ligament** connects the fibula and talus on the posterolateral side. The **calcaneofibular ligament** connects the fibula and calcaneus. The medial side of the ankle bones is held together by the **deltoid ligament**. The deltoid ligament is triangular shaped, hence the name, and attaches from the medial malleolus to various locations on the talus and calcaneus.

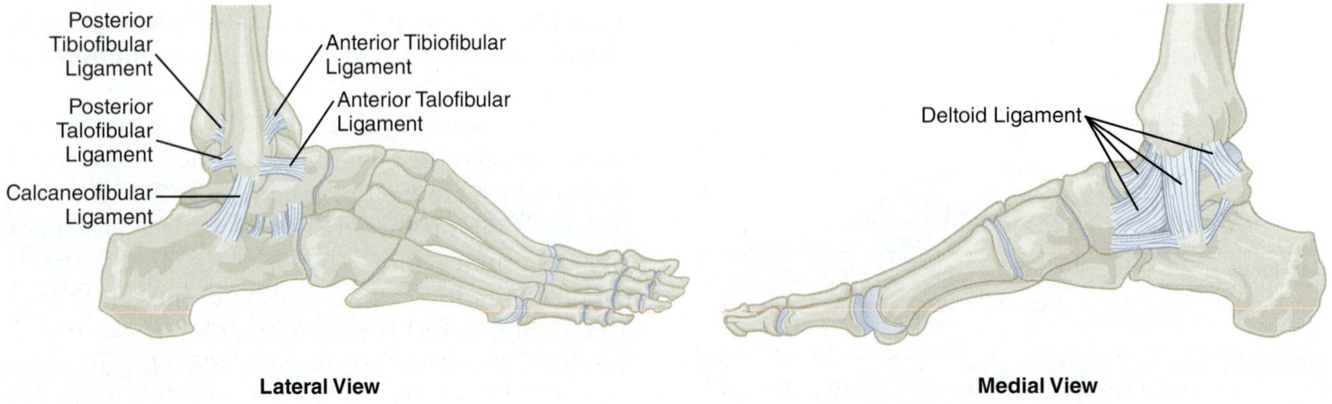

FIGURE 9.4 The Ankle Ligaments.

The Motions of the Ankle

The ankle moves in four directions (**FIGURE 9.5**). They are plantar flexion, dorsiflexion, inversion, and eversion. **Plantar flexion** and **dorsiflexion** straighten and bend the ankle in the sagittal plane rotating about the frontal axis. **Inversion** and **eversion** rotate the ankle in the frontal plane around the sagittal axis.

The average ranges of motions of the ankle motions are summarized in **TABLE 9.1**. The average plantar flexion range of motion is 50 degrees. Average dorsiflexion range of motion is 20 degrees. Average inversion range of motion is 20 degrees while average eversion range of motion is 5 degrees.

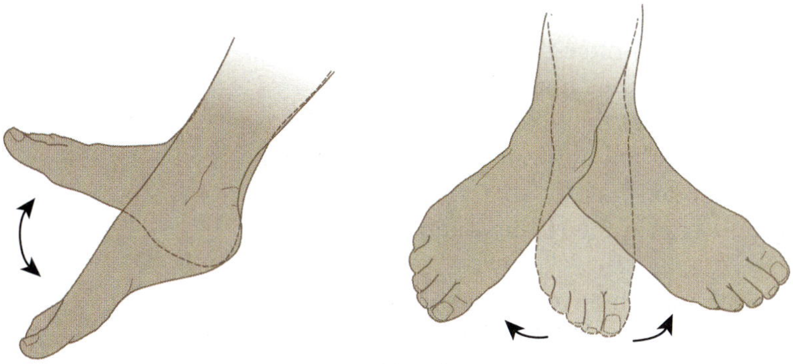

FIGURE 9.5 Plantar Flexion & Dorsiflexion (left) and Inversion & Eversion (right).

TABLE 9.1 Average Ankle Ranges of Motions

Motion	Average Degrees of Motion
Dorsiflexion	20°
Plantar Flexion	50°
Combined Dorsiflexion – Plantar Flexion	70°
Eversion	5°
Inversion	20°
Combined Eversion – Inversion	25°

PAUSE TO CHECK FOR UNDERSTANDING

1. List the joints of the ankle.
2. Then, tell which bones form each joint.
3. What are the four motions of the ankle?

The Muscles that Move the Ankle

The rest of this chapter is dedicated to learning the ankle muscles; their attachment sites; the *primary*, main motions they produce; and the nerves that innervate them. Muscles work in groups and so each muscle within the group supplies a certain percentage of the force needed to move a body segment and that amount (percentage) is not the same for each muscle. Furthermore, the force each muscle contributes varies depending on the starting position of the body segment and changes throughout the range of motion. This can be confusing to the beginning kinesiology student. So, this text has made certain choices in an attempt to simplify the muscle groups based on initial motion from anatomical position.

The muscle groups are based on the motion caused by their *concentric* action (i.e., The ankle dorsiflexors dorsiflex the ankle when they concentrically act.). However, it should not be forgotten that all muscles can and will act *eccentrically* and *isometrically*. The reader should also notice that the table for each individual muscle has a section called "Other Motions." *Other Motions* means that this muscle aides, to some degree, in other motion(s) besides the group it has been placed in. Some muscles may appear in more than one group because their role is too great to ignore.

Since there are four ankle motions, there are four groups. The muscle groups are *dorsiflexors, plantar flexors, evertors*, and *invertors*. Also, the ankle muscles reside within compartments of the lower leg (crural region). Listed in a logical order around the crural region, they are the anterior, lateral, posterior, and finally deep posterior compartments. The **anterior compartment** contains the muscles that dorsiflex the ankle. The **lateral compartment** contains the muscles that evert the ankle. The **posterior** (also called superficial posterior) **compartment** contains the muscles that plantar flex the ankle. The **deep posterior compartment** contains the muscles that invert the ankle.

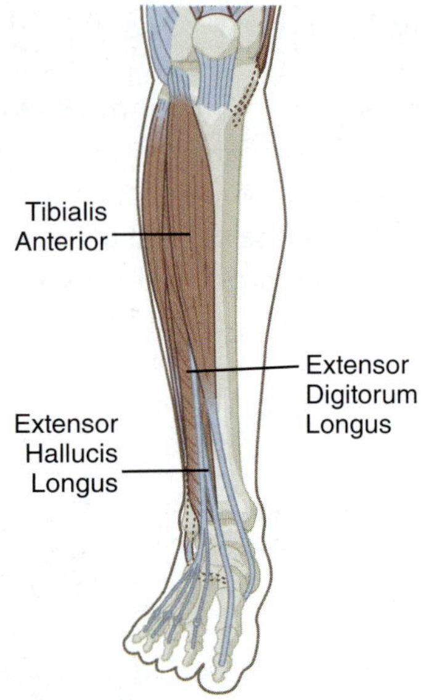

FIGURE 9.6 The Dorsiflexors.

The Ankle Dorsiflexors (Anterior Compartment Muscles)

There are three muscles in the anterior compartment that dorsiflex the ankle. They are the *tibialis anterior*, the *extensor hallucis longus*, and the *extensor digitorum longus* (**FIGURE 9.6**). The dorsiflexors are summarized at the end of this section in **TABLE 9.2**.

The **tibialis (TIB-ee-al-us) anterior** passes longitudinally through the anterior crural region from the proximal tibia across the ankle to the foot (**FIGURE 9.7**). It originates from the proximal two-thirds

FIGURE 9.7 The Tibialis Anterior Muscle.

ORIGIN	
proximal 2/3 of lateral tibia	
INSERTIONS	
medial cuneiform	
base of metatarsal I	
OTHER MOTIONS	
inversion	
NERVE	
Deep Fibular	

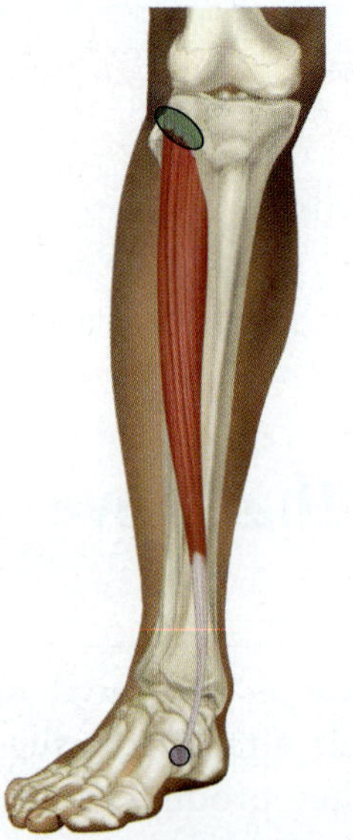

of the lateral side of the tibia and then inserts on two locations. One is on the medial cuneiform and the other is on the base of the first metatarsal.

The **extensor hallucis (ha-LOO-sis) longus** resides in the anterior compartment inferior to and between the tibialis anterior and the extensor digitorum longus (**FIGURE 9.8**). It originates along the middle two-thirds of the anterior side of the diaphysis of the fibula and inserts on the dorsal side of the base of the first distal phalange.

The **extensor digitorum (DIDJ-ih-TOE-rum) longus** is the most lateral of the three anterior compartment muscles (**FIGURE 9.9**). It has three origins: (1) the fibular head, (2) the lateral side of the lateral tibial condyle, and (3) the proximal two-thirds of the anterior side of the fibula. It inserts on the dorsal side of the second through fifth middle phalanges and on the dorsal side of the second through fifth distal phalanges.

FIGURE 9.8 The Extensor Hallucis Longus Muscle.

ORIGIN
middle 2/3 of anterior fibula
INSERTION
dorsal base of distal phalange I
OTHER MOTIONS
1st toe extension
NERVE
Deep Fibular

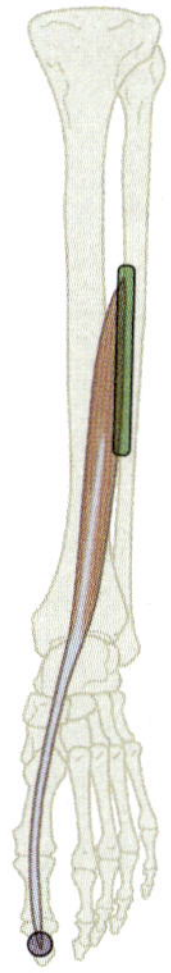

FIGURE 9.9 The Extensor Digitorum Longus Muscle.

ORIGINS
fibular head
lateral tibial condyle
proximal 2/3 of anterior fibula
INSERTIONS
dorsal side of middle phalanges II–V
dorsal side of distal phalanges II–V
OTHER MOTIONS
2nd–5th toe extension
NERVE
Deep Fibular

TABLE 9.2 The Muscles that Dorsiflex the Ankle

Muscle	Origins	Insertions	Nerve
Tibialis Anterior	proximal 2/3 of lateral tibia	medial cuneiform & the base of metatarsal I	
Extensor Hallucis Longus	middle 2/3 of anterior fibula	dorsal base of distal phalange I	Deep Fibular
Extensor Digitorum Longus	fibular head, lateral tibial condyle, & proximal 2/3 of anterior fibula	dorsal side of middle & distal phalanges II–V	

PAUSE TO CHECK FOR UNDERSTANDING

Table 9.2 should serve as a good review for much of the material, but here are a few other checks for understanding

1. How many muscles are in the ankle dorsiflexion group? List them.
2. What motion occurs if these muscles act eccentrically?
3. Which compartment does this muscle group reside in?

The Ankle Evertors (Lateral Compartment Muscles)

There are three muscles residing in the lateral compartment that evert the ankle. They are the *fibularis longus*, the *fibularis brevis*, and the *fibularis tertius* (**FIGURE 9.10**). For many years, the first part of their name was known as peroneus rather than fibularis. Peroneus is a Latin word referring to the fibula bone. The modern version now refers to them as fibularis which obviously means fibula as well. The evertors are summarized at the end of this section in **TABLE 9.3**.

The **fibularis (FIB-you-LARE-iss) longus** passes longitudinally on the lateral lower leg (**FIGURE 9.11**). It originates from the lateral side of the fibular head. Its tendon curves behind the lateral malleolus before turning anteriorly, "running" on the plantar side of the foot, and finally inserting onto the plantar side of the first metatarsal and the medial cuneiform.

The **fibularis (FIB-you-LARE-iss) brevis** lies deep to the fibularis longus on the lateral lower leg (**FIGURE 9.12**). It originates more distally than the fibularis longus, hence the word "brevis," from the lateral mid shaft of the fibula. Its tendon "runs" similarly to the tendon of the longus behind the lateral malleolus, but inserts into the base of the fifth metatarsal.

The **fibularis tertius** (FIB-you-LARE-iss TUR-she-us) resides just anterior to the fibularis brevis, but is missing from much of the population (**FIGURE 9.13**). It originates from the medial side of the distal one-third of the fibula and inserts on the base of the fifth metatarsal superior to the fibularis brevis attachment.

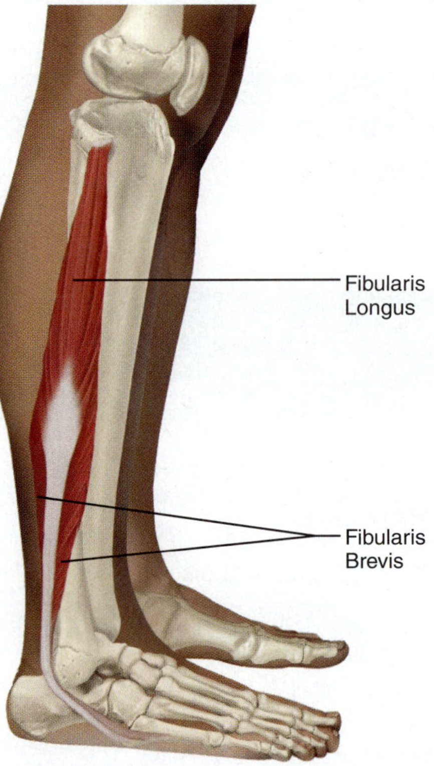

FIGURE 9.10 The Evertors.

FIGURE 9.11 The Fibularis Longus Muscle.

ORIGIN	
lateral side of fibular head	
INSERTIONS	
plantar side of metatarsal I	
plantar side of medial cuneiform	
OTHER MOTIONS	
plantar flexion	
NERVE	
Superficial Fibular	

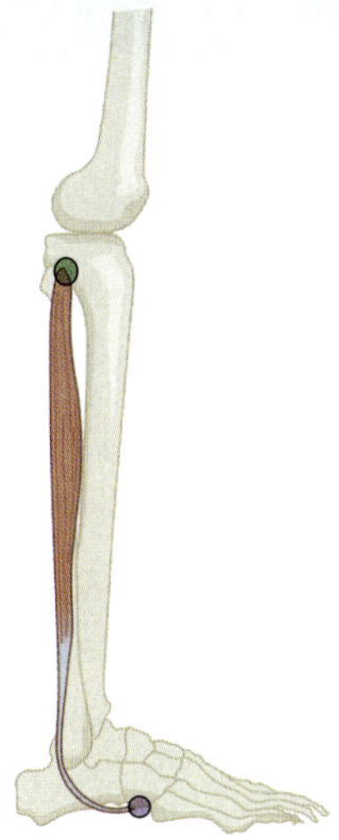

FIGURE 9.12 The Fibularis Brevis Muscle.

ORIGIN	
lateral side of mid fibula	
INSERTION	
base of metatarsal V	
OTHER MOTIONS	
plantar flexion	
NERVE	
Superficial Fibular	

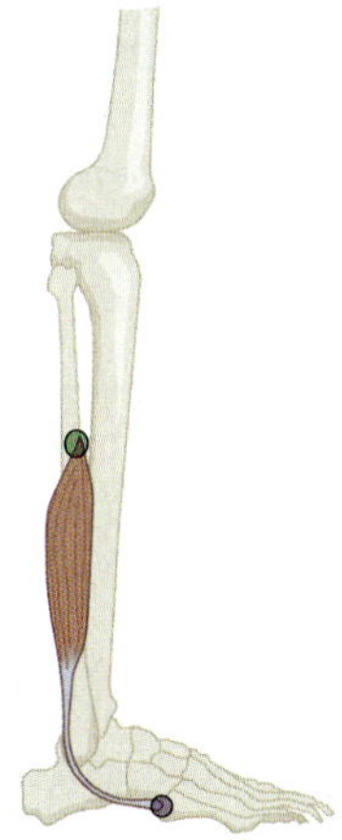

FIGURE 9.13 The Fibularis Tertius Muscle.

ORIGIN	
distal 1/3 of medial fibula	
INSERTION	
base of metatarsal V	
OTHER MOTIONS	
dorsiflexion	
NERVE	
Deep Fibular	

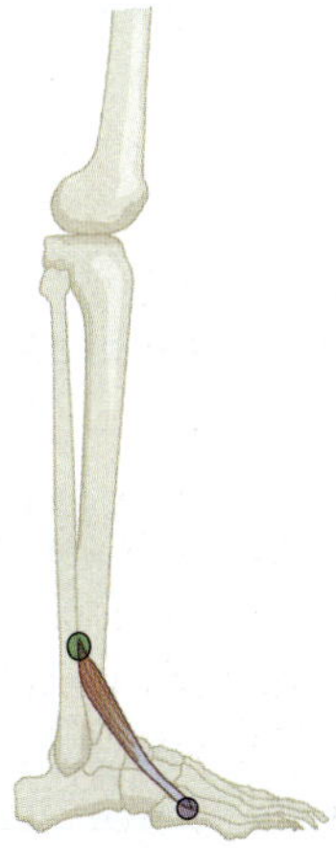

TABLE 9.3 Summary of the Evertors

Muscle	Origins	Insertions	Nerve
Fibularis Longus	lateral side of fibular head	plantar sides of metatarsal I & medial cuneiform	Superficial Fibular
Fibularis Brevis	lateral side of mid fibula	base of metatarsal V	Deep Fibular
Fibularis Tertius	distal 1/3 of medial fibula		

PAUSE TO CHECK FOR UNDERSTANDING

Table 9.3 should serve as a good review for much of the material, but here are a few other checks for understanding

1. How many muscles are in the ankle eversion group? List them.
2. What motion occurs if these muscles act eccentrically?
3. Which compartment does this muscle group reside in?

The Ankle Plantar Flexors (Posterior Compartment Muscles)

There are two muscles residing in the posterior compartment that are mainly responsible for plantar flexing the ankle. They are the *gastrocnemius* and *soleus* (**FIGURE 9.14**). Because of their common insertion, the gastrocnemius and the soleus are sometimes called the **triceps surae** which means three headed calf. A third muscle, the *plantaris*, is located here as well, but it offers little to no action in most of the population. These muscles are summarized at the end of this section in **TABLE 9.4**.

The **gastrocnemius** (GAS-trock-NEE-me-us) is the most superficial muscle in the posterior compartment (**FIGURE 9.15**). It is referred to as a two-jointed muscle

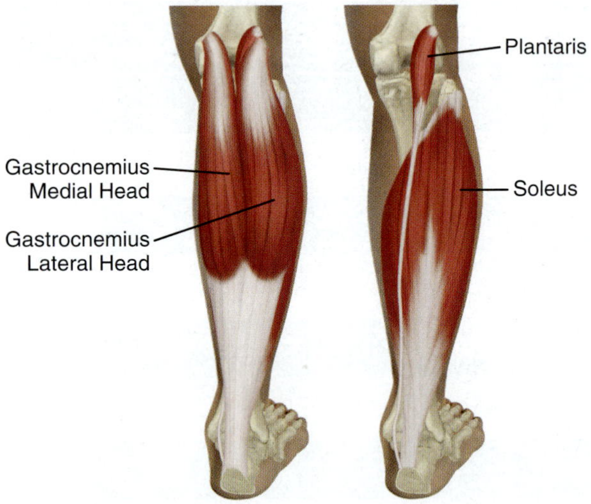

FIGURE 9.14 The Plantar Flexors.

FIGURE 9.15 The Gastrocnemius Muscle.

ORIGINS
posterior side of medial femoral condyle
posterior side of lateral femoral condyle
INSERTION
posterior calcaneus
OTHER MOTIONS
knee flexion
NERVE
Tibial

because it crosses both the ankle and the knee and therefore acts upon both. Its two origins are on the posterior side of the medial and lateral condyles of the femur and it inserts on the posterior calcaneus via the Achilles tendon.

The **soleus** (SO-lee-us) lies deep to the gastrocnemius muscle in the posterior compartment (**FIGURE 9.16**). It originates below the knee from the posterior side of the fibular head, the proximal one-fourth of the fibula, and the middle one-third of the tibia. It joins with the gastrocnemius "using" the Achilles tendon to insert on the posterior calcaneus.

The **plantaris** (PLAN-tear-us) is a long thin muscle originating from the posterior lateral femoral condyle and inserts on the posterior calcaneus (**FIGURE 9.17**). However, it offers very little to motion. In fact, about 10% of the population does not have this muscle.

FIGURE 9.16 The Soleus Muscle.

ORIGINS
posterior side of fibular head
proximal 1/4 of fibula
middle 1/3 of tibia
INSERTION
posterior calcaneus
OTHER MOTIONS
none
NERVE
Tibial

FIGURE 9.17 The Plantaris Muscle.

ORIGIN
posterior side of lateral femoral condyle
INSERTION
posterior calcaneus
OTHER MOTIONS
knee flexion
NERVE
Tibial

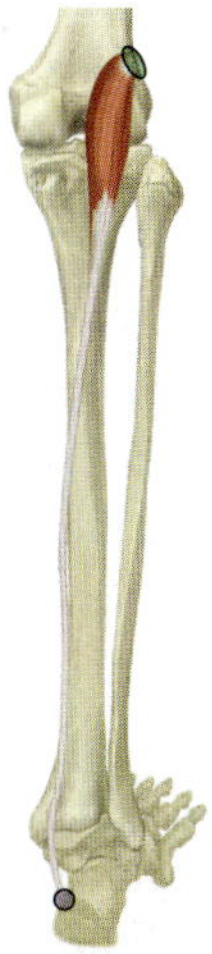

TABLE 9.4 Summary of the Plantar Flexors

Muscle	Origins	Insertions	Nerve
Gastrocnemius	posterior side of the medial & lateral femoral condyles	posterior calcaneus	Tibial
Soleus	posterior fibular head, proximal 1/4 fibula, & middle 1/3 tibia		
Plantaris	posterior side of lateral femoral condyle		

PAUSE TO CHECK FOR UNDERSTANDING

Table 9.4 should serve as a good review for much of the material, but here are a few other checks for understanding

1. How many muscles are in the ankle plantar flexion group? List them.
2. What motion occurs if these muscles act eccentrically?
3. Which compartment does this muscle group reside in?

The Ankle Invertors (Deep Posterior Compartment Muscles)

There are three muscles in the deep posterior compartment that invert the ankle. They are the *posterior tibialis*, the *flexor hallucis longus*, and the *flexor digitorum longus* (**FIGURE 9.18**). The invertors are summarized at the end of this section in **TABLE 9.5**.

The **tibialis (TIB-ee-al-us) posterior** lies somewhat on the lateral side of the posterior compartment (**FIGURE 9.19**). It originates on the posterior side of the proximal one-half of the tibia and the fibula. It inserts on the plantar side of 1) the navicular, 2) the medial cuneiform, and 3) the bases of the second through fourth metatarsals.

The **flexor hallucis (ha-LOO-sis) longus** originates on the posterior side of the distal two-thirds of the fibular diaphysis (**FIGURE 9.20**). From there, its belly runs obliquely toward the middle of the posterior lower leg (crural region) before turning to pass over the ankle inserts on the plantar side of the base of the first distal phalange. Its tendon then turns behind the medial malleolus before inserting on plantar side of the base of the first distal phalange.

The **flexor digitorum (DIDJ-ih-TOE-rum) longus** rests just medial to the tibialis posterior (**FIGURE 9.21**). It originates on the posterior side of the middle one-third of the tibia behind the medial malleolus before splitting into four tendons which insert on the plantar side of the bases of the second through fifth distal phalanges.

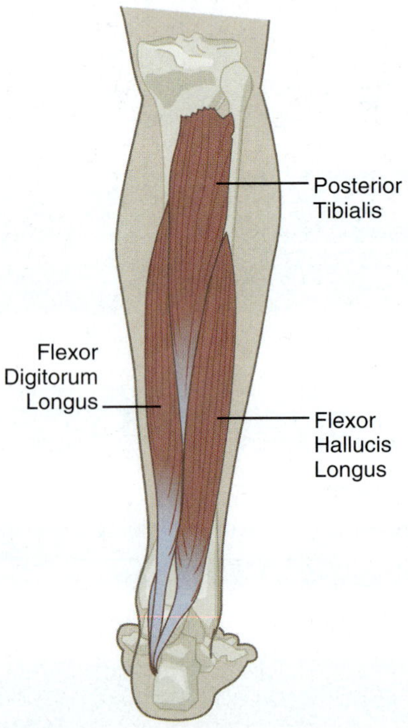

FIGURE 9.18 The Invertors.

FIGURE 9.19 The Posterior Tibialis Muscle.

ORIGINS
proximal 1/2 of posterior tibia
proximal 1/2 of posterior fibula
INSERTIONS
navicular bone
medial cuneiform
bases of metatarsals II–V
OTHER MOTIONS
plantar flexion
NERVE
Tibial

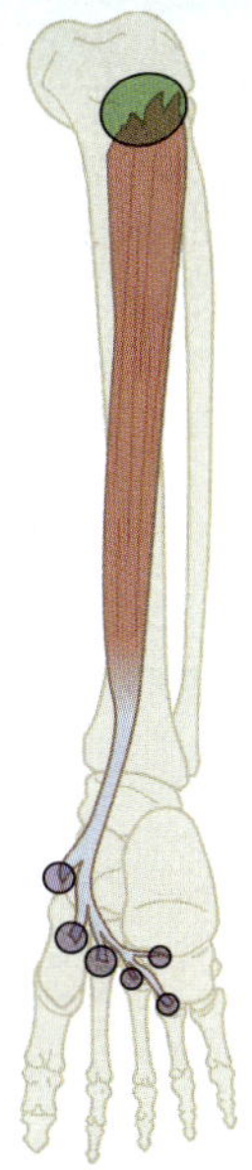

FIGURE 9.20 The Flexor Hallucis Longus Muscle.

ORIGIN
distal 2/3 of posterior fibula
INSERTION
plantar base of distal phalange I
OTHER MOTIONS
1st toe flexion
NERVE
Tibial

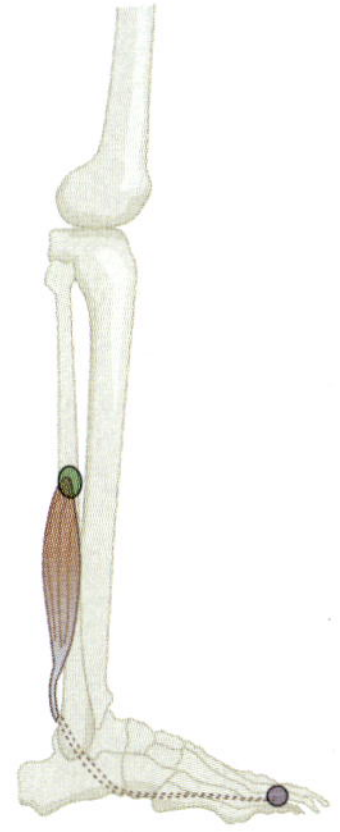

FIGURE 9.21 The Flexor Digitorum Longus Muscle.

ORIGIN
middle 1/3 of posterior tibia
INSERTION
plantar base of distal phalanges II–V
OTHER MOTIONS
2nd–5th toe flexion
NERVE
Tibial

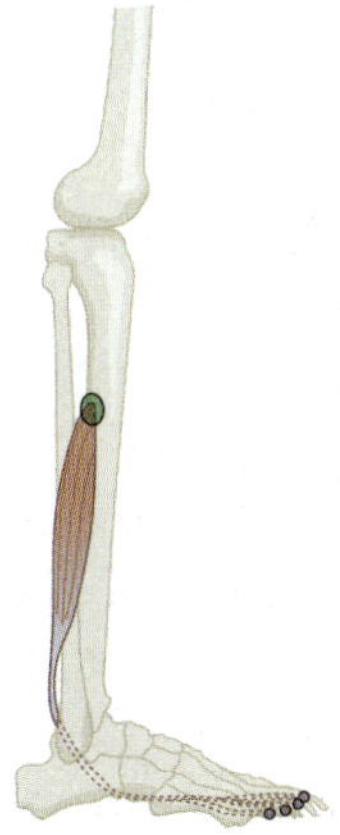

TABLE 9.5 Summary of the Invertors

Muscle	Origins	Insertions	Nerve
Tibialis Posterior	proximal 1/2 of posterior tibia & of the posterior fibula	navicular bone, medial cuneiform, & the bases of metatarsals II–V	Tibial
Flexor Hallucis Longus	distal 2/3 of posterior fibula	plantar base of distal phalange I	
Flexor Digitorum Longus	middle 1/3 of posterior tibia	plantar base of distal phalanges II–V	

PAUSE TO CHECK FOR UNDERSTANDING

Table 9.5 should serve as a good review for much of the material, but here are a few other checks for understanding

1. How many muscles are in the ankle inversion group? List them.
2. What motion occurs if these muscles act eccentrically?
3. Which compartment does this muscle group reside in?

CHAPTER 10

The Knee

CHAPTER OBJECTIVES

After completing this chapter, the student will be able to:

1. summarize the joints of the knee region;
2. comprehend the knee motions;
3. identify the names and locations of the knee muscles;
4. recall the attachment sites and nerves of the knee muscles; and
5. apply the motions produced by the actions of the knee muscles.

The knee is a region of the body where the thigh and lower leg (crural region) meet (**FIGURE 10.1**). Within the knee region are three joints between the femur, patella, tibia, and fibula. The arrangement of the bones and the actions of the knee muscles produce two main motions (flexion and extension) and two lesser motions.

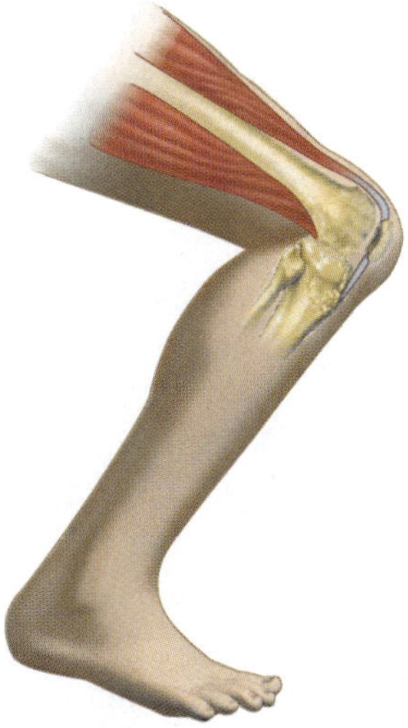

FIGURE 10.1 The Knee.

The Joints of the Knee

The knee contains three joints. They are the *tibiofemoral joint*, the *patellofemoral joint*, and *proximal tibiofibular joint*. The **tibiofemoral joint** (**FIGURE 10.2**) is the largest of the body and is an articulation between

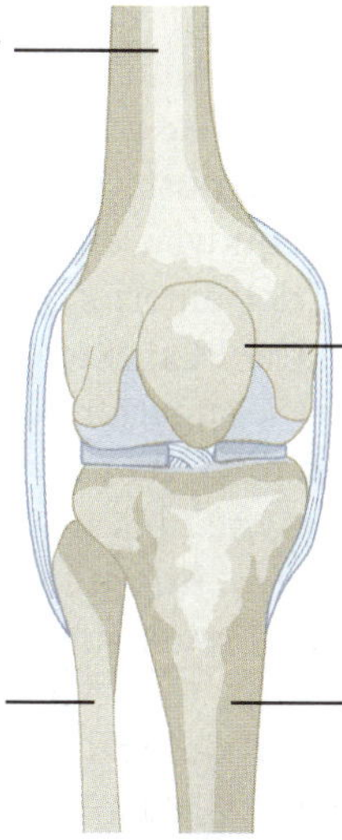

FIGURE 10.2 The Tibiofemoral Joint.

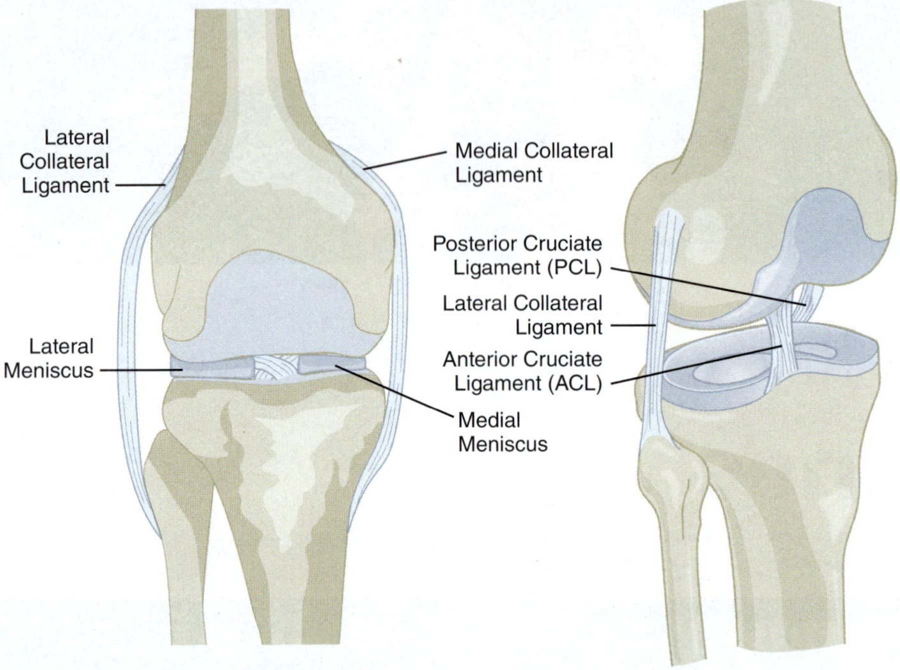

FIGURE 10.3 The Main Ligaments of the Tibiofemoral Joint.

the proximal tibia and the distal femur. The superior edge of the proximal tibia is called the tibial plateau which has two shallow concavities that articulate with the two femoral condyles. The tibial plateau also has two relatively shallow C-shaped concave menisci (fibrocartilage tissue). The **lateral meniscus** is on the lateral depression and a **medial meniscus** is on the medial depression. These two menisci deepen the tibial concavities and act as "shock absorbers" to compression forces as well as adding in the protection of the ends of the bones.

There are four main ligaments that stabilize the bones of the tibiofemoral joint (**FIGURE 10.3**). The **anterior cruciate ligament** (ACL) and **posterior cruciate ligament** (PCL) criss-cross obliquely through the middle of the joint. Cruciate means cross and the two ligaments are like a twisted "X." The ACL extends from the inferior posterior femur and to the superior anterior intercondylar eminence of the tibia. Its purpose is to restrict anterior translation of the tibia from the femur. The PCL extends from the inferior anterior femur to the superior posterior intercondylar eminence of the tibia. Its purpose is to restrict posterior translation of the tibia from the femur. The **medial (tibial) collateral ligament** and the **lateral (fibular) collateral ligament** stabilize their respective sides of the knee.

The **patellofemoral joint** (**FIGURE 10.4**) is the articulation between the patella and the femur. The patella glides back and forth in the femoral groove which is between the femoral condyles during extension and flexion of the knee.

The **proximal tibiofibular joint**, as the name implies, is the articulation between the proximal ends of the tibia and fibula (**FIGURE 10.5**). More specifically, it is where the fibular head articulates with the lateral side of the lateral tibial condyle. The ligaments that connect the bones are the **proximal anterior tibiofibular ligament** and the **proximal posterior tibiofibular ligament**. The tibia and

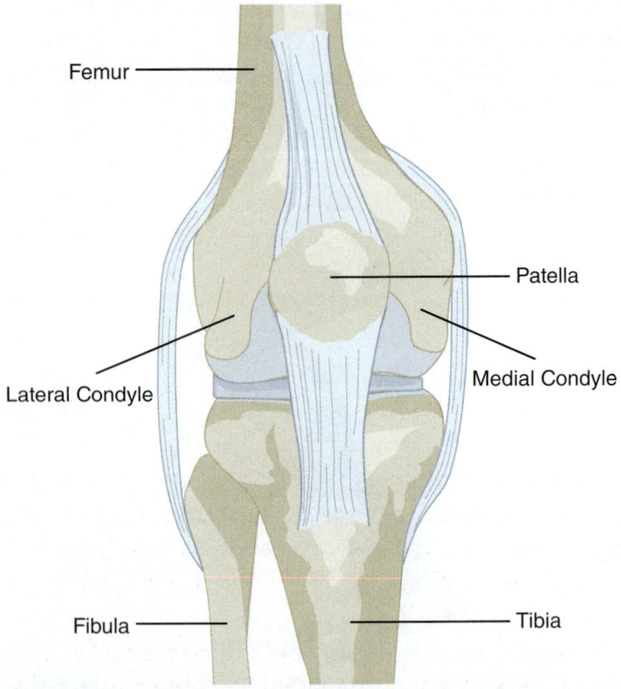

FIGURE 10.4 The Patellofemoral Joint.

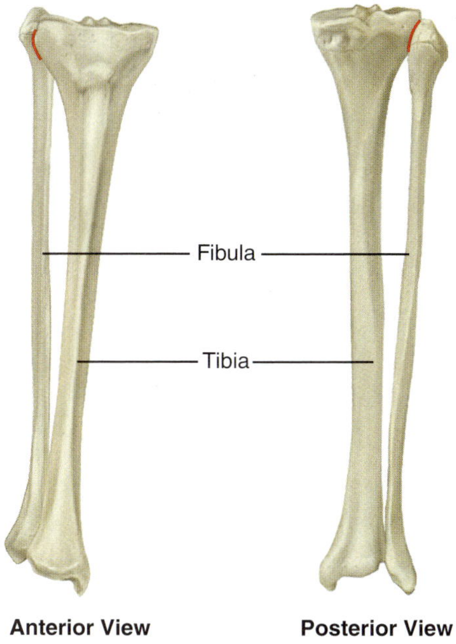

FIGURE 10.5 The Proximal Tibiofibular Joint.

fibula are also held together by a band of tissue called the **interosseous membrane** that runs the length of the bones.

The Motions of the Knee

Four motions occur at the knee. However, two of the motions are referred to as independent while the other two are known as dependent or accessory motions. Independent means that these motions are free to occur without additional motions. Dependent motions, also called accessory motions, are not free and only occur during the act of another motion.

Knee extension and **flexion** are the two independent motions (**FIGURE 10.6**). They straighten and bend the knee in the sagittal plane rotating about the frontal axis. **Hyperextension** is sometimes listed as another motion and is the extension beyond 0 degrees. Average knee ranges of motions are presented in **TABLE 10.1**.

Knee **internal rotation** and **external rotation** are the two dependent motions. During *non-weight bearing* activities, the tibia rotates while the femur remains relatively motionless. Interestingly enough during *weight bearing* activities, it is the femur that rotates and the tibia remains relatively motionless. In ***non-weight bearing*** activities during the last few degrees of knee extension, the tibia externally rotates a few degrees. Then, during the first few degrees of knee flexion, the tibia internally rotates a few degrees.

During ***weight bearing*** activities during the last few degrees of knee extension, the femur internally rotates a few degrees. This accessory motion during knee extension is sometimes called the "screw home" motion or mechanism (**FIGURE 10.7**). Also, it has been called "locking" the knee. Conversely, the knee needs to be "unlocked" during the first few degrees of knee flexion. During weight bearing activities, the

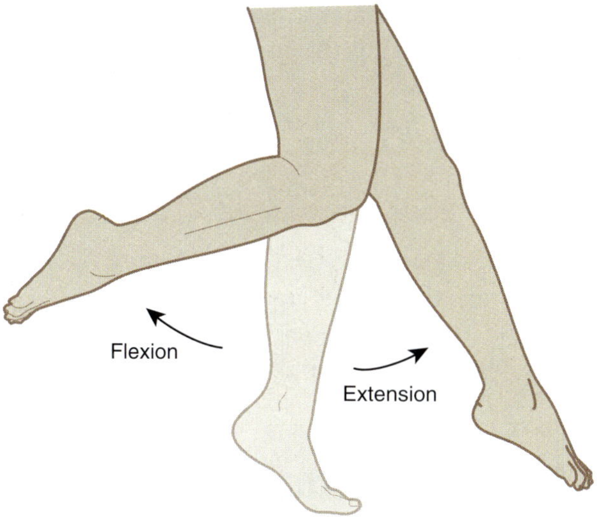

FIGURE 10.6 Knee Extension and Flexion.

TABLE 10.1 Average Knee Ranges of Motions

Motion	Average Degrees of Motion
Extension Hyperextension	0° 0–10°
Flexion	135–145°
Combined Hyper/ Extension-Flexion	135–155°

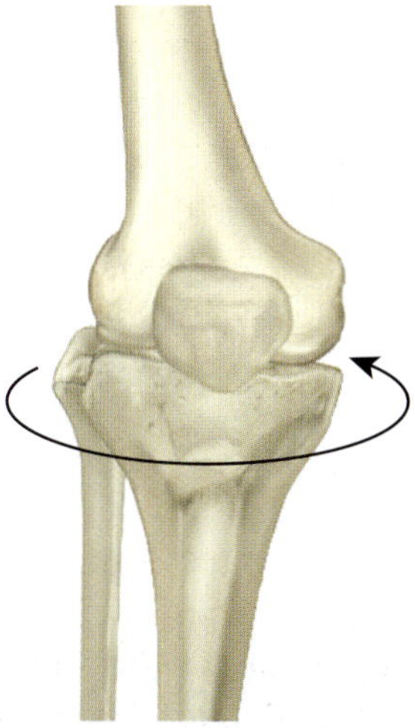

FIGURE 10.7 Screw Home Mechanism During Weight Bearing Knee Extension.

femur externally rotates a few degrees back in line with the tibia during the first few degrees of knee flexion.

PAUSE TO CHECK FOR UNDERSTANDING

1. List the joints of the knee.
2. Then, tell which bones form each joint.
3. What are the two motions of the knee? Also, what are the two accessory motions?

The Muscles that Move the Knee

The rest of this chapter is dedicated to learning the knee muscles; their attachment sites; the *primary*, main motions they produce; and the nerves that innervate them. Muscles work in groups and so each muscle within the group supplies a certain percentage of the force needed to move a body segment and that amount (percentage) is not the same for each muscle. Furthermore, the force each muscle contributes varies depending on the starting position of the body segment and changes throughout the range of motion. This can be confusing to the beginning kinesiology student. So, this text has made certain choices in an attempt to simplify the muscle groups based on initial motion from anatomical position.

The muscle groups are based on the motion caused by their *concentric* action (i.e., The knee flexors flex the knee when acting concentrically.). However, it should not be forgotten that all muscles can and will act *eccentrically* and *isometrically*. The reader should also notice that the table for each individual muscle has a section called "Other Motions." *Other Motions* means that this muscle aides, to some degree, in other motion(s) besides the group it has been placed in.

Only the independent knee motions will be considered. So, there are two groups because there are two motions. They are the *knee flexors* and *knee extensors*.

The Knee Flexors

Three main muscles are primarily responsible for flexing the knee and they reside in the posterior thigh. This group of three muscles is sometimes called the **hamstrings** and they are the *biceps femoris*, the *semitendinosus*, and the *semimembranosus* (**FIGURE 10.8**). Additionally, there are three other muscles that assist in knee flexion. They are the *gastrocnemius*, *plantaris*, and the *popliteus*. At the end of this section, the knee flexors are summarized in **TABLE 10.2**.

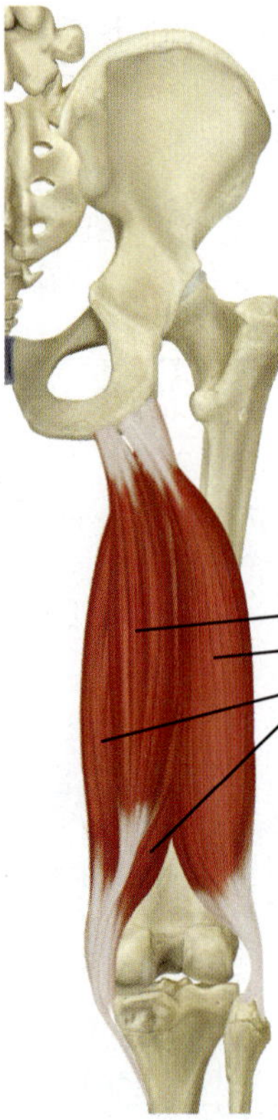

FIGURE 10.8 The Knee Flexors (Hamstrings).

The **biceps femoris** (FAH-more-us) resides on the lateral side of the posterior thigh (**FIGURE 10.9**). It has two heads and therefore two origins. The **long head** originates from the ischial tuberosity and the **short head** from the linea aspera. The fibers of the long head cross the hip and merge with the fibers of the short head in the posterior thigh. They then eventually insert on the fibular head and the lateral side of the lateral condyle of the tibia.

The belly of the **semitendinosus** (SEM-ee-TEN-din-OH-sus) is rod shaped (like a tendon) and it resides in the medial side of the posterior thigh superficial to the semimembranosus (**FIGURE 10.10**). It originates from the tibial tuberosity and inserts on the anteromedial side of the proximal tibia just below the medial condyle.

The **semimembranosus** (SEM-ee-MEM-bran-OH-sus) is a relatively flat, membrane-like muscle that lies beneath (deep) the semitendinosus on the medial side of the posterior thigh (**FIGURE 10.11**). It originates from the ischial tuberosity and inserts on the posterior side of the medial condyle of the tibia.

The **gastrocnemius** (GAS-trock-NEE-me-us) is a two jointed muscle crossing both the knee and ankle (**FIGURE 10.12**). It originates from the posterior side of both femoral condyles and inserts into the posterior calcaneus via the Achilles tendon.

The **plantaris** (PLAN-tear-us) is a long thin muscle that also crosses both the knee and ankle (**FIGURE 10.13**). However, its function offers little to no action and is missing from about 10% of the population. It has a single head that originates from the posterior side of the lateral femoral condyle and inserts on the posterior calcaneus.

The **popliteus** (pop-lah-TEE-us) muscle is a relatively short muscle that obliquely crosses the knee and therefore is not involved in any motion of any other joint (**FIGURE 10.14**). It originates from the lateral condyle of the femur and its insertion is on the posterior side of the medial condyle of the tibia.

FIGURE 10.9 The Biceps Femoris Muscle.

ORIGINS	
LH: ischial tuberosity	
SH: lateral side of linea aspera	
INSERTIONS	
fibular head	
lateral tibial condyle	
OTHER MOTIONS	
LH: hip extension & knee "locking"	
SH: knee "locking"	
NERVES	
LH: Sciatic	
SH: Common Fibular	

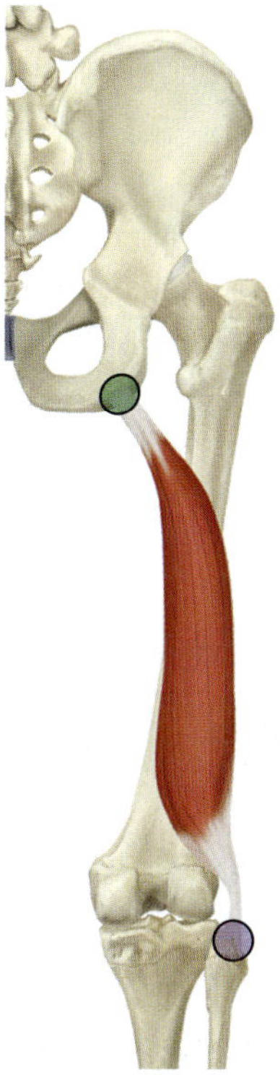

FIGURE 10.10 The Semitendinosus Muscle.

ORIGIN	
ischial tuberosity	
INSERTION	
anteromedial side of proximal tibia	
OTHER MOTIONS	
hip extension	
NERVE	
Sciatic	

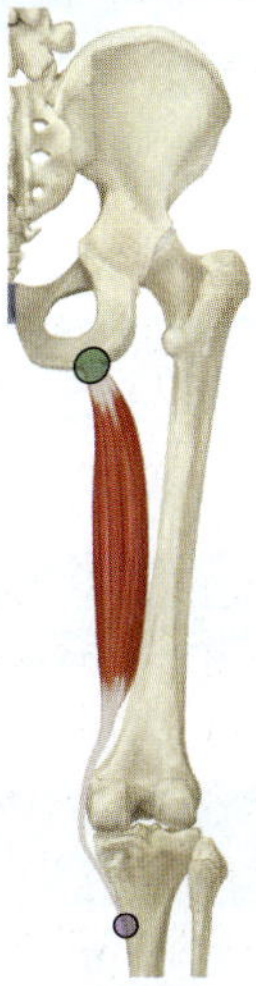

FIGURE 10.11 The Semimembranosus Muscle.

ORIGIN	
ischial tuberosity	
INSERTION	
posterior side of medial tibial condyle	
OTHER MOTIONS	
hip extension	
NERVE	
Sciatic	

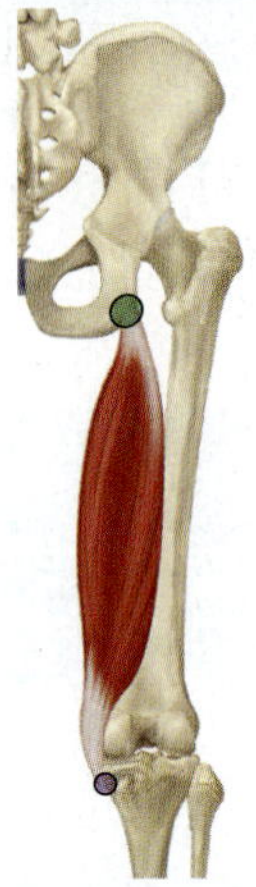

FIGURE 10.12 The Gastrocnemius Muscle.

ORIGINS	
posterior side of medial femoral condyle	
posterior side of lateral femoral condyle	
INSERTION	
posterior calcaneus	
OTHER MOTIONS	
plantar flexion	
NERVE	
Tibial	

FIGURE 10.13 The Plantaris Muscle.

ORIGIN	
posterior side of lateral femoral condyle	
INSERTION	
posterior calcaneus	
OTHER MOTIONS	
plantar flexion	
NERVE	
Tibial	

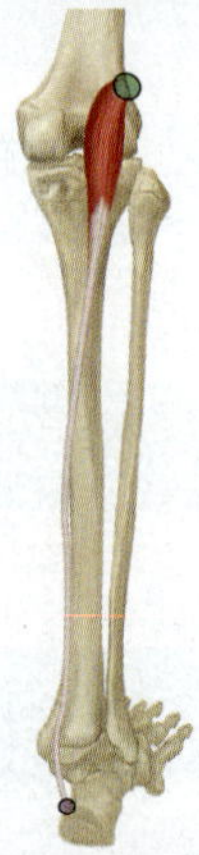

FIGURE 10.14 The Popliteus Muscle.

ORIGIN

lateral side of lateral femoral condyle

INSERTION

posterior side of medial tibial condyle

OTHER MOTIONS

knee "unlocking"

NERVE

Tibial

TABLE 10.2 Summary of the Knee Flexors

Muscle	Origins	Insertions	Nerve
Biceps Femoris (SH)	lateral side of linea aspera	fibular head & lateral condyle of tibia	Common Fibular
Biceps Femoris (LH)			
Semitendinosus	ischial tuberosity	anteromedial side of proximal tibia	Sciatic
Semimembranosus		posterior side of medial condyle of tibia	
Gastrocnemius	posterior sides of medial and lateral condyles of tibia	posterior calcaneus	Sciatic
Plantaris	posterior side of lateral femoral condyle		Tibial
Popliteus	lateral side of lateral condyle of femur	posterior side of medial condyle of tibia	

PAUSE TO CHECK FOR UNDERSTANDING

Table 10.2 should serve as a good review for much of the material, but here are a few other checks for understanding

1. How many muscles are in the knee flexor group? List them.
2. What motion occurs if these muscles act eccentrically?
3. Where is the general location for the muscle group?

The Knee Extensors

There are four muscles in the anterior thigh that extend the knee (**FIGURE 10.15**). They are the *rectus femoris*, *vastus lateralis*, *vastus intermedius*, and *vastus medialis*. The knee extensors are sometimes called the quadriceps (four heads) femoris because they all have different origins, but share a common insertion. These muscles converge distally into a common tendon that passes across the anterior patella and eventually inserts on the tibial tuberosity. Furthermore, as this tendon passes across the patella, fibers extend from the tendon attaching all along the superior pole, anterior surface, and inferior pole of the patella. The section of the tendon from the muscles to the superior pole of the patella is called the **quadriceps tendon**. The section of the tendon from the inferior pole to the tibial tuberosity is called the **patella tendon**. At the end of this section, the knee flexors are summarized in **TABLE 10.3**.

The **rectus femoris** (REK-tus FAH-more-us) is superficial and passes longitudinally through the center of the anterior thigh (**FIGURE 10.16**). It is

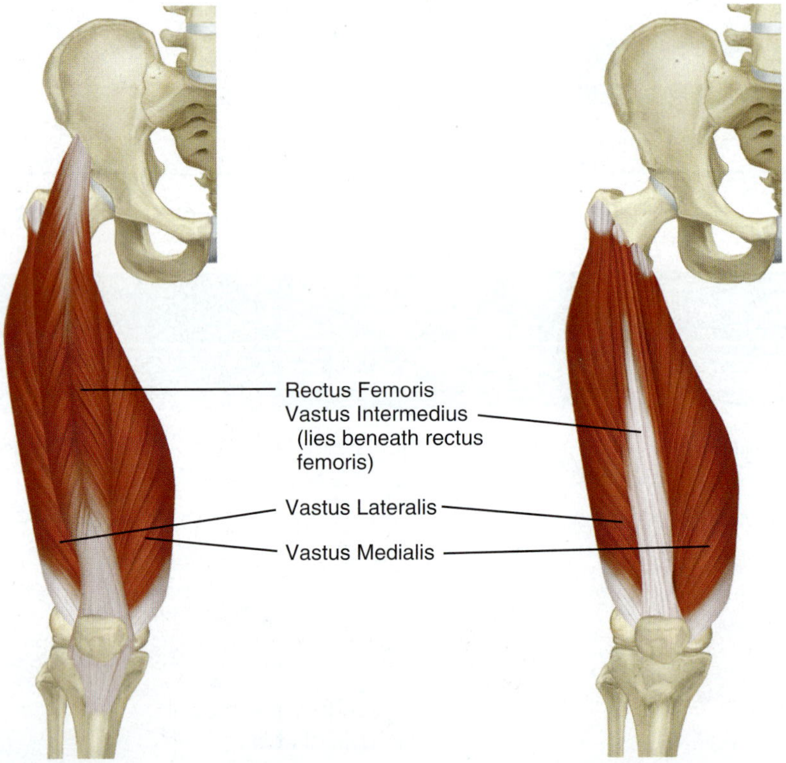

FIGURE 10.15 The Knee Extensors.

FIGURE 10.16 The Rectus Femoris Muscle.

ORIGIN	
anterior inferior iliac spine (AIIS)	
INSERTION	
tibial tuberosity	
OTHER MOTIONS	
hip flexion	
NERVE	
Femoral	

a double jointed muscle in that it crosses both the hip and the knee and therefore acts upon both. It originates from the anterior inferior iliac spine (AIIS) and inserts on the tibial tuberosity via the patellar tendon.

As part of their names, the other three muscles share the word "vastus" which literally means "great." The **vastus lateralis** (lah-TUR-al-us) is the most lateral of the three muscles (**FIGURE 10.17**). It originates from the intertrochanteric line and the linea aspera before inserting on the tibial tuberosity.

The **vastus intermedius** (inter-ME-dee-us) lies deep to the rectus femoris and between the other two vasti muscles (**FIGURE 10.18**). It originates on the anterior, proximal femur just inferior to the level of the trochanters and inserts on the tibial tuberosity.

The **vastus medialis** (VAST-us ME-dee-al-us) lies on the medial side of the anterior thigh (**FIGURE 10.19**). It originates from three locations: the intertrochanteric line, the linea aspera, and the medial supracondylar line. It inserts on the tibial tuberosity.

FIGURE 10.17 The Vastus Lateralis Muscle.

ORIGINS	
intertrochanteric line	
linea aspera	
INSERTION	
tibial tuberosity	
OTHER MOTIONS	
none	
NERVE	
Femoral	

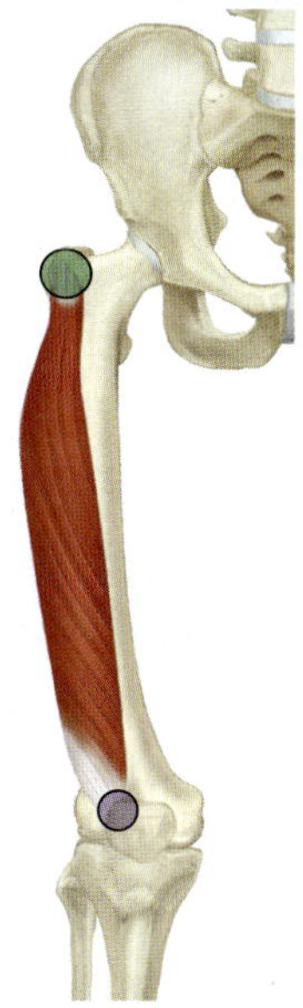

FIGURE 10.18 The Vastus Intermedius Muscle.

ORIGIN	
anterior side of proximal femur	
INSERTION	
tibial tuberosity	
OTHER MOTIONS	
none	
NERVE	
Femoral	

FIGURE 10.19 The Vastus Medialis Muscle.

ORIGIN
intertrochanteric line
linea aspera
medial supracondylar line
INSERTION
tibial tuberosity
OTHER MOTIONS
none
NERVE
Femoral

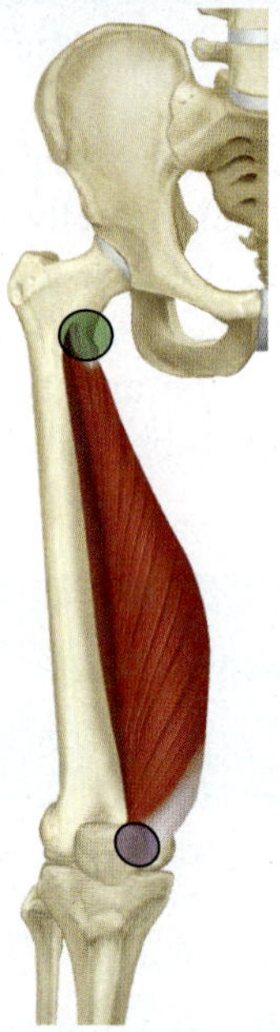

TABLE 10.3 Summary of the Knee Extensors

Muscle	Origins	Insertions	Nerve
Rectus Femoris	AIIS	tibial tuberosity	Femoral
Vastus Lateralis	intertrochanteric line & linea aspera		
Vastus Intermedius	anterior side of proximal femur		
Vastus Medialis	intertrochanteric line, linea aspera, & medial supracondylar line		

PAUSE TO CHECK FOR UNDERSTANDING

Table 10.3 should serve as a good review for much of the material, but here are a few other checks for understanding

1. How many muscles are in the knee extensor group? List them.
2. What motion occurs if these muscles act eccentrically?
3. Where is the general location for the muscle group?

CHAPTER 11

The Hip

CHAPTER OBJECTIVES

After completing this chapter, the student will be able to:

1. summarize the joints of the hip region;
2. comprehend the hip motions;
3. identify the names and locations of the hip muscles;
4. recall the attachment sites and nerves of the hip muscles; and
5. apply the motions produced by the actions of the hip muscles.

The hip is a region of the body where the thigh and pelvis intersect (**FIGURE 11.1**). Only two bones intersect here. Therefore, the hip is comprised by just one joint which has been classified as a ball and socket joint. Its muscles and joint structure is able to produce six different motions.

The Joints of the Hip

The one joint in the hip is formally known as the **coxal joint** (**FIGURE 11.2**). The two bones that intersect and form the coxal joint are the femur and coxal bone. More specifically, the head (ball) of the femur articulates with the acetabulum (socket) of the coxal bone. The joint is strengthened by a **labrum** (cartilage) that deepens the acetabulum.

Several strong ligaments stabilize the coxal joint (**FIGURE 11.3**). On the interior of the coxal joint is the **round ligament**. The main ligament on the anterior side is the **pubiofemoral ligament**. The main ligament on the superior side is the **iliofemoral ligament** and the main ligament on the posterior side is the **ischiofemoral ligament**.

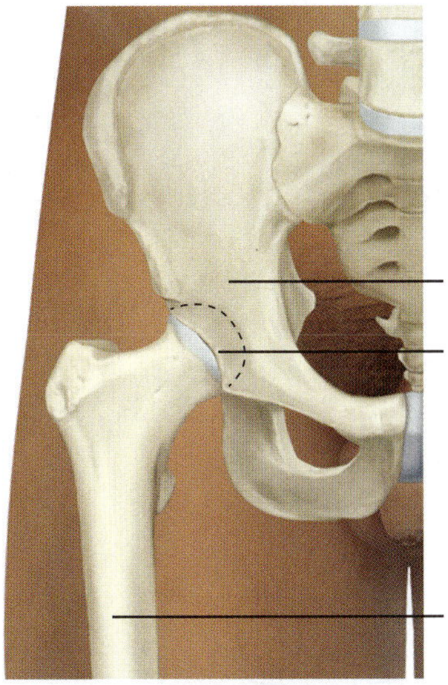

FIGURE 11.1 The Hip.

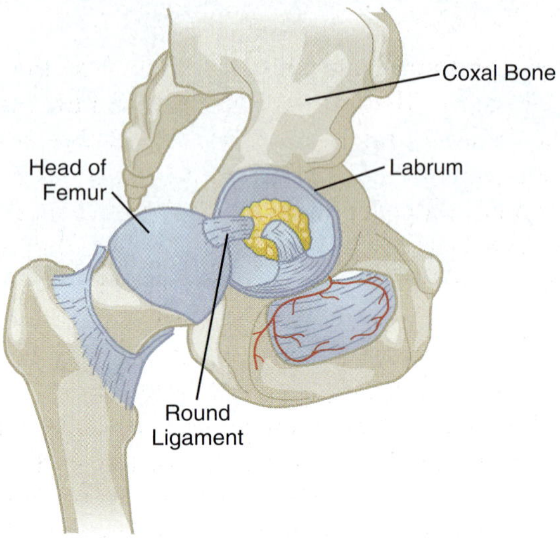

FIGURE 11.2 The Coxal Joint (lateral view).

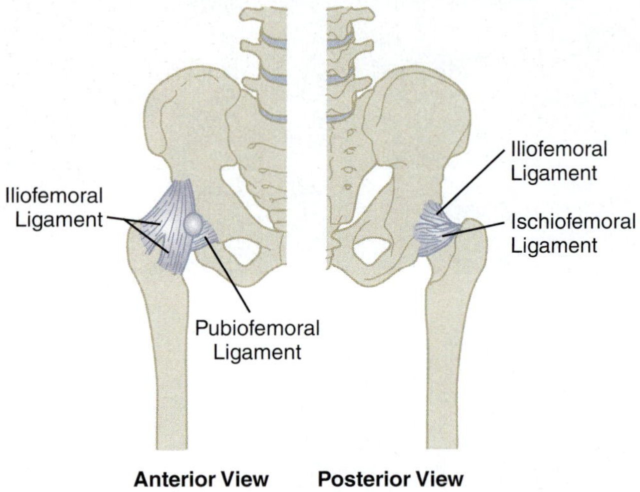

FIGURE 11.3 The Ligaments of the Coxal Joint.

PAUSE TO CHECK FOR UNDERSTANDING

1. List the joints of the hip.
2. Then, tell which bones form each joint.
3. What are the six motions of the hip?

The Motions of the Hip

Six motions occur at the coxal joint (**FIGURE 11.4**). They are extension, flexion, abduction, adduction, external (lateral) rotation, and internal (medial) rotation. **Hip extension** and **flexion** straighten and bend the hip in the sagittal plane about the frontal axis. **Hip adduction** and **abduction** move the hip in the frontal plane rotating about the sagittal axis. **Hip external** and **internal rotation** rotate the hip in the transverse plane about the longitudinal axis.

TABLE 11.1 summarizes the average ranges of motions of the hip. Average hip extension is 10–20° and for flexion is 120–130°. Average hip adduction is 20–30° and for abduction is 45°. Average hip internal rotation is 45° and for external rotation is 40–50°.

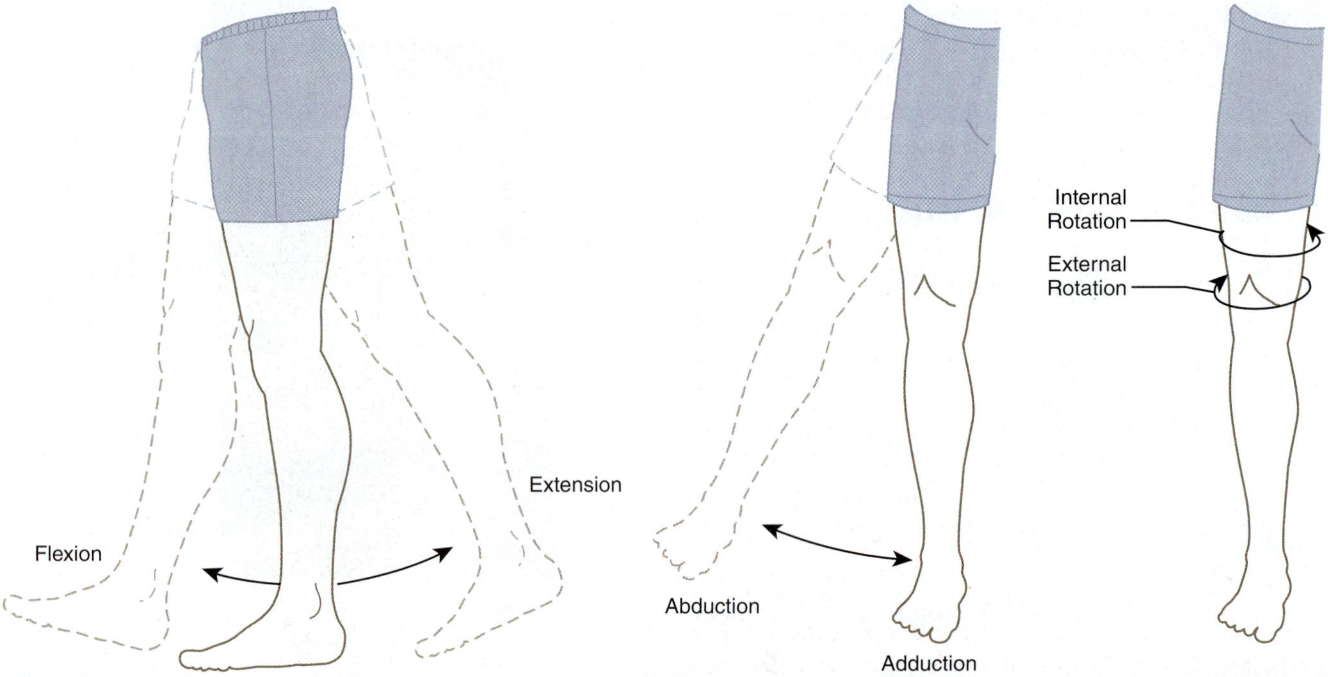

FIGURE 11.4 Hip Motions.

TABLE 11.1 Average Hip Range of Motions

Motion	Average Degrees of Motion
Extension	10–20°
Flexion	120–130°
Combined Extension-Flexion	130–150°
Adduction	20–30°
Abduction	45°
Combined Adduction-Abduction	65–75°
Internal Rotation	45°
External Rotation	40–50°
Combined Internal-External Rotation	85–95°

The Muscles that Move the Hip

The rest of this chapter is dedicated to learning the hip muscles; their attachment sites; the *primary*, main motions they produce; and the nerves that innervate them. Muscles work in groups and so each muscle within the group supplies a certain percentage of the force needed to move a body segment and that amount (percentage) is not the same for each muscle. Furthermore, the force each muscle contributes varies depending on the starting position of the body segment and changes throughout the range of motion. This can be confusing to the beginning kinesiology student. So, this text has made certain choices in an attempt to simplify the muscle groups based on initial motion from anatomical position.

The muscle groups are based on the motion caused by their *concentric* action (i.e., The hip flexors flex the hip when acting concentrically.). However, it should not be forgotten that all muscles can and will act *eccentrically* and *isometrically*. The reader should also notice that the table for each individual muscle has a section called "Other Motions." *Other Motions* means that this muscle aides, to some degree, in other motion(s) besides the group it has been placed in. Some muscles may appear in more than one group because their role is too great to ignore.

There are six muscle groups of the hip because there are six motions. They are the *hip flexors, hip extensors, hip abductors, hip adductors, hip external rotators*, and *hip internal rotators*.

The Hip Flexors

There are four main muscles that flex the hip (**FIGURE 11.5**). They are the *iliacus* (ill-LEE-a-cus), *psoas* (SO-as) *major*, *sartorius* (SAR-tore-ee-us), and *rectus femoris*. The iliacus and the psoas major are sometimes collectively called the **iliopsoas** (ill-LEE-oh-SO-as). This is because although they have different origins and nerve innervations, they have the same insertion like the gastrocnemius and soleus muscles that move the ankle. At the end of this section, the hip flexors are summarized in **TABLE 11.2.**

The **psoas** (SO-as) **major** muscle originates from the lateral side of the bodies of vertebrae T12–L5 (**FIGURE 11.6**). It merges with the iliacus muscle and they insert on the lesser trochanter and just distally to that on the medial side of the diaphysis.

The **iliacus** (ill-LEE-a-cus) **muscle** lies deep within the pelvis (**FIGURE 11.7**). It originates from the medial edge of the iliac spine, the iliac fossa, and the anterior side of superior end of the lateral sacrum. It merges with the psoas major muscle and they insert

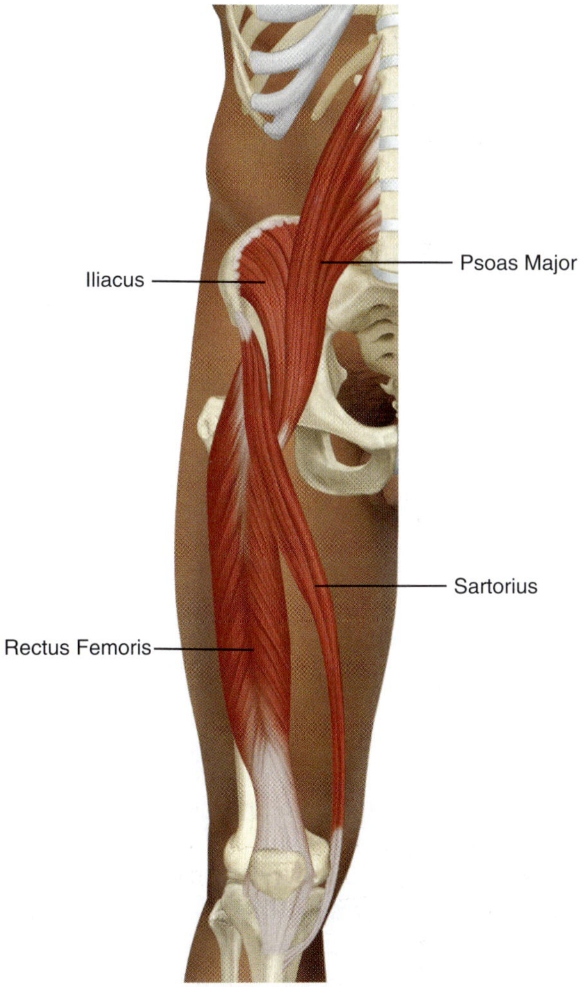

FIGURE 11.5 The Hip Flexors.

FIGURE 11.6 The Psoas Major Muscle of the Iliopsoas.

ORIGINS
lateral side of vertebral bodies T12–L5
INSERTIONS
lesser trochanter
medial side of proximal femoral diaphysis
OTHER MOTIONS
none
NERVES
Lumbar Nerve Roots

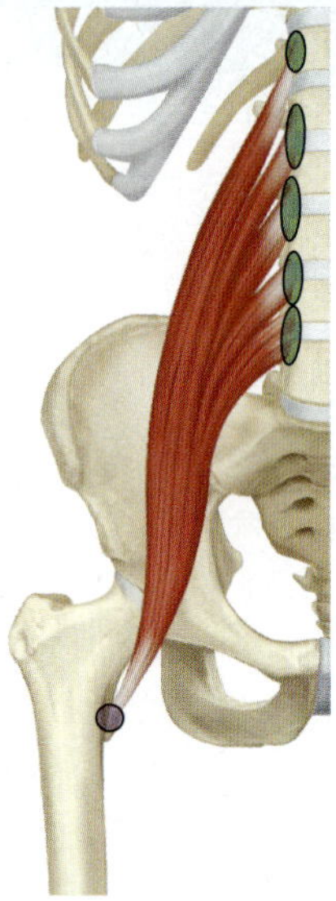

FIGURE 11.7 The Iliacus Muscle of the Iliopsoas.

ORIGINS
medial edge of the iliac spine
iliac fossa
anterior side of superior end of lateral sacrum
INSERTION
lesser trochanter
medial side of proximal femur
OTHER MOTIONS
none
NERVE
Femoral

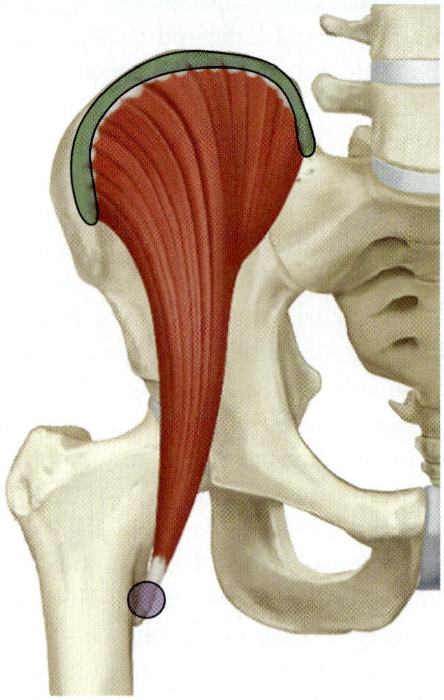

on the lesser trochanter and just distally to that on the medial side of the diaphysis.

The **sartorius** (SAR-tore-ee-us) **muscle** is the longest in the body (**FIGURE 11.8**). It originates from the anterior superior iliac spine (ASIS), crosses the hip, then obliquely crosses the thigh, and inserts just below the medial tibial condyle on the medial side.

The **rectus femoris** (REK-tus FAH-more-us) is superficial and passes longitudinally through the center of the anterior thigh (**FIGURE 11.9**). It is a double jointed muscle in that it crosses both the hip and the knee and therefore acts upon both. It originates from the anterior inferior iliac spine (AIIS) and inserts on the tibial tuberosity via the patellar tendon.

FIGURE 11.8 The Sartorius Muscle.

ORIGIN

ASIS

INSERTION

medial side of proximal tibia

OTHER MOTIONS

hip lateral rotation

hip abduction

knee flexion

NERVE

Femoral

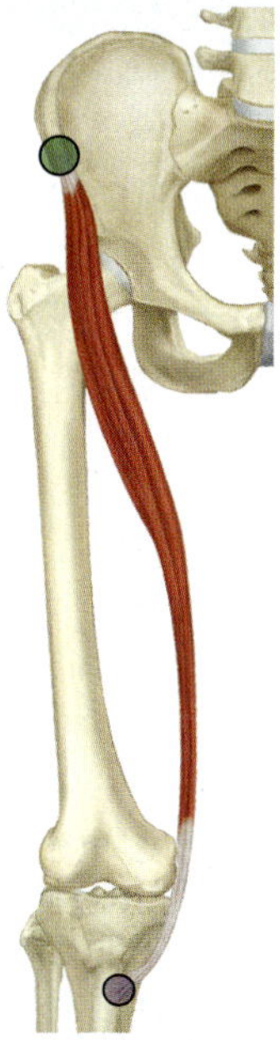

FIGURE 11.9 The Rectus Femoris Muscle.

ORIGIN

AIIS

INSERTION

tibial tuberosity

OTHER MOTIONS

knee extension

NERVE

Femoral

TABLE 11.2 Summary of the Hip Flexors

<table>
<tr><th>Muscle</th><th>Origins</th><th>Insertions</th><th>Nerve</th></tr>
<tr><td>Psoas Major</td><td>lateral side of vertebral bodies T12–L5</td><td rowspan="2">lesser trochanter & medial side of proximal femur</td><td>Lumbar Nerve Roots</td></tr>
<tr><td>Iliacus</td><td>medial edge of the iliac spine, iliac fossa, & anterior side of superior end of lateral sacrum</td><td rowspan="3">Femoral</td></tr>
<tr><td>Sartorius</td><td>ASIS</td><td>medial side of proximal tibia</td></tr>
<tr><td>Rectus Femoris</td><td>AIIS</td><td>tibial tuberosity</td></tr>
</table>

PAUSE TO CHECK FOR UNDERSTANDING

Table 11.2 should serve as a good review for much of the material, but here are a few other checks for understanding:

1. How many muscles are in the hip flexor group? List them.
2. What motion occurs if these muscles act eccentrically?
3. Where is the general location for the muscle group?

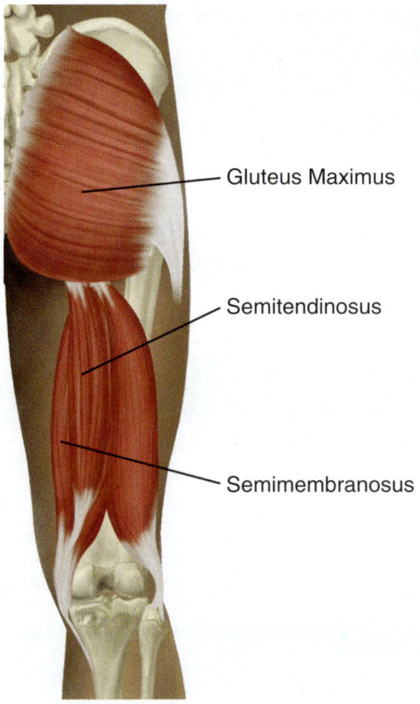

FIGURE 11.10 The Hip Extensors.

The Hip Extensors

There are four main muscles that extend the hip (**FIGURE 11.10**). They are the *gluteus maximus*, the *biceps femoris*, the *semitendinosus*, and the *semimembranosus*. However, the biceps femoris has two sections called the long head and the short head. Only the long head originates proximal to the coxal joint and therefore acts upon it. At the end of this section, the hip extensors are summarized in **TABLE 11.3.**

The **gluteus** (GLUE-tee-us) **maximus** is the largest and strongest muscles in the body (**FIGURE 11.11**). It originates from four locations. One origin is on the lateral edge of the iliac spine from the posterior portion to the posterior superior iliac spine (PSIS). It also originates from the posterior gluteal line and the lateral edges of the posterior sacrum and coccyx. Arising from these locations, it obliquely covers the majority of the gluteal region and inserts on the gluteal tuberosity of

FIGURE 11.11 The Gluteus Maximus Muscle.

ORIGINS
lateral edge of posterior portion of iliac spine
posterior gluteal line
posterior side of lateral sacrum
posterior side of lateral coccyx
INSERTIONS
gluteal tuberosity of femur
lateral tibial condyle
OTHER MOTIONS
hip external rotation
hip abduction
NERVE
Inferior Gluteal

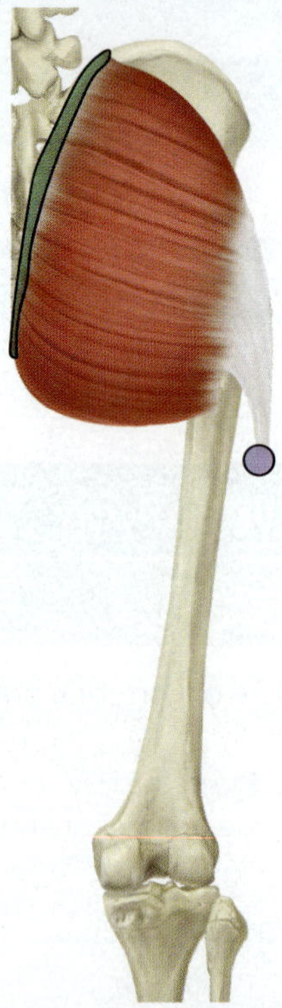

the femur. Also, some of the muscle fibers merge with the iliotibial (IT) tract (or band). The IT band is connective tissue that passes longitudinally down the lateral thigh eventually inserting on the lateral side of the lateral tibial condyle and so is a second insertion site.

The **biceps femoris** is another hip extensor (**FIGURE 11.12**). However, only the fibers of its long head are involved in hip extension. The long head originates on the ischial tuberosity of the coxal bone and passes along the posterior mid-lateral thigh. The origin of the short head has been omitted because it is not involved in this motion. After merging with the fibers of the short head, the muscle inserts on the fibular head and the lateral tibial condyle.

The belly of the **semitendinosus** (SEM-ee-TEN-din-OH-sus) is rod-shaped (like a tendon) residing in the medial side of the posterior thigh superficial to the semimembranosus (**FIGURE 11.13**). It originates from the tibial tuberosity and inserts on the anteromedial side of the proximal tibia just below the medial condyle.

The **semimembranosus** (SEM-ee-MEM-bran-OH-sus) is a relatively flat, membrane-like muscle that lies beneath (deep) to the semitendinosus on the medial side of the posterior thigh (**FIGURE 11.14**). It originates from the ischial tuberosity and inserts on the posterior side of the medial condyle of the tibia.

FIGURE 11.12 The Long Head of the Biceps Femoris Muscle.

ORIGIN	
ischial tuberosity	
INSERTIONS	
fibular head	
lateral tibial condyle	
OTHER MOTIONS	
knee flexion	
NERVE	
Sciatic	

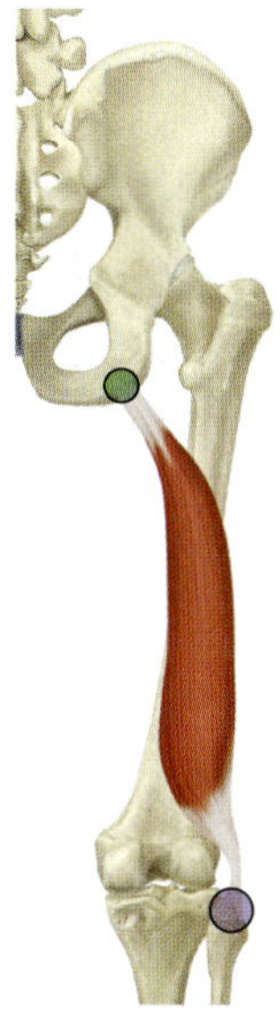

FIGURE 11.13 The Semitendinosus Muscle.

ORIGIN	
ischial tuberosity	
INSERTIONS	
anteromedial side of proximal tibia	
OTHER MOTIONS	
knee flexion	
NERVE	
Sciatic	

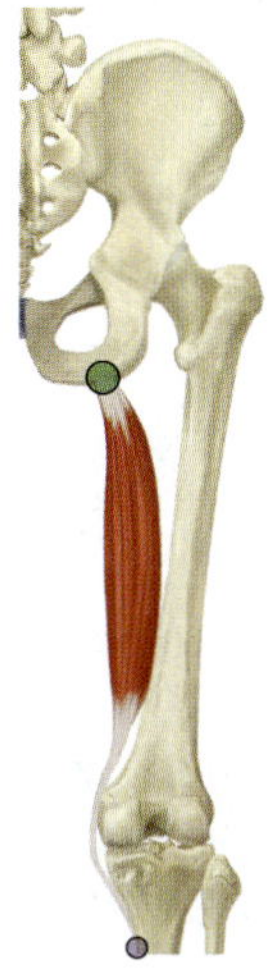

FIGURE 11.14 The Semimembranosus Muscle.

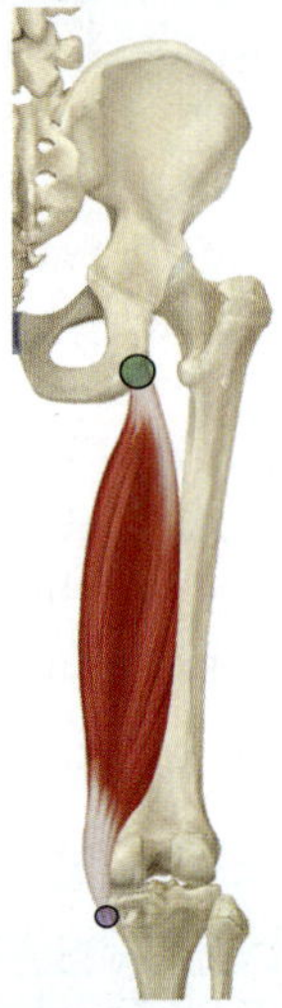

ORIGIN	ischial tuberosity
INSERTIONS	posterior side of medial tibial condyle
OTHER MOTIONS	knee flexion
NERVE	Sciatic

TABLE 11.3 Summary of the Hip Extensors

Muscle	Origins	Insertions	Nerve
Gluteus Maximus	posterior sides of lateral edge of iliac spine, of gluteal line, of lateral sacrum, & of lateral coccyx	gluteal tuberosity of femur & lateral tibial condyle	Inferior Gluteal
Biceps Femoris (Long Head)	ischial tuberosity	fibular head & lateral condyle of tibia	Sciatic
Semitendinosus		anteromedial side of proximal tibia	
Semimembranosus		posterior side of medial condyle of tibia	

PAUSE TO CHECK FOR UNDERSTANDING

Table 11.3 should serve as a good review for much of the material, but here are a few other checks for understanding:

1. How many muscles are in the hip extensor group? List them.
2. What motion occurs if these muscles act eccentrically?
3. Where is the general location for the muscle group?

The Hip Abductors

There are two muscles that abduct the hip (**FIGURE 11.15**). They are the gluteus medius and the tensor fasciae latae. The gluteus maximus also plays a fairly significant role in hip abduction. However, it will not be repeated since it has already been described in the Hip Flexor ection. The hip abductors are summarized in **TABLE 11.4** at the end of this section.

The **gluteus medius** (ME-dee-us) lies deep beneath the gluteus maximus in the gluteal region

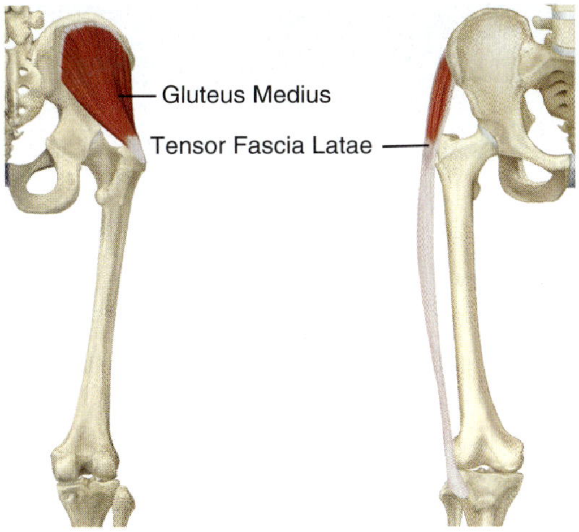

FIGURE 11.15 The Hip Abductors.

(**FIGURE 11.16**). It originates from the ilium just inferior to the lateral side of the iliac spine. Its belly extends over the lateral coxal bone and inserts on the greater trochanter.

The **tensor fasciae latae** (TEN-sur FASH-ee-ee LAY-tee) resides in the anterolateral thigh (**FIGURE 11.17**). It originates from the anterior superior iliac spine (ASIS) of the coxal bone. Its insertion is via the iliotibial (IT) tract (or band). As stated earlier, the IT band is a connective tissue that passes longitudinally down the lateral thigh eventually terminating (inserting) on the lateral side of the lateral tibial condyle. All of the fibers of the tensor fasciae latae merge into the IT band at a similar junction as a portion of the gluteus maximus.

FIGURE 11.16 The Gluteus Medius Muscle.

ORIGIN
lateral side of ilium just inferior to iliac spine
INSERTION
greater trochanter
OTHER MOTIONS
none
NERVE
Superior Gluteal

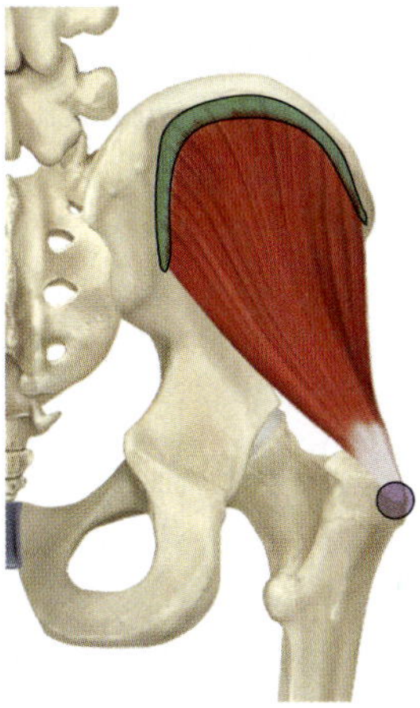

FIGURE 11.17 The Tensor Fasciae Latae Muscle.

ORIGINS
ASIS
INSERTIONS
lateral tibial condyle
OTHER MOTIONS
hip flexion
hip internal rotation
NERVE
Superior Gluteal

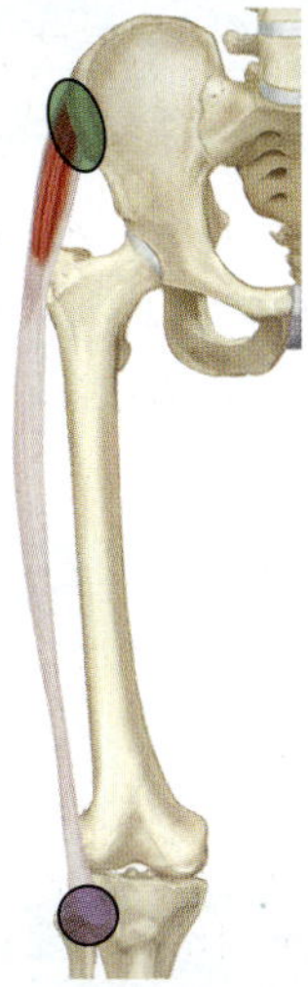

TABLE 11.4 Summary of the Hip Abductors

Muscle	Origins	Insertions	Nerve
Gluteus Medius	lateral side of ilium just inferior to iliac spine	greater trochanter	Superior Gluteal
Tensor Fasciae Latae	ASIS	lateral tibial condyle	

PAUSE TO CHECK FOR UNDERSTANDING

Table 11.4 should serve as a good review for much of the material, but here are a few other checks for understanding

1. How many muscles are in the hip abductor group? List them.
2. What motion occurs if these muscles act eccentrically?
3. Where is the general location for the muscle group?

The Hip Adductors

Five muscles adduct the hip (**FIGURE 11.18**). They are the *pectineus*, the *adductor brevis*, the *adductor longus*, the *gracilis*, and the *adductor magnus*. The hip abductors are summarized in **TABLE 11.5** at the end of this section.

The **pectineus** (PECK-tee-knee-us) is the most superior of this group of five muscles (**FIGURE 11.19**). It originates from the superior ramus of the pubis and passes through the groin. It then inserts on the pectineal line of the femur.

The **adductor brevis** lies inferior to the pectineus in the groin (**FIGURE 11.20**). It originates on the body and inferior ramus of the pubis and inserts on the pectineal line and the proximal end of the linea aspera.

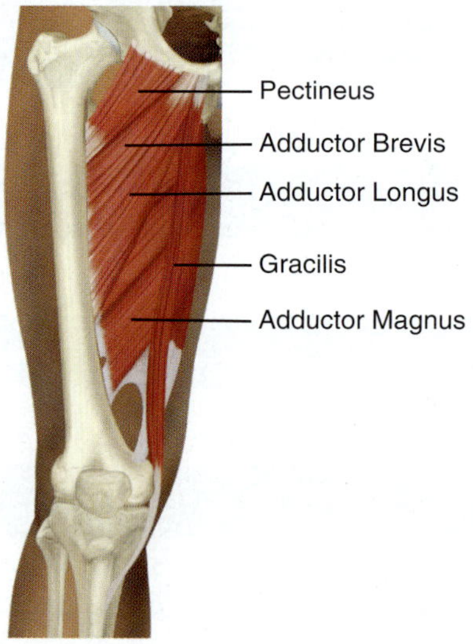

FIGURE 11.18 The Hip Adductors.

FIGURE 11.19 The Pectineus Muscle.

ORIGIN

superior ramus of pubis

INSERTION

pectineal line

OTHER MOTIONS

hip flexion

minimal assistance in other motions

NERVE

Femoral

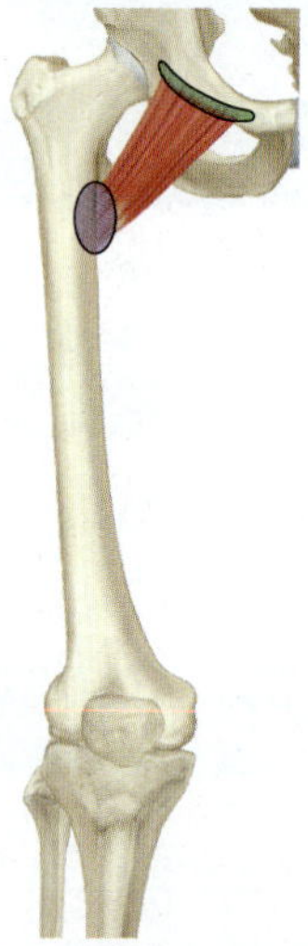

The **adductor longus** lies just inferior to the adductor brevis muscle (**FIGURE 11.21**). It originates on the body and inferior ramus of the pubis just inferior to the adductor brevis origin. It then passes through the groin and inserts on the middle one-third of the linea aspera.

The **gracilis** is the most medial of the other hip adductors and spans the length of the medial thigh (**FIGURE 11.22**). It originates from the superior ramus of the pubis and inserts just below the medial tibial condyle on the anteromedial side of the diaphysis.

The **adductor magnus** lies deep to the rest of the hip adductors as it passes along the medial thigh (**FIGURE 11.23**). It originates from three locations: the inferior ramus of the pubis, the ischial ramus, and the ischial tuberosity. It inserts along the linea aspera and on the adductor tubercle of the femur.

FIGURE 11.20 The Adductor Brevis Muscle.

ORIGIN
body of pubis
inferior ramus of pubis
INSERTION
pectineal line of femur
proximal linea aspera
OTHER MOTIONS
hip flexion
minimal assistance in other motions
NERVE
Obturator

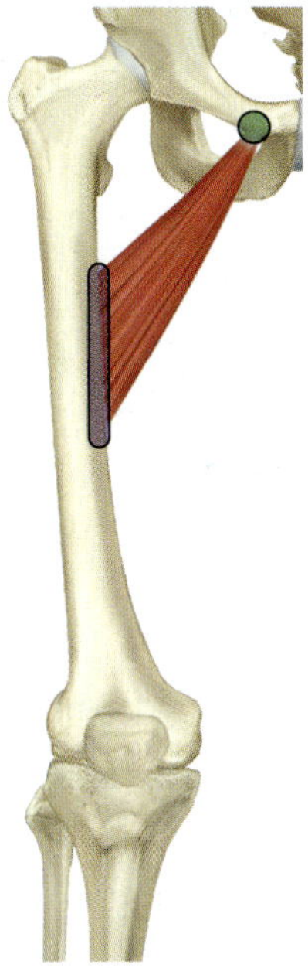

FIGURE 11.21 The Adductor Longus Muscle.

ORIGINS
body of pubis
inferior ramus of pubis
INSERTION
middle 1/3 of linea aspera
OTHER MOTIONS
hip flexion
minimal assistance in other motions
NERVE
Obturator

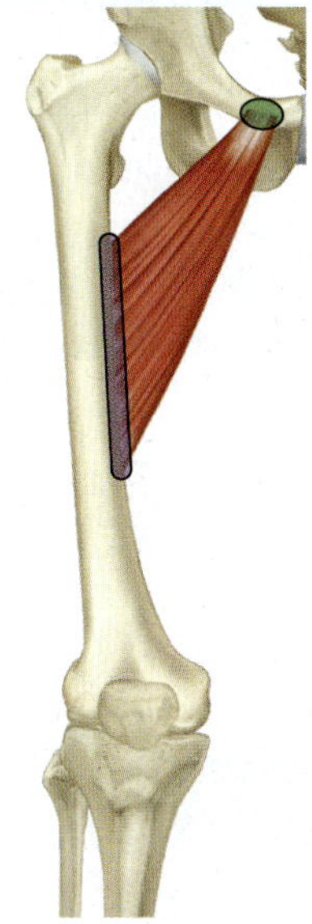

FIGURE 11.22 The Gracilis Muscle.

ORIGINS

superior ramus of pubis

INSERTION

anteromedial side of proximal tibia

OTHER MOTIONS

minimal assistance in other motions

NERVE

Obturator

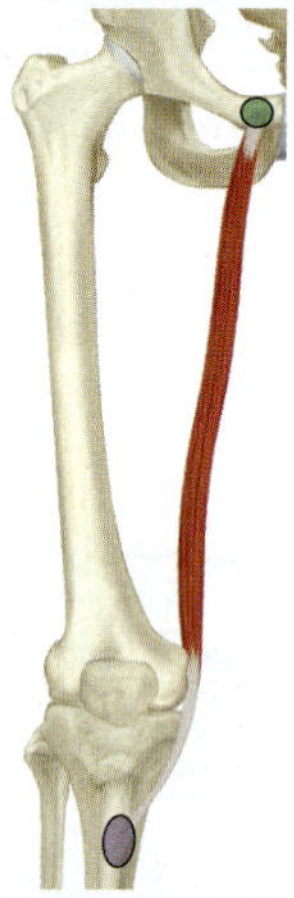

FIGURE 11.23 The Adductor Magnus Muscle.

ORIGINS

inferior ramus of pubis

ischial ramus

ischial tuberosity

INSERTIONS

linea aspera

adductor tubercle of femur

OTHER MOTIONS

minimal assistance in other motions

NERVES

Obturator & Tibial

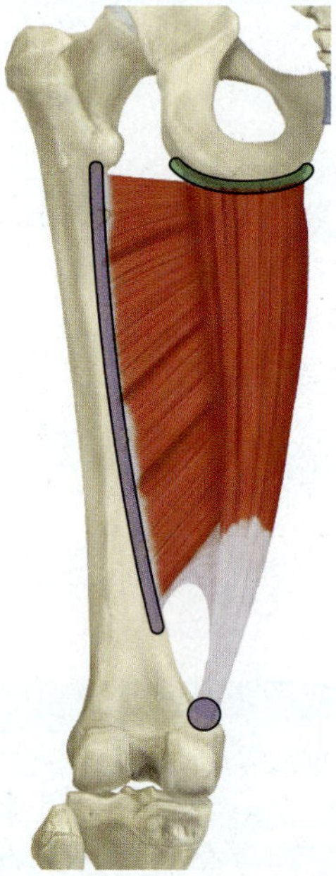

TABLE 11.5 Summary of the Hip Adductors

Muscle	Origins	Insertions	Nerve
Pectineus	superior ramus of pubis	pectineal line	Femoral
Adductor Brevis	body & inferior ramus of pubis	pectineal line of femur & proximal linea aspera	Obturator
Adductor Longus		middle 1/3 of linea aspera	
Gracilis	superior ramus of pubis	anteromedial side of proximal tibia	
Adductor Magnus	inferior ramus, ischium ramus, & ischial tuberosity	linea aspera & adductor tubercle of femur	Obturator & Tibial

PAUSE TO CHECK FOR UNDERSTANDING

Table 11.5 should serve as a good review for much of the material, but here are a few other checks for understanding

1. How many muscles are in the hip adductor group? List them.
2. What motion occurs if these muscles act eccentrically?
3. Where is the general location for the muscle group?

The Hip External Rotators

Six muscles externally rotate the hip (**FIGURE 11.24**). They are the piriformis, the gemellus superior, the gemellus inferior, the obturator internus, the obturator externus, and the quadratus femoris. All the hip external rotators run relatively transverse across the gluteal region originating in various locations on the pelvic bones and inserting on the proximal femur. At the end of this section, the hip abductors are summarized in **TABLE 11.6.**

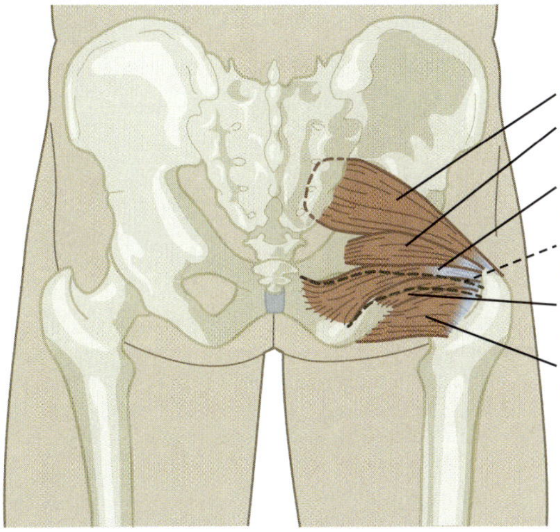

FIGURE 11.24 The Hip External Rotators.

The **piriformis** (PIER-eh-for-mis) is the most superior of the hip lateral rotators originating from the anterior side of the sacrum (**FIGURE 11.25**). It then passes across the gluteal region and inserts on the posterolateral side of the greater trochanter of the femur.

The **gemellus superior** (gah-MEL-lus) originates from the ischial spine (**FIGURE 11.26**). Its distal fibers run deep to the pectineus and obturator internus inserting on the posterolateral side of the greater trochanter of the femur.

The **gemellus inferior** lies inferior to the superior gemellus (**FIGURE 11.27**). It originates from the ischial tuberosity and inserts on the posterolateral side of the greater trochanter of the knee.

The **obturator internus** lies superficial to the gemelli muscles originating from the ischial ramus (**FIGURE 11.28**). It inserts superficially on the posterolateral side of the greater trochanter of the femur.

Most of the **obturator externus** is deep to the other hip external rotator muscles (**FIGURE 11.29**). It originates from the ischial ramus and inserts on the posterolateral side of the greater trochanter of the femur.

The most inferior muscle in this group is the **quadratus femoris** (**FIGURE 11.30**). Its belly is rectangular, hence its name quadratus. It originates from the ischial tuberosity and inserts on the intertrochanteric crest of the femur.

FIGURE 11.25 The Piriformis Muscle.

ORIGIN	
anterior side of sacrum	
INSERTION	
posterolateral side of greater trochanter	
OTHER MOTIONS	
minimal assistance in other motions	
NERVES	
L5–S2 Nerve Roots	

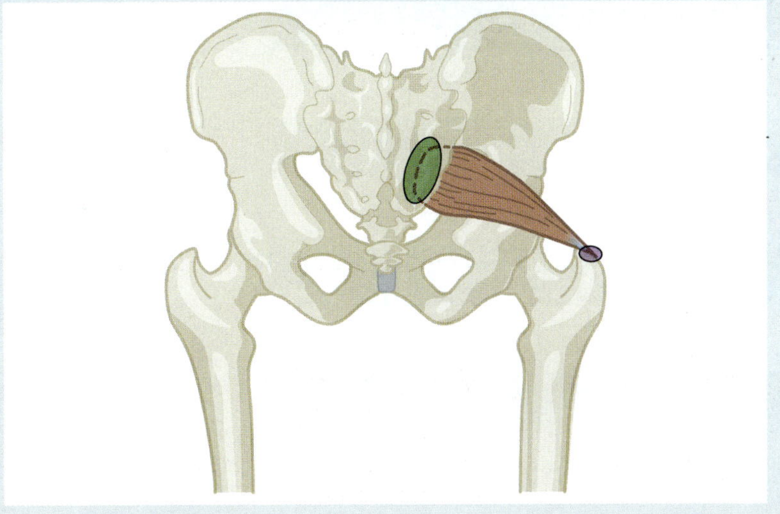

FIGURE 11.26 The Gemellus Superior Muscle.

ORIGIN
ischial spine
INSERTION
posterolateral side of greater trochanter
OTHER MOTIONS
minimal assistance in other motions
NERVES
L5–S2 Nerve Roots

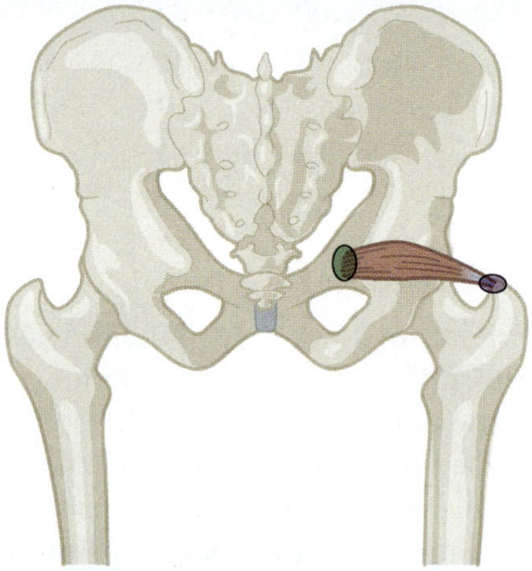

FIGURE 11.27 The Gemellus Inferior Muscle.

ORIGIN
ischial tuberosity
INSERTION
posterolateral side of greater trochanter
OTHER MOTIONS
minimal assistance in other motions
NERVES
L4–S2 Nerve Roots

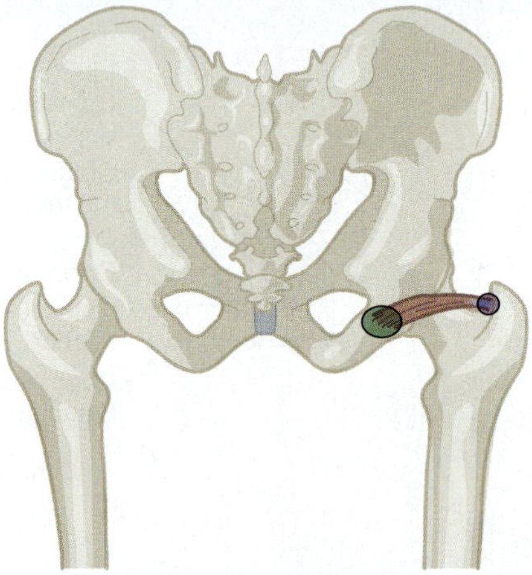

FIGURE 11.28 The Obturator Internus Muscle.

ORIGIN
ischial ramus
INSERTION
posterolateral side of greater trochanter
OTHER MOTIONS
minimal assistance in other motions
NERVES
L4–S2 Nerve Roots

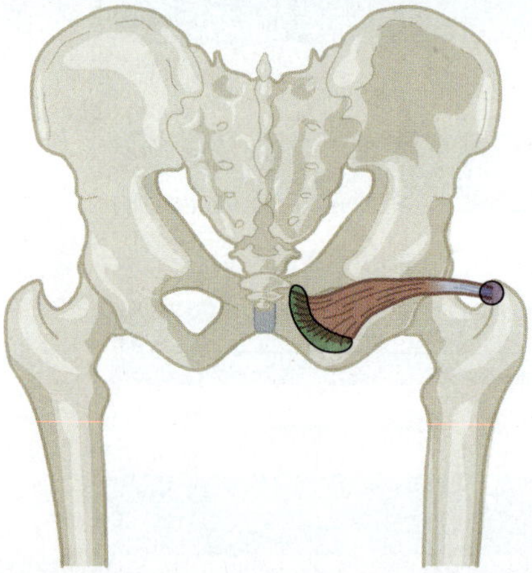

FIGURE 11.29 The Obturator Externus Muscle.

ORIGIN

ischial ramus

INSERTION

posterolateral side of greater trochanter

OTHER MOTIONS

minimal assistance in other motions

NERVE

Obturator

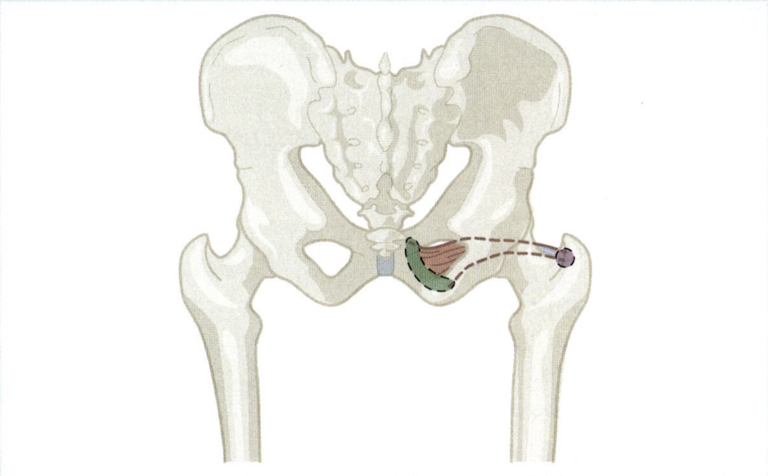

FIGURE 11.30 The Quadratus Femoris Muscle.

ORIGIN

ischial tuberosity

INSERTION

intertrochanteric crest

OTHER MOTIONS

minimal assistance in other motions

NERVES

L4–S1 Nerve Roots

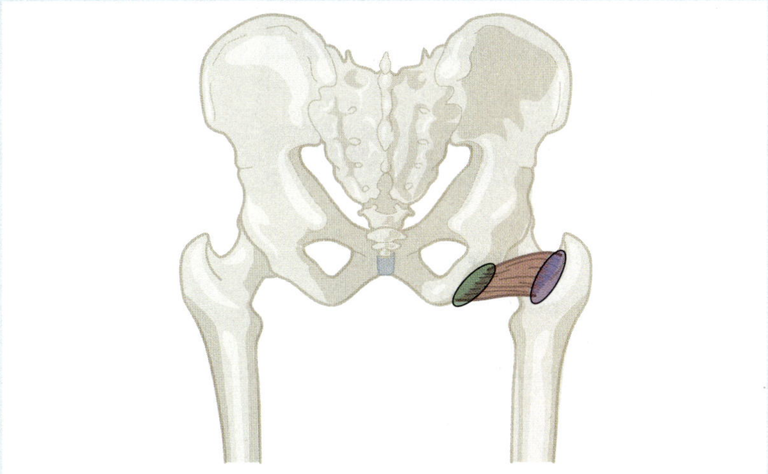

TABLE 11.6 Summary of the Hip External Rotators

Muscle	Origins	Insertions	Nerve
Piriformis	anterior side of sacrum	posterolateral side of greater trochanter	L5–S2 Nerve Roots
Gemellus Superior	ischial spine	posterolateral side of greater trochanter	L5–S2 Nerve Roots
Gemellus Inferior	ischial tuberosity	posterolateral side of greater trochanter	L4–S2 Nerve Roots
Obturator Internus	ischial ramus	posterolateral side of greater trochanter	L4–S2 Nerve Roots
Obturator Externus	ischial ramus	posterolateral side of greater trochanter	Obturator Nerve
Quadratus Femoris	ischial tuberosity	intertrochanteric crest	L4–S1 Nerve Roots

PAUSE TO CHECK FOR UNDERSTANDING

Table 11.6 should serve as a good review for much of the material, but here are a few other checks for understanding

1. How many muscles are in the hip external rotator group? List them.
2. What motion occurs if these muscles act eccentrically?
3. Where is the general location for the muscle group?

The Hip Internal Rotators

Although some muscles assist in hip internal rotation, only one muscle is primarily responsible for this motion. The **gluteus minimus** originates from the lateral side of the ilium inferior to the spine and deep to, as well as inferior to, the origin of the gluteus medius (**FIGURE 11.31**). Passing deep to the gluteus medius across the lateral ilium, it inserts on the anterior side of the greater trochanter.

FIGURE 11.31 The Gluteus Minimus Muscle.

ORIGIN	lateral ilium inferior to gluteus medius origin
INSERTION	anterior side of greater trochanter
OTHER MOTIONS	minimal assistance in other motions
NERVE	Superior Gluteal

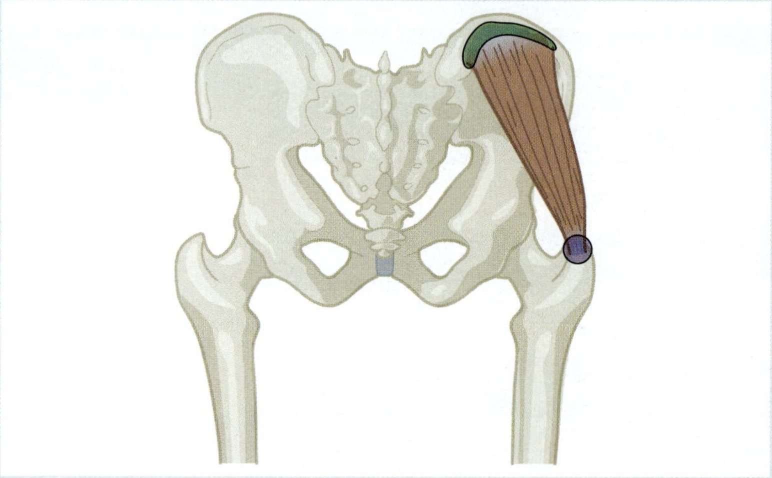

© BraunS/E+/Getty Images

SECTION 4

The Axial Region

CHAPTER 12

The Trunk

CHAPTER OBJECTIVES

After completing this chapter, the student will be able to:

1. summarize the joints of the trunk region;
2. comprehend the trunk motions;
3. identify the names and locations of the trunk muscles;
4. recall the attachment sites and nerves of the trunk muscles; and
5. apply the motions produced by the actions of the trunk muscles.

The trunk is a region of the body inferior to the neck that ends where the upper and lower extremities begin (**FIGURE 12.1**). The trunk contains numerous joints between the vertebrae of the spine as well as between the sacrum and fifth lumbar vertebra. The arrangement of the bones and the actions of the muscles produce six motions.

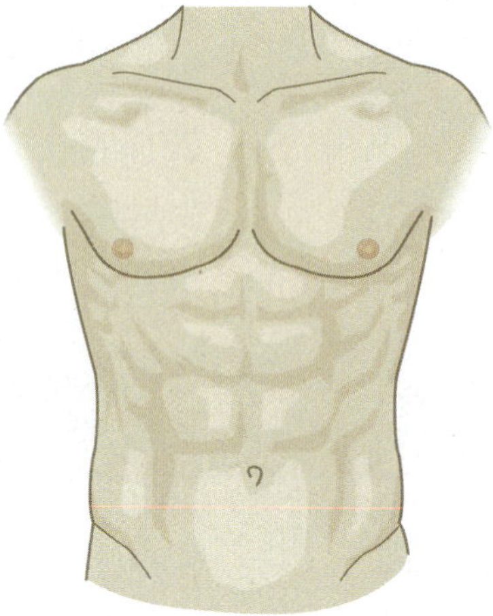

FIGURE 12.1 The Trunk.

The Joints of the Trunk

Numerous intervertebral joints and the lumbosacral joint are involved in the trunk motions. An **intervertebral joint** is the articulation between one vertebrae and the next (**FIGURE 12.2**). Between each vertebra is an intervertebral disc which protects the ends of the bones and absorbs compression forces.

Six main ligaments stabilize the intervertebral joints. The **anterior longitudinal ligament** attaches on the anterior surface of each vertebral body as it runs the entire length of the column. The **posterior longitudinal ligament** spans the entire length of the column stabilizing the posterior bodies of each vertebra as it passes through the vertebral foramina. The **ligamentum flavum** attaches the lamina of each vertebra. Superficially, the **supraspinous ligament** spans the vertebral column from spinous process to spinous process. Deep to that ligament, the **interspinous ligaments** also connect the spinous processes. Finally, the **intertransverse ligaments** stabilize all the transverse processes.

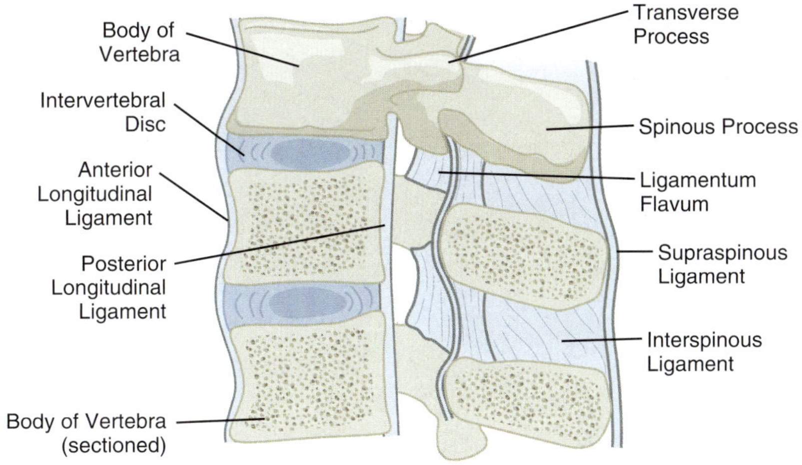

FIGURE 12.2 Intervertebral Joints (A Section of the Vertebral Column).

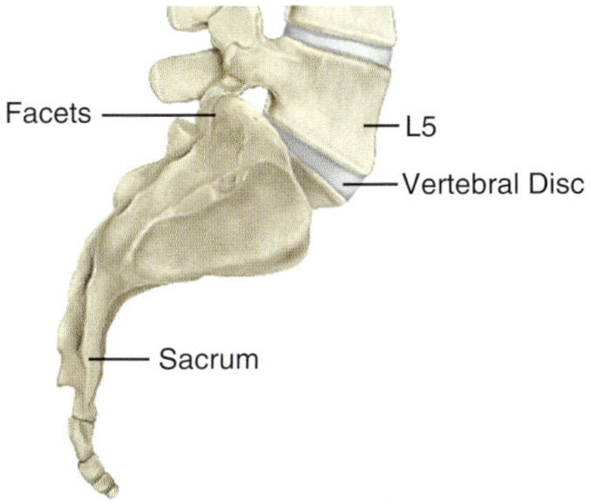

FIGURE 12.3 The Lumbosacral Joint.

The **lumbosacral joint** is the articulation between the distal end of the vertebral column (fifth lumbar vertebra) and the sacrum (**FIGURE 12.3**). At the superior end of the sacrum is a concavity (slight depression) that is called the base. The inferior side of the body of the fifth lumbar vertebra sits on this base. A vertebral disc resides in between the body of the fifth vertebra and the base of the sacrum. Additionally, the sacrum has two superior facets on the posterosuperior side that articulate with two inferior articular facets on the posterior aspect of the fifth lumbar vertebra.

The Motions of the Trunk

The trunk is capable of six motions (**FIGURE 12.4**). They are extension, flexion, right lateral flexion, left lateral flexion, right rotation, and left rotation. **Extension** and **flexion** straighten and bend the trunk in the sagittal plane about the frontal axis. **Right** and **left lateral flexion** bend the trunk to each side in the frontal plane rotating around the sagittal axis. **Right** and **left rotation** turn the trunk in the transverse plane around the longitudinal axis.

The average trunk ranges of motions are summarized in **TABLE 12.1**. There are no established averages for trunk extension and flexion. Trunk lateral flexion averages 20–45° in both directions while rotation averages 20–40° in both directions.

TABLE 12.1 Average Trunk Ranges of Motions

Motion	Average Degrees of Motion
Extension Flexion Combined Extension-Flexion	There are no established norms.
Right Lateral Flexion Left Lateral Flexion Combined Lateral Flexion	20–45° 20–45° 40–90°
Right Rotation Left Rotation Combined Rotation	20–40° 20–40° 40–80°

PAUSE TO CHECK FOR UNDERSTANDING

1. List the joints of the trunk.
2. Then, tell which bones form each joint.
3. What are the six motions of the trunk?

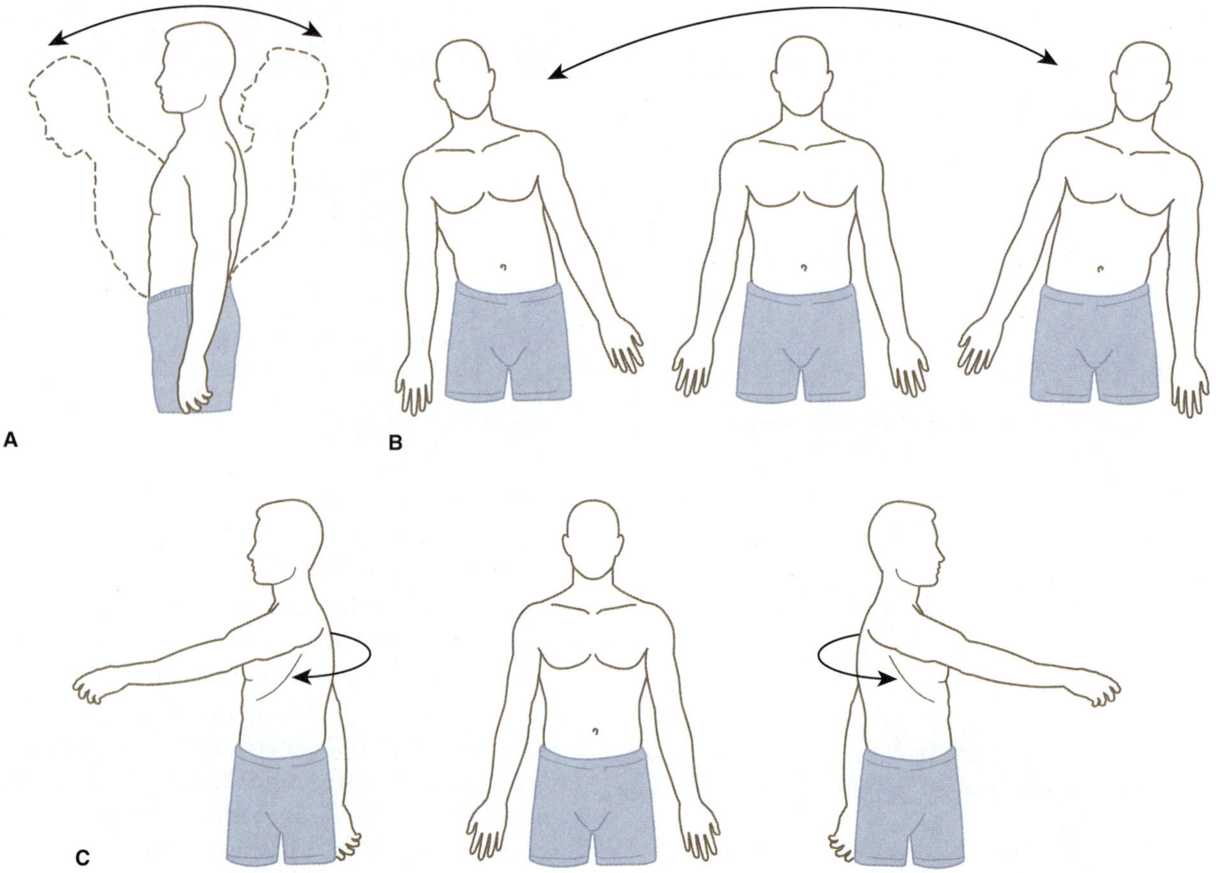

FIGURE 12.4 Trunk Motions. **(A)** extension & flexion **(B)** right & left lateral flexion **(C)** right & left rotation

The Muscles that Move the Trunk

The rest of this chapter is dedicated to learning the trunk muscles; their attachment sites; the *primary*, main motions they produce; and the nerves that innervate them. Muscles work in groups and so each muscle within the group supplies a certain percentage of the force needed to move a body segment and that amount (percentage) is not the same for each muscle. Furthermore, the force each muscle contributes varies depending on the starting position of the body segment and changes throughout the range of motion. This can be confusing to the beginning kinesiology student. So, this text has made certain choices in an attempt to simplify the muscle groups based on initial motion from anatomical position.

The muscle groups are based on the motion caused by their *concentric* action (i.e., The trunk flexors flex the trunk when they act concentrically). However, it should not be forgotten that all muscles can and will act *eccentrically* and *isometrically*. The reader should also notice that the table for each individual muscle has a section called "Other Motions." *Other Motions* means that this muscle aides, to some degree, in other motion(s) besides the group it has been placed in. Some muscles may appear in more than one group because their role is too great to ignore.

Unlike the other regions studied thus far, there are only four groups despite the fact that there are six motions. This is because four of the motions use the same term. For instance, right and left trunk rotation are two different motions, but only one group is needed to describe rotation. The same is of right and left lateral flexion of the trunk. Therefore, the muscle groups of the trunk are the *extensors, flexors, rotators*, and *lateral flexors*.

The Trunk Extensors

Only one muscle is primarily responsible for extension of the trunk when it acts *bilaterally* (**FIGURE 12.5**). That muscle is called the **erector spinae**. Sometimes the erector spinae is referred to as a muscle group of three individual muscles. However, this text will refer to it as one muscle with three sections. The sections of the erector spinae are the *spinalis, longissimus*, and *iliocostalis*. The erector spinae and its three portions run the length of the posterior trunk close to the vertebral column. The most medial of the sections of this muscle is the spinalis. Next to it is the

FIGURE 12.5 The Erector Spinae Muscle.

ORIGINS
posterior sacrum
posterior iliac spine
spinous processes of lower vertebrae
transverse processes of lower vertebrae
INSERTIONS
posterior side of ribs
spinous processes of upper vertebrae
transverse processes of upper vertebrae
mastoid process
OTHER UNILATERAL MOTIONS
ipsilateral trunk lateral flexion
NERVES
Cervical – lumbar nerve roots

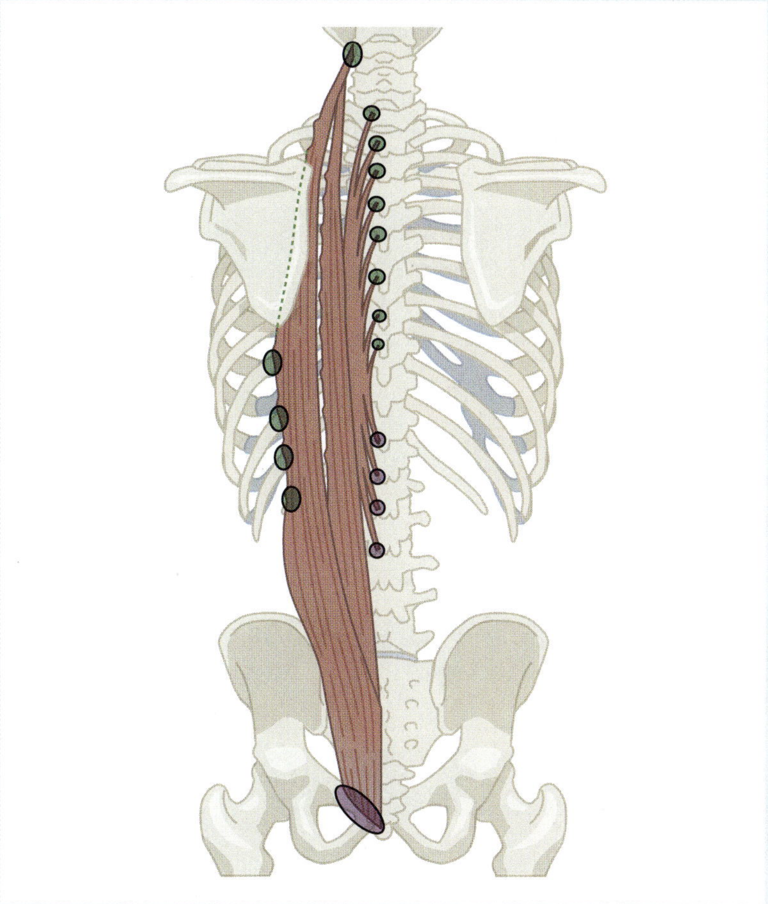

longissimus and finally the iliocostalis is the most lateral of the three.

The erector spinae originates in several locations. It arises from the posterior sacrum and the posterior iliac crest in addition to various spinous and transverse processes of the lower vertebrae. It inserts on the posterior side of various ribs, various spinous and transverse processes of the upper vertebrae, and the mastoid process.

PAUSE TO CHECK FOR UNDERSTANDING

1. How many sections are of the muscle that extends the trunk? List them.
2. What motion occurs if this muscle acts eccentrically?
3. Where is the general location for the muscle?

The Trunk Flexors

There are three muscles that are primarily responsible for flexing the trunk. However, this only occurs when the muscles bilaterally (both sides) act. They are the *rectus abdominis, external abdominal oblique*, and *internal abdominal oblique* (**FIGURE 12.6**). The trunk flexors are summarized in **TABLE 12.2** at the end of this section.

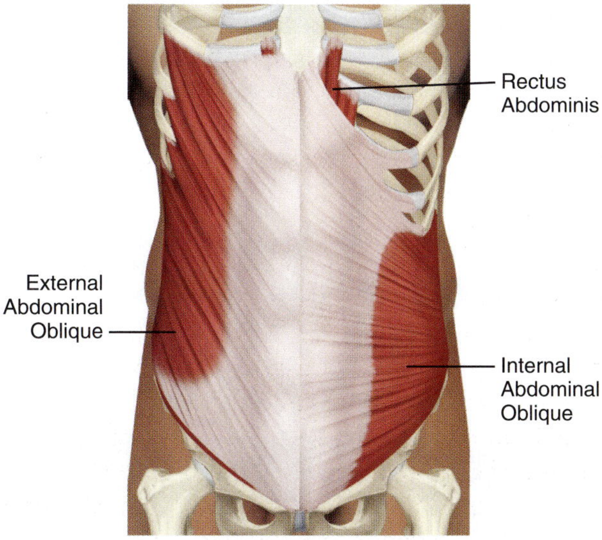

FIGURE 12.6 The Trunk Flexors.

The **rectus abdominis** resides vertically in the middle of the abdomen (**FIGURE 12.7**). It originates from the superior side of the pubis and inserts on the xiphoid process as well as the costal cartilage of ribs 5–7.

The **external abdominal oblique** runs obliquely from the lateral side of the abdomen toward the midline of the body (**FIGURE 12.8**). The external abdominal oblique originates from the lateral side of ribs 5–12. It inserts in three locations which are the anterior half of the iliac spine, the superior side of the pubis, and the linea alba. The **linea alba** which literally means "white line" is a cord-like connective tissue that runs vertically through the length of the mid-abdomen from the xiphoid process to the symphysis pubis.

FIGURE 12.7 The Rectus Abdominis Muscle.

ORIGIN
superior side of pubis
INSERTIONS
xiphoid process
costal cartilage of ribs 5–7
OTHER MOTIONS
None
NERVES
T6–T12 Nerve Roots

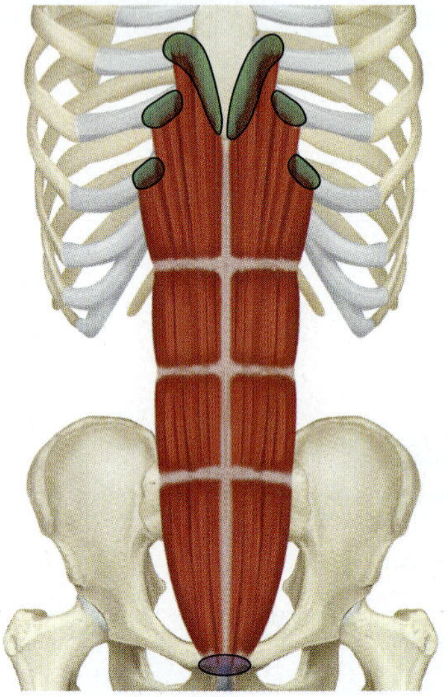

FIGURE 12.8 The External Abdominal Oblique Muscle.

ORIGINS
lateral side of ribs 5–12
INSERTIONS
anterior 1/2 of iliac spine
superior side of pubis
linea alba
OTHER UNILATERAL MOTIONS
contralateral trunk rotation
ipsilateral trunk lateral flexion
NERVES
T7–T12 Nerve Roots

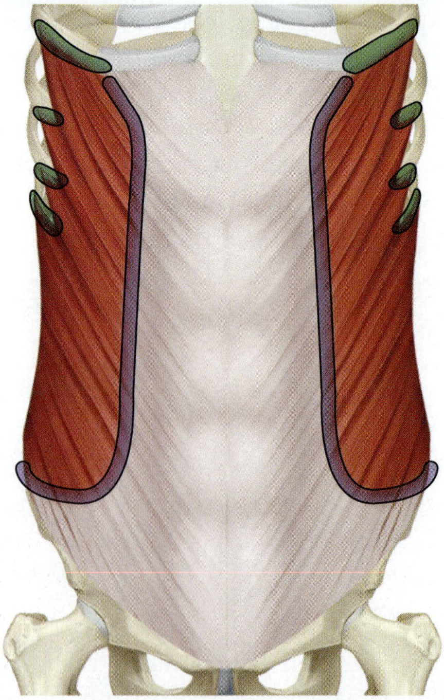

FIGURE 12.9 The Internal Abdominal Oblique Muscle.

ORIGINS
iliac spine
inguinal ligament
INSERTIONS
ribs 10–12
costal cartilage of rib 7–10
linea alba
OTHER MOTIONS
ipsilateral trunk rotation
ipsilateral trunk lateral flexion
NERVE
T7–L1 Nerve Roots

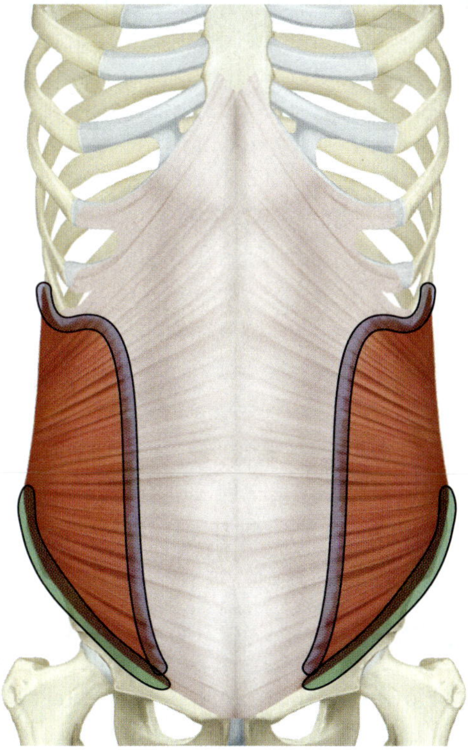

TABLE 12.2 Summary of the Trunk Flexors

Muscle	Origins	Insertions	Nerve
Rectus Abdominis	superior side of pubis	xiphoid process & the costal cartilage of ribs 5–7	T6–T12 Nerve Roots
External Abdominal Oblique	lateral side of ribs 5–12	anterior half of iliac spine, superior side of pubis, & linea alba	T7–T12 Nerve Roots
Internal Abdominal Oblique	iliac spine & inguinal ligament	ribs 10–12, costal cartilage of ribs 7–10, & linea alba	T7–L1 Nerve Roots

The **internal abdominal oblique** also passes obliquely across the abdomen from the midline to the coxal bone (**FIGURE 12.9**). This is in the opposite direction of the external abdominal oblique. To illustrate the difference in fiber direction between the two muscles, imagine only seeing one fiber from each muscle. If this were done, it would resemble an "X." Specifically, the internal abdominal oblique originates from the iliac spine and the inguinal ligament. It inserts on ribs 10–12, the costal cartilage of ribs 7–12, and the linea alba.

PAUSE TO CHECK FOR UNDERSTANDING

Table 12.2 should serve as a good review for much of the material, but here are a few other checks for understanding.

1. How many muscles are in the trunk flexor group? List them.
2. What motion occurs if these muscles act eccentrically?
3. What does bilateral action mean?

The Trunk Rotators

Two muscles are primarily responsible for trunk rotation. Both sets (sides) of muscles only rotate the trunk during unilateral (one side) action. The **external abdominal oblique** (Figure 12.8) causes contralateral (opposite side) trunk rotation when unilaterally (one side) acting. The **internal abdominal oblique** (Figure 12.9) causes ipsilateral (same side) trunk rotation during unilateral (one side) action. So, when the left external abdominal oblique and the right internal abdominal oblique unilaterally contract, they cause trunk rotation to the right. And, trunk rotation to the left occurs when the right external abdominal oblique and left internal abdominal oblique act. Their origins, insertions, other actions, and the nerves that innervate them will not be repeated since they were indicated in the previous section.

PAUSE TO CHECK FOR UNDERSTANDING

1. How many muscles are in the trunk rotator group? List them.
2. What motion occurs if these muscles act eccentrically?
3. What does it mean for a muscle to unilaterally act?
4. What does contralateral and ipsilateral mean?

The Trunk Lateral Flexors

The **quadratus lumborum** is the main muscle that laterally flexes the trunk when it acts unilaterally (**FIGURE 12.10**). However, the internal and external abdominal oblique and the erector spinae muscles assist in lateral flexion. The quadratus lumborum passes obliquely on each side of the low back region originating from the iliac crest. It inserts on rib 12 and the transverse processes of all of the lumbar vertebrae. It performs no other actions and is innervated by nerve roots T12–L4.

FIGURE 12.10 The Quadratus Lumborum Muscle.

ORIGIN	
iliac spine	
INSERTIONS	
rib 12	
transverse processes of L1–L5 vertebrae	
OTHER MOTIONS	
none	
NERVES	
T12–L4 Nerve Roots	

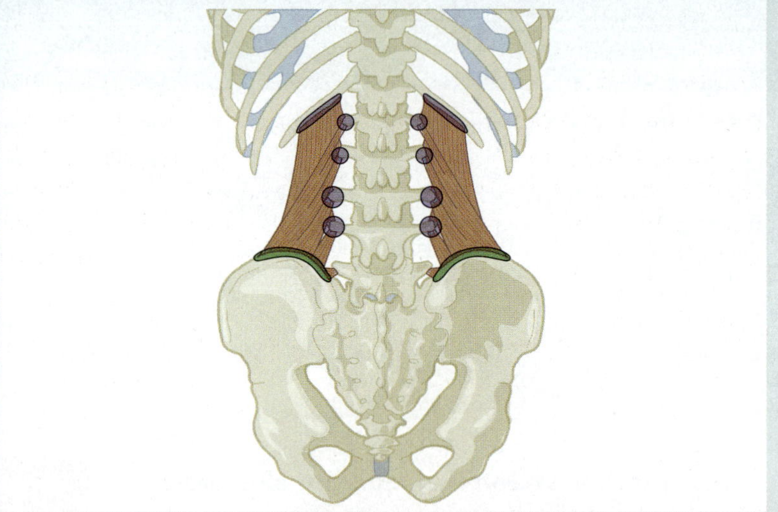

PAUSE TO CHECK FOR UNDERSTANDING

1. How many muscles are in the trunk lateral flexor group? List them.
2. Where is the general location for the muscle group?
3. What does it mean for a muscle to unilaterally act?

CHAPTER 13

The Neck

CHAPTER OBJECTIVES

After completing this chapter, the student will be able to:

1. summarize the joints of the neck region;
2. comprehend the neck motions;
3. identify the names and locations of the neck muscles;
4. recall the attachment sites and nerves of the neck muscles; and
5. apply the motions produced by the actions of the neck muscles.

The neck is a region of the body between the trunk and the head (**FIGURE 13.1**). Similar to the waist, the neck motions actually occur as a result of the combined motions of the intervertebral joints in the cervical spine as well as the articulation between atlas (C1) and the occipital bone. The arrangement of the bones and the actions of the neck muscles produce six motions.

The Joints of the Neck

There are seven joints in the neck. The most superior is the **atlanto-occipital joint** which is the articulation between the occipital bone and the first cervical vertebra (C1) which is also named atlas (**FIGURE 13.2**). This joint is stabilized by three main structures: the **atlanto-occipital joint capsule**, the **posterior**

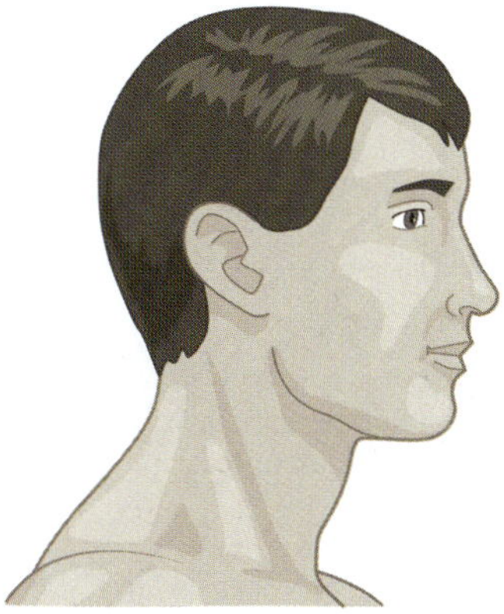

FIGURE 13.1 The Neck.

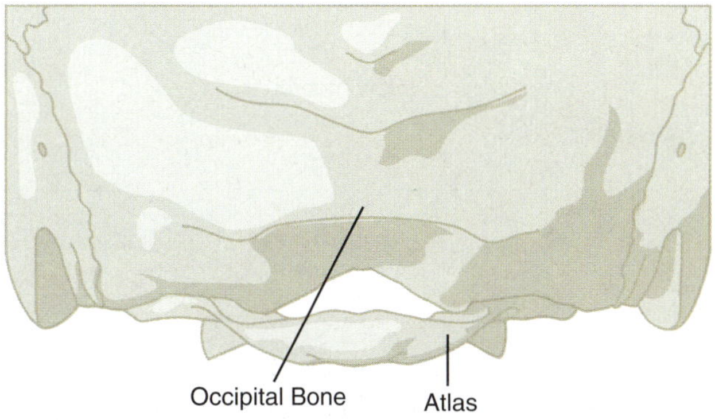

FIGURE 13.2 The Atlanto-Occipital Joint.

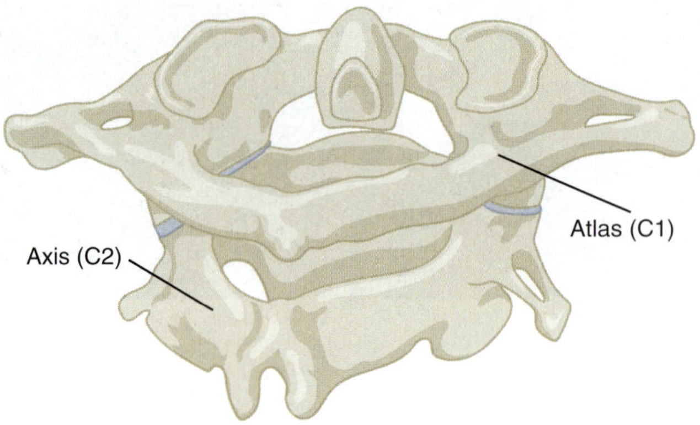

FIGURE 13.3 The Atlantoaxial Joint.

atlanto-occipital membrane (ligament), and the **anterior atlanto-occipital membrane** (ligament).

The other six are **intervertebral joints** between the seven cervical vertebrae. The most superior intervertebral joint between atlas (C1) and axis (C2) is called the **atlantoaxial joint** (**FIGURE 13.3**).

Six main ligaments stabilize the bones of the intervertebral joints (**FIGURE 13.4**). The **anterior longitudinal ligament** attaches on the anterior surface of each vertebral body as it runs the entire length of the column. The **posterior longitudinal ligament** attaches to the posterior body of each vertebra as it pass through the vertebral foramen. The **ligamentum flavum** attaches the lamina of each vertebra. Superficially, the **supraspinous ligament** spans the vertebral column from spinous process to spinous process. Deep to that ligament, the **interspinous ligaments** also connect the spinous processes. Finally, the **intertransverse ligaments** stabilize all the transverse processes.

The Motions of the Neck

The neck is capable of six motions (**FIGURE 13.5**). They are extension, flexion, right and left lateral flexion, and right and left rotation. **Neck extension** and **flexion** are straightening and bending in the sagittal plane rotating about the frontal axis. **Neck lateral flexion** bends the head to each side through the frontal plane about the sagittal axis. **Neck rotation** turns the head from side to side through the transverse plane about the longitudinal axis.

Average neck ranges of motions are summarized in **TABLE 13.1**. There are no established range of motion averages for neck extension and flexion. Lateral flexion to either side averages 20–45°. And, neck rotation averages 70–90° to either side.

TABLE 13.1 Average Neck Range of Motions

Motion	Average Degrees of Motion
Extension Flexion Combined	There are no established norms.
Right Lateral Flexion Left Lateral Flexion Combined	20–45° 20–45° 40–90°
Right Rotation Left Rotation Combined	70–90° 70–90° 140–180°

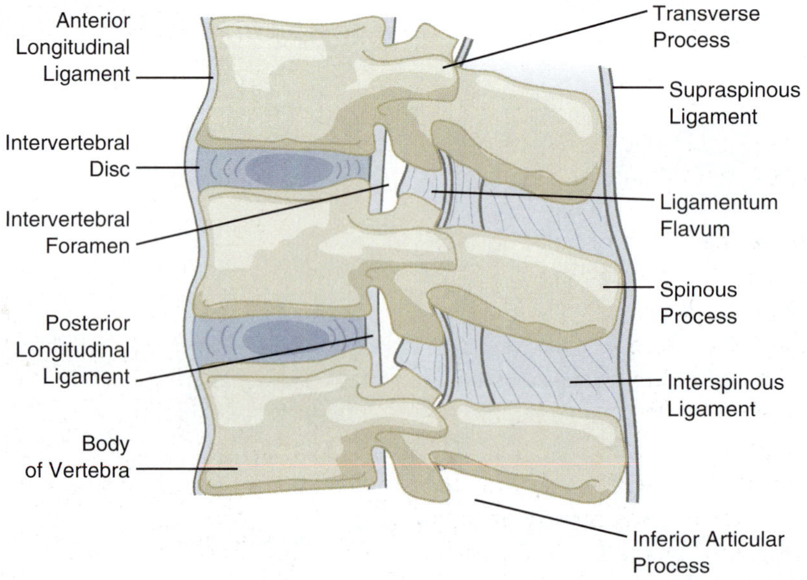

FIGURE 13.4 A Section of the Vertebral Column Showing a Few Intervertebral Joints.

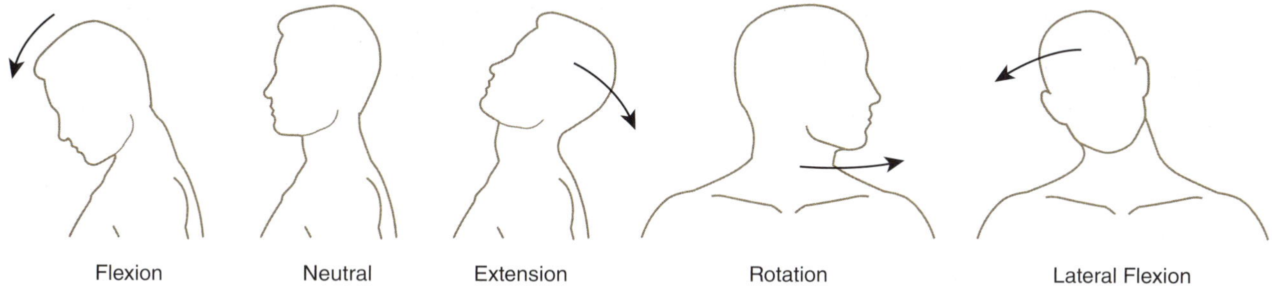

FIGURE 13.5 The Motions of the Neck.

PAUSE TO CHECK FOR UNDERSTANDING

1. List the joints of the neck.
2. Then, tell which bones form each joint.
3. What are the six motions of the neck?

The Muscles That Move the Neck

The rest of this chapter is dedicated to learning the neck muscles; their attachment sites; the *primary*, main motions they produce; and the nerves that innervate them. Muscles work in groups and so each muscle within the group supplies a certain percentage of the force needed to move a body segment and that amount (percentage) is not the same for each muscle. Furthermore, the force each muscle contributes varies depending on the starting position of the body segment and changes throughout the range of motion. This can be confusing to the beginning kinesiology student. So, this text has made certain choices in an attempt to simplify the muscle groups based on initial motion from anatomical position.

The muscle groups are based on the motion caused by their *concentric* action (i.e., The neck flexors flex the neck when they act concentrically). However, it should not be forgotten that all muscles can and will act *eccentrically* and *isometrically*. The reader should also notice that the table for each individual muscle has a section called "Other Motions." *Other Motions* means that this muscle aides, to some degree, in other motion(s) besides the group it has been placed in. Some muscles may appear in more than one group because their role is too great to ignore.

For the same reason as the waist muscles in the previous chapter, there are fewer neck muscle groups (four) than there are neck motions (six). The muscle groups are the *neck flexors*, *neck extensors*, *neck rotators*, and *neck lateral flexors*.

The Neck Flexors

Four muscles flex the neck when they ***bilaterally*** (both sides) act (**FIGURE 13.6**). They are the *sternocleidomastoid*, the *anterior scalenes*, the *middle scalenes*, and the *posterior scalenes*. The anterior, middle, and posterior scalenes are commonly grouped together and simply called the *scalenes*. At the end of this section, the neck flexors will be summarized in **TABLE 13.2**.

The **sternocleidomastoid** runs through the anterolateral neck originating from the manubrium and proximal clavicle. It inserts into the mastoid process of the temporal bone (**FIGURE 13.7**).

The **scalenes** (anterior, middle, and posterior muscles) lie deep in the posterolateral neck (**FIGURE 13.8**). The muscle group originates from the transverse processes of all the cervical vertebrae (C1–C7). The **anterior** and **middle scalene** insert on the first rib while the **posterior scalene** inserts on the second rib.

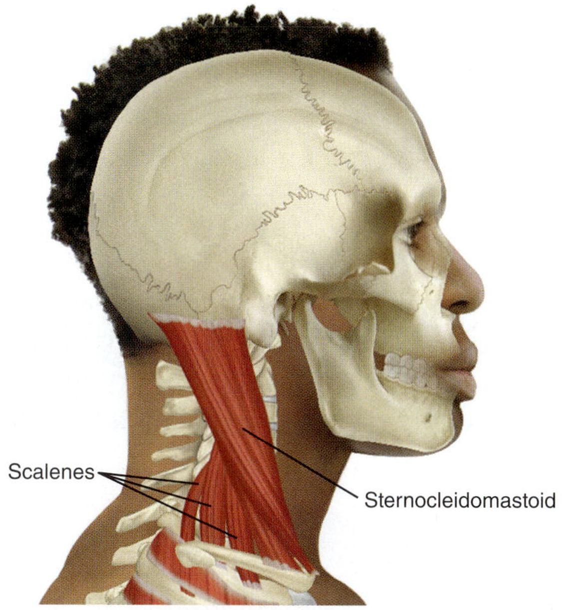

FIGURE 13.6 The Neck Flexors.

FIGURE 13.7 The Sternocleidomastoid Muscle.

ORIGINS

manubrium

proximal clavicle

INSERTION

mastoid process

OTHER UNILATERAL MOTIONS

contralateral neck rotation

NERVES

C2–C3 Nerve Roots

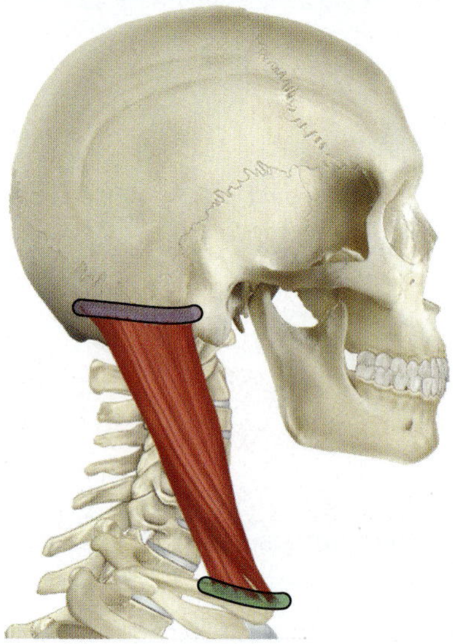

FIGURE 13.8 The Scalenes Muscles.

ORIGINS

transverse processes of C1–C7

INSERTIONS

first and second ribs

OTHER UNILATERAL MOTIONS

ipsilateral neck lateral flexion

NERVES

Accessory Nerve & C3–C8 Nerve Roots

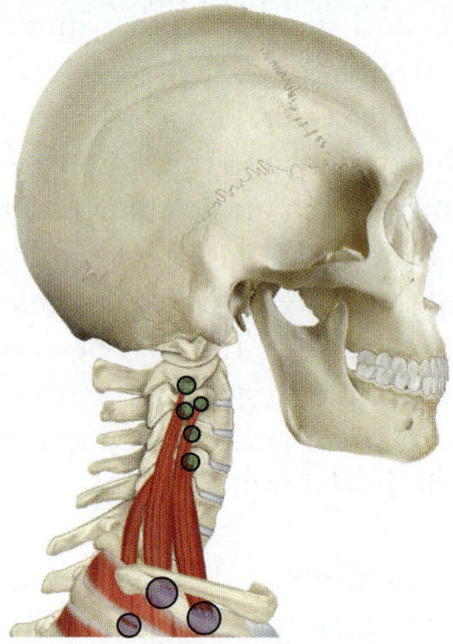

TABLE 13.2 Summary of the Neck Flexors

Muscle	Origins	Insertions	Nerve
Sternocleidomastoid	manubrium & proximal clavicle	mastoid process	C2–C3 Nerve Roots
Scalenes	transverse processes of C1–C7	first and second ribs	Accessory Nerve & C3–C8 Nerve Roots

PAUSE TO CHECK FOR UNDERSTANDING

Table 13.2 should serve as a good review for much of the material, but here are a few other checks for understanding:

1. How many muscles are in the neck flexor group? List them.
2. What motion occurs if these muscles act eccentrically?
3. Where is the general location for the muscle group?

The Neck Extensors

There are five main muscles that extend the neck when they bilaterally act (**FIGURE 13.9**). They are the *trapezius*, *splenius capitis*, *splenius cervicis*, *semispinalis capitis*, and the *semispinalis cervicis* muscles. In regards to the trapezius muscle, it is commonly divided into three sections (upper, middle, and lower) because each section causes a different motion. The *upper section* of this muscle extends the neck. The neck extensors will be summarized in **TABLE 13.3** at the end of this section.

The **upper trapezius** runs obliquely through the neck to the superior shoulder (**FIGURE 13.10**).

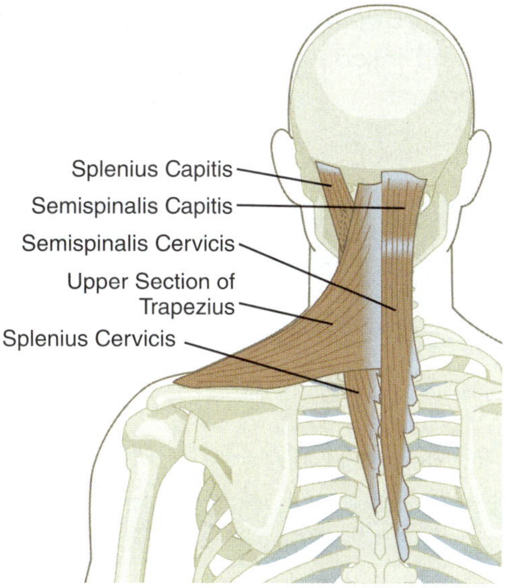

FIGURE 13.9 The Muscles That Extend the Neck.

It originates from the occipital protuberance and the spinous processes of the upper cervical vertebrae. It then inserts on three locations on the superior side of the scapular spine, the distal one-third of the clavicle, and the acromion.

The **splenius capitis** originates from the distal half of the nuchal ligament, spinous processes of the last cervical vertebra (C7), and the first three or

FIGURE 13.10 The Upper Section of the Trapezius Muscle.

ORIGINS	
occipital protuberance	
spinous processes of upper C-spine	
INSERTIONS	
superior side of scapular spine	
distal 1/3 clavicle	
acromion	
OTHER UNILATERAL MOTIONS	
ipsilateral neck lateral flexion	
NERVE	
Accessory	

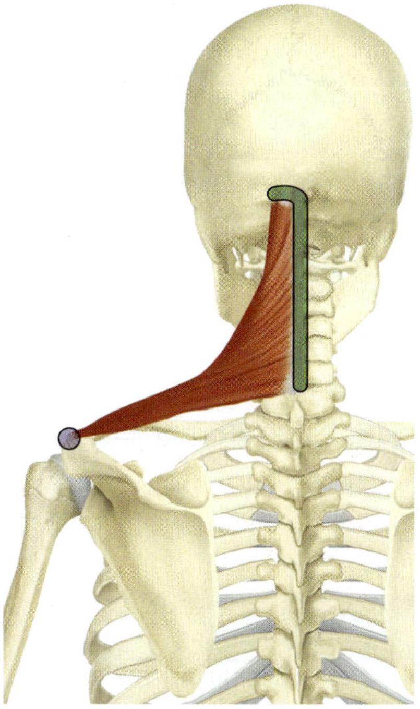

four thoracic vertebrae (**FIGURE 13.11**). It passes up the neck and inserts on the occipital bone and the mastoid process.

The **splenius cervicis** originates lower on the spine than the splenius capitis from the spinous processes of T3–T6 vertebrae (**FIGURE 13.12**). After

FIGURE 13.11 The Splenius Capitis Muscle.

ORIGINS
distal half of nuchal ligament
spinous processes of C7–T3 or T4
INSERTIONS
occipital bone
mastoid process
OTHER UNILATERAL MOTIONS
ipsilateral neck lateral flexion
ipsilateral neck rotation
NERVES
C3–C8 Nerve Roots

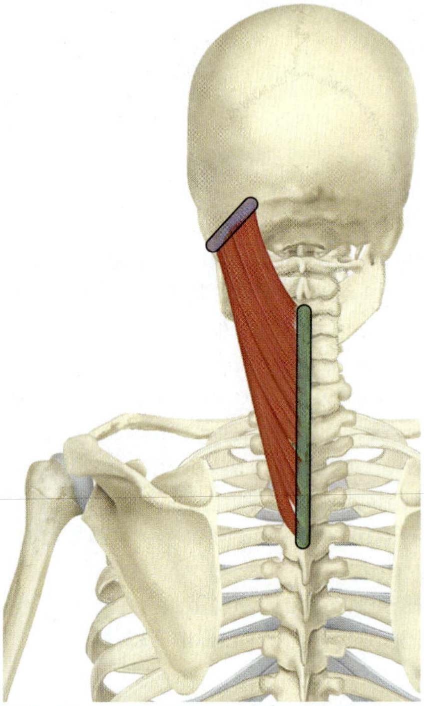

FIGURE 13.12 The Splenius Cervicis Muscle.

ORIGINS
spinous processes of T3–T6
INSERTIONS
transverse processes of C1–C3
OTHER UNILATERAL MOTIONS
ipsilateral neck lateral flexion
ipsilateral neck rotation
NERVES
C4–C8 Nerve Roots

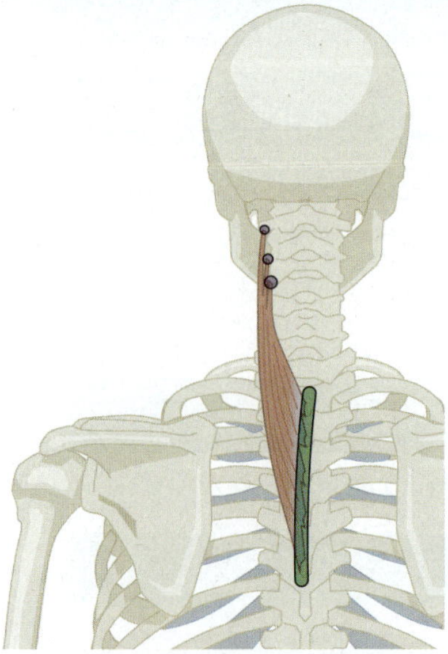

passing through the neck, it inserts on the transverse processes of C1–C3 vertebrae.

The **semispinalis capitis** originates from the transverse processes of C4–T4 vertebrae, passes through the posterior neck region, and inserts on the occipital bone between the superior and inferior nuchal lines (**FIGURE 13.13**).

The **semispinalis cervicis** originates more distally than the semispinalis capitis from the transverse processes of the upper thoracic spine (T-spine) (**FIGURE 13.14**). From there, it runs longitudinally into the neck and inserts on the spinous processes of the cervical spine (C-spine).

FIGURE 13.13 The Semispinalis Capitis Muscle.

ORIGINS	
transverse processes of C4–T4	
INSERTION	
between superior & inferior nuchal lines	
OTHER UNILATERAL MOTIONS	
ipsilateral neck lateral flexion	
ipsilateral neck rotation	
NERVE	
Greater Occipital	

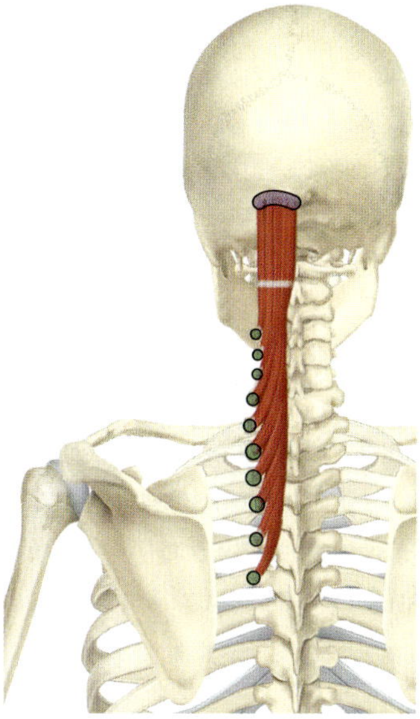

FIGURE 13.14 The Semispinalis Cervicis Muscle.

ORIGINS	
transverse processes of upper T-spine	
INSERTIONS	
spinous processes of C-spine	
OTHER UNILATERAL MOTIONS	
ipsilateral neck lateral flexion	
ipsilateral neck rotation	
NERVES	
C3–C5 Nerve Roots	

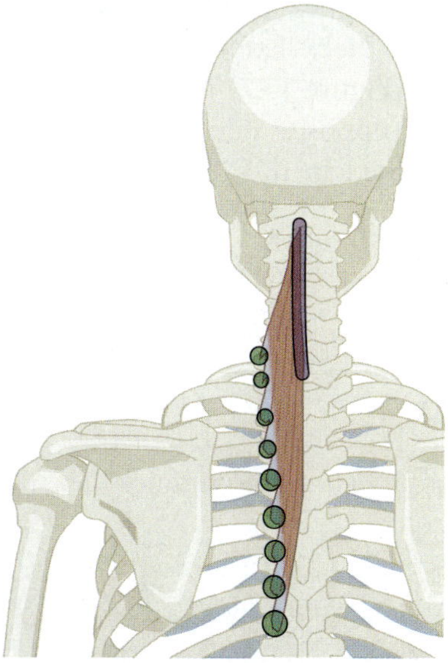

TABLE 13.3 Summary of the Neck Extensors

Muscle	Origins	Insertions	Nerve
Upper Trapezius	occipital protuberance & spinous processes of the upper C-spine	distal 1/3 clavicle & acromion	Accessory
Splenius Capitis	distal half of nuchal ligament & spinous processes of C7–T3 or T4	occipital bone & mastoid process	C3–C8 Nerve Roots
Splenius Cervicis	spinous processes of T3–T6	transverse processes of C1–C3	C4–C8 Nerve Roots
Semispinalis Capitis	transverse processes of C4–T4	occipital bone between the superior & inferior nuchal lines	Greater Occipital Nerve
Semispinalis Cervicis	transverse processes of the upper T-spine	spinous processes of the C-spine	C3–C5 Nerve Roots

PAUSE TO CHECK FOR UNDERSTANDING

Table 13.3 should serve as a good review for much of the material, but here are a few other checks for understanding:

1. How many muscles are in the neck extensor group? List them.
2. What motion occurs if these muscles act eccentrically?
3. Where is the general location for the muscle group?
4. What does it mean for a muscle to bilaterally act?

The Neck Rotators

Five muscles ***unilaterally*** (one side) act to rotate the neck. Four of them ***ipsilaterally*** (same side) rotate the neck. The list of these muscles includes the splenius capitis (Figure 13.11), the splenius cervicis (Figure 13.12), the semispinalis capitis (Figure 13.13), and the semispinalis cervicis (Figure 13.14). These four are described in the "The Neck Extensors" section. The sternocleidomastoid (Figure 13.7) is the fifth muscle, but it contralaterally (opposite side) rotates the neck. This muscle is described in the "The Neck Flexors" section.

PAUSE TO CHECK FOR UNDERSTANDING

1. How many muscles are in the neck rotator group? List them.
2. What motion occurs if these muscles act eccentrically?
3. Where is the general location for the muscle group?
4. What does it mean for a muscle to unilaterally act?
5. What does ipsilateral and contralateral refer to?

The Neck Lateral Flexors

Eight muscles laterally flex the neck when they ***unilaterally*** (one side) act. Five of them are the **trapezius** (upper section) (Figure 13.10), splenius capitis (Figure 13.11), the splenius cervicis (Figure 13.12), the semispinalis capitis (Figure 13.13), and the semispinalis cervicis (Figure 13.14). These five muscles were described in section titled "The Neck Extensors." The other three are the anterior, middle, and posterior scalenes (Figure 13.8). The description of the scalenes is found in the section related to "The Neck Flexors."

PAUSE TO CHECK FOR UNDERSTANDING

1. How many muscles are in the neck lateral flexor group? List them.
2. What motion occurs if these muscles act eccentrically?
3. Where is the general location for the muscle group?
4. What does it mean for a muscle to unilaterally act?

CHAPTER 14

The Shoulder Girdle

CHAPTER OBJECTIVES

After completing this chapter, the student will be able to:

1. summarize the joints of the shoulder girdle;
2. comprehend the shoulder girdle motions;
3. identify the names and locations of the shoulder girdle muscles;
4. recall the attachment sites and nerves of the shoulder girdle muscles; and
5. apply the motions produced by the actions of the shoulder girdle muscles.

This text considers the shoulder girdle as part of the axial skeleton. It is not to be confused with the shoulder region covered in the next chapter. The shoulder girdle is a region of the upper trunk that includes the sternal, pectoral, and scapular regions (**FIGURE 14.1**). The arrangement of the bones and actions of the muscles of the shoulder girdle cause four motions.

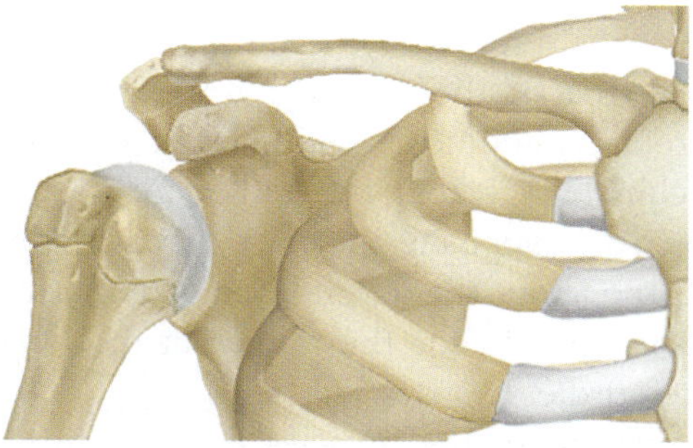

FIGURE 14.1 The Shoulder Girdle.

The Joints of the Shoulder Girdle

Two cartilaginous joints (slightly movable) and one articulation comprise the shoulder girdle. The two joints are the *sternoclavicular joint* and the *acromioclavicular joint*. The articulation is the *scapulothoracic articulation*.

The scapulothoracic articulation is called this rather than a joint because it does not fit the definition of the intersection of two bones. The **scapulothoracic articulation** is the relationship of the scapula moving back and forth over the posterior thorax; more specifically, ribs 2–7 (**FIGURE 14.2**). There are no ligaments connecting the scapula and ribs. Rather, superficial soft tissues as well as several muscles keep the scapula, relatively speaking, hovering over the ribs.

The **sternoclavicular (SC) joint** (**FIGURE 14.3**) is the intersection between the superolateral end of the

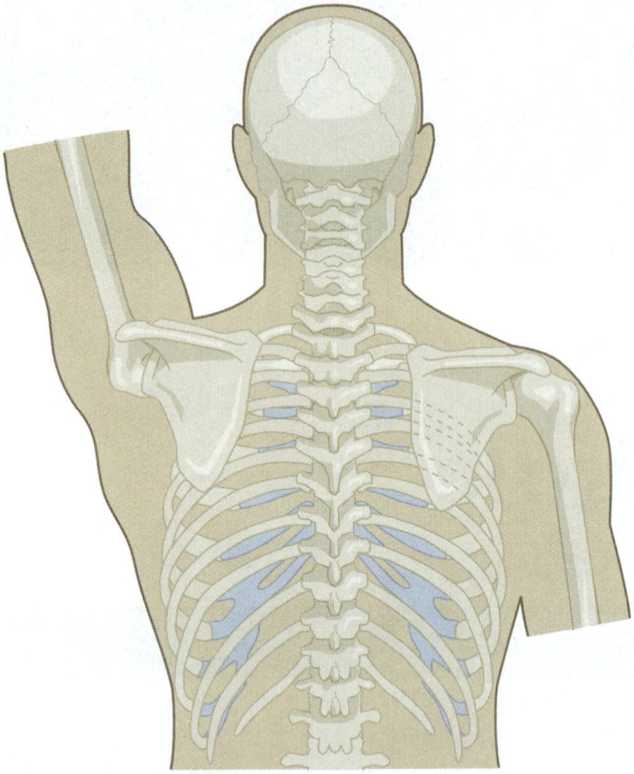

FIGURE 14.2 The Scapulothoracic Articulation.

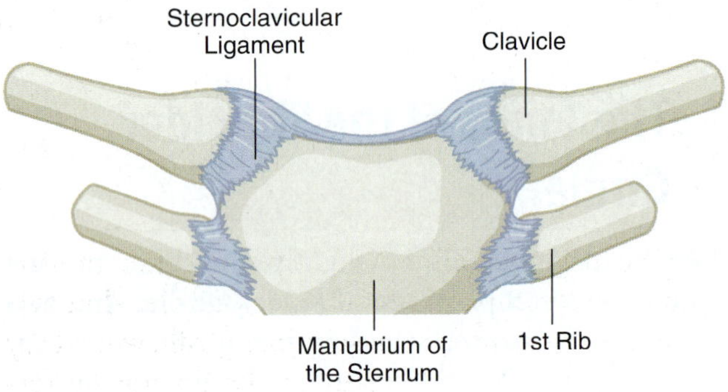

FIGURE 14.3 The Sternoclavicular Joint.

sternum and the proximal end of the clavicle. This joint is primarily stabilized by the **sternoclavicular ligament**.

The distal end of the clavicle intersects with the acromial process (acromion) of the scapula to form the **acromioclavicular (AC) joint** (**FIGURE 14.4**). One main ligament, the **acromioclavicular ligament**, connects these two bone parts. However, two other main ligaments are part of the stability of the joint. The **coracoclavicular ligament** attaches the distal end of the clavicle and the coracoid process of the scapula.

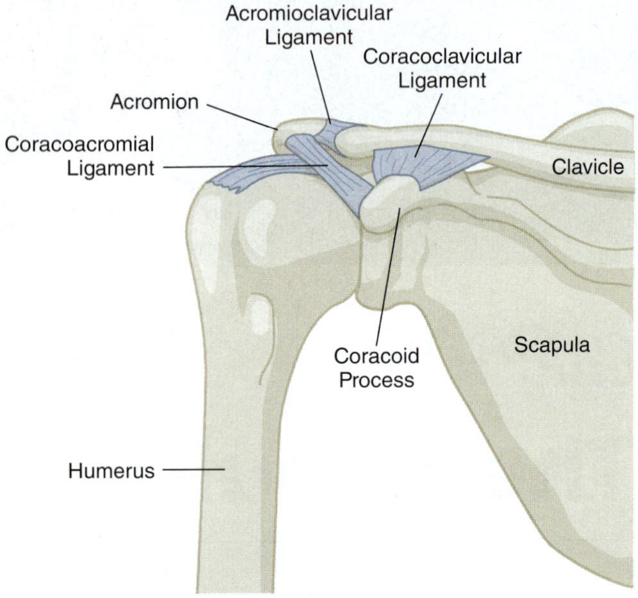

FIGURE 14.4 The Acromioclavicular Joint.

The **coracoacromial ligament** runs from the acromion to the coracoid process.

> **PAUSE TO CHECK FOR UNDERSTANDING**
>
> 1. List the joints of the shoulder girdle.
> 2. Then, tell which bones form each joint.
> 3. What are the four motions of the shoulder girdle?

The Motions of the Shoulder Girdle

The shoulder girdle is capable of four motions (**FIGURE 14.5**). These motions include *elevation*, *depression*, *protraction*, and *retraction*. **Elevation** and **depression**, respectively, is lifting (shoulder shrug) and lowering of the shoulder girdle in the frontal plane rotating about the sagittal axis. **Retraction**, as in pinching the scapulae together in the back, and **protraction** (opposite motion) occur in the transverse plane about the longitudinal axis. There are no established norms for the ranges of motions of the shoulder girdle.

A number of the muscles listed in this chapter assist in *scapular rotation*. However, these are accessory motions mostly to shoulder abduction and adduction and will be described in Chapter 15.

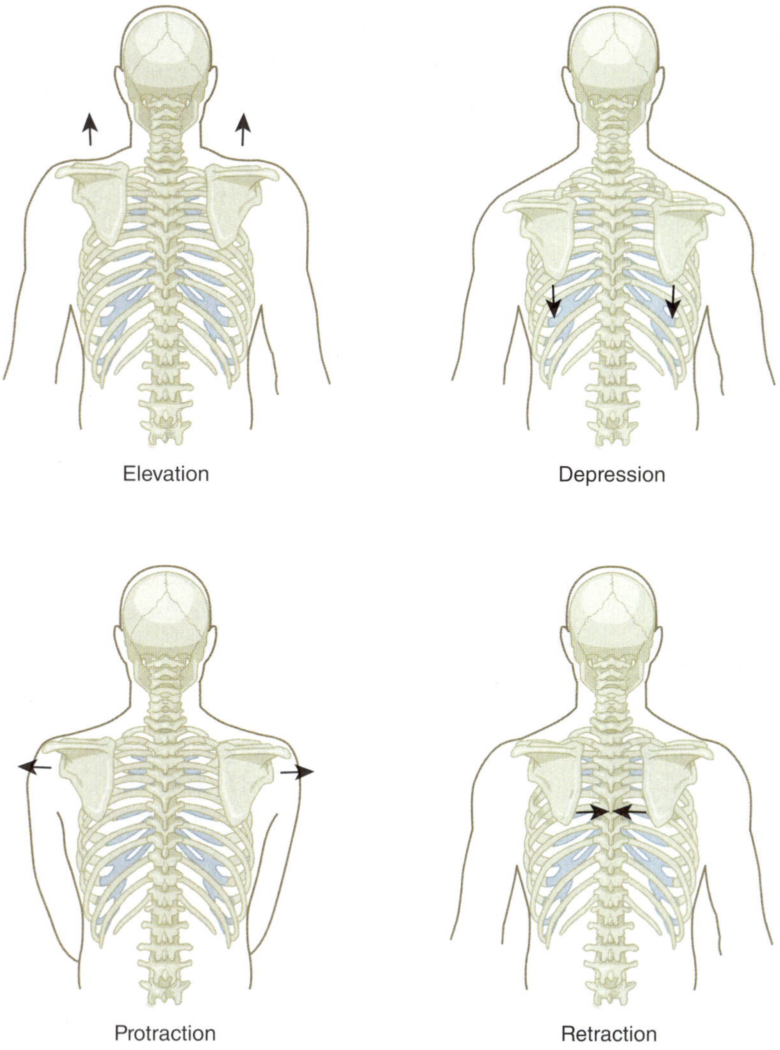

FIGURE 14.5 Motions of the Shoulder Girdle.

The Muscles That Move the Shoulder Girdle

The rest of this chapter is dedicated to learning the muscles of the shoulder girdle; their attachment sites; the *primary*, main motions they produce; and the nerves that innervate them. Muscles work in groups and so each muscle within the group supplies a certain percentage of the force needed to move a body segment and that amount (percentage) is not the same for each muscle. Furthermore, the force each muscle contributes varies depending on the starting position of the body segment and changes throughout the range of motion. This can be confusing to the beginning kinesiology student. So, this text has made certain choices in an attempt to simplify the muscle groups based on initial motion from anatomical position.

The muscle groups are based on the motion caused by their *concentric* action (i.e., The shoulder sirdle elevators elevate the shoulder girdle when they act concentrically). However, it should not be forgotten that all muscles can and will act *eccentrically* and *isometrically*. The reader should also notice that the table for each individual muscle has a section called "Other Motions." *Other Motions* means that this muscle aides, to some degree, in other motion(s) besides the group it has been placed in.

There are four muscle groups of the shoulder girdle because there are four motions. They are the *shoulder girdle elevators, depressors, retractors*, and *protractors*. Several muscles are involved in shoulder girdle motions. One of them is the trapezius. The *trapezius* muscle has commonly been divided into three sections (upper, middle, and lower) because each section causes different motions (**FIGURE 14.6**). Therefore, the three sections of this muscle will be

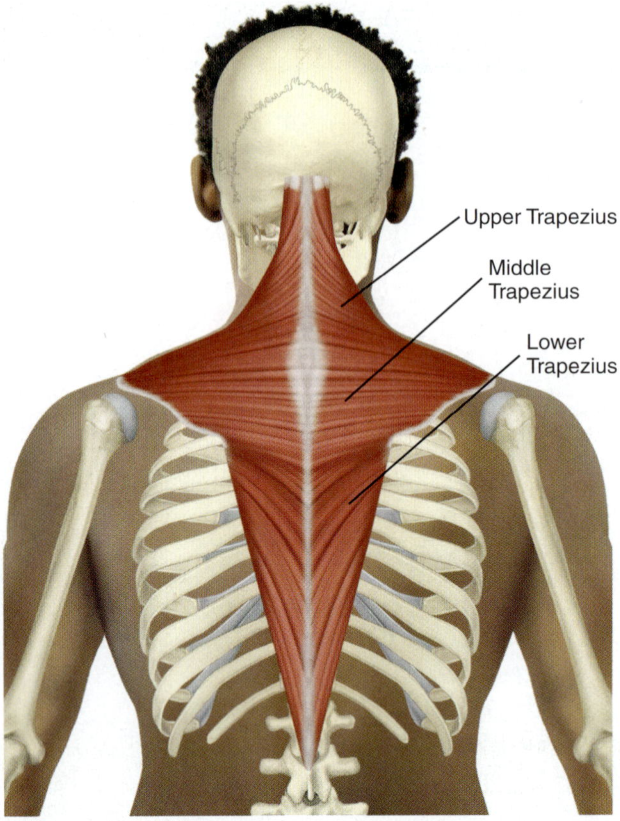

FIGURE 14.6 The Sections of the Trapezius Muscle.

described separately in the appropriate section based on its action.

The Shoulder Girdle Elevators

Two muscles are primarily responsible for shoulder girdle elevation. They are the *trapezius* (*upper section*) and the *levator scapula*. The shoulder girdle elevators will be summarized in **TABLE 14.1** at the end of this section.

The **upper trapezius** section of the trapezius muscle originates from the occipital protuberance and the spinous processes of the upper C-spine (**FIGURE 14.7**). The fibers of the upper trapezius pass through the neck and across the superior end of the upper back toward the shoulder. The upper trapezius inserts on three locations: the superior side of the scapular spine, the distal one-third of the clavicle, and the acromion.

The **levator** (leh-VAY-tur) **scapula** passes through the neck region originating from the transverse processes of the first four cervical vertebrae (C1–C4) (**FIGURE 14.8**). It inserts along the vertebral (medial) border of the scapula from the superior angle to the level of the base of the scapular spine.

FIGURE 14.7 The Upper Section of the Trapezius Muscle.

ORIGINS
occipital protuberance
spinous processes of upper C-spine
INSERTIONS
superior side of scapular spine
distal 1/3 clavicle
acromion
OTHER UNILATERAL MOTIONS
ipsilateral neck lateral flexion
scapular upward rotation
NERVE
Accessory

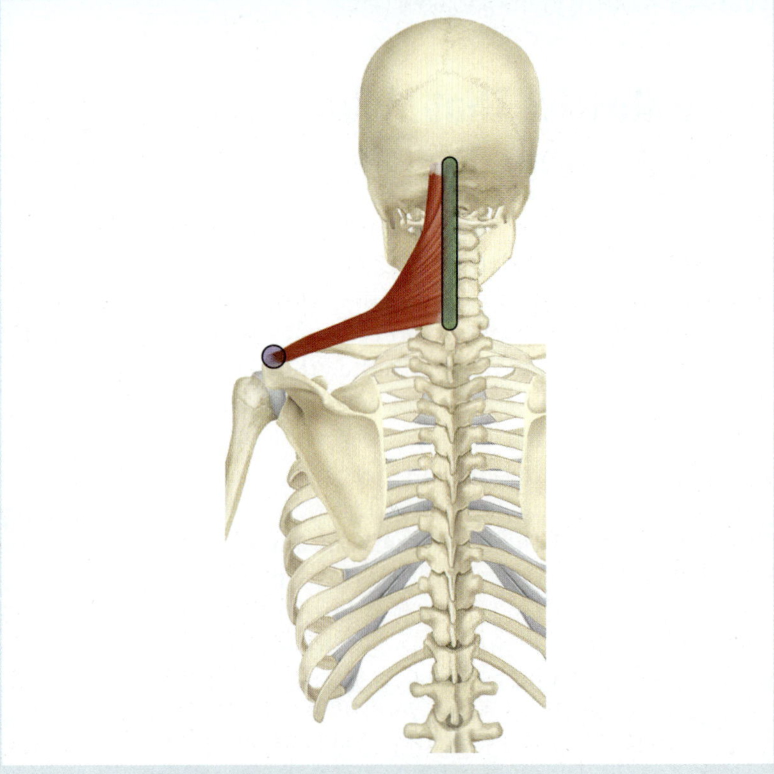

FIGURE 14.8 The Levator Scapula Muscle.

ORIGINS	
transverse processes of C1–C4	
INSERTION	
vertebral border of scapula from superior angle to base of scapular spine	
OTHER MOTIONS	
scapular downward rotation	
NERVE	
Dorsal Scapular	

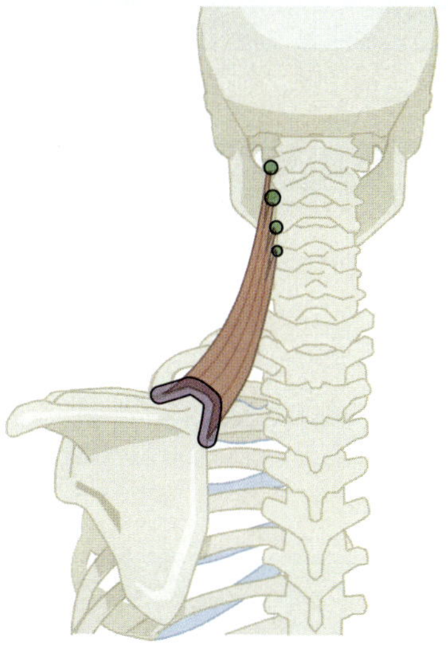

TABLE 14.1 Summary of the Shoulder Girdle Elevators

Muscle	Origins	Insertions	Nerves
Upper Trapezius	occipital protuberance & spinous processes of the upper C-spine	superior side of scapular spine, distal 1/3 clavicle, & acromion	Accessory
Levator Scapula	transverse processes of C1–C4	vertebral border of scapula from superior angle to base of scapular spine	Dorsal Scapular

PAUSE TO CHECK FOR UNDERSTANDING

Table 14.1 should serve as a good review for much of the material, but here are a few other checks for understanding

1. How many muscles are in the shoulder girdle elevator group? List them.
2. What motion occurs if these muscles act eccentrically?
3. Where is the general location for the muscle group?

The Shoulder Girdle Depressors

The *trapezius* (lower section) muscle is the only muscle primarily responsible for shoulder girdle depression (**FIGURE 14.9**). The **lower trapezius** passes from the vertebral column relatively obliquely through the mid back to the shoulder. It originates from the spinous processes of the last nine thoracic vertebrae (T4–T12) and inserts on the inferior side of the spine of the scapula and the acromion (acromial process).

PAUSE TO CHECK FOR UNDERSTANDING

1. How many muscles are in the shoulder girdle depressor group? List them.
2. What motion occurs if these muscles act eccentrically?
3. Where is the general location for the muscle group?

FIGURE 14.9 The Lower Section of the Trapezius Muscle.

ORIGINS	spinous processes of T4–T12
INSERTIONS	inferior side of scapular spine; acromion
OTHER MOTIONS	scapular upward rotation
NERVE	Accessory

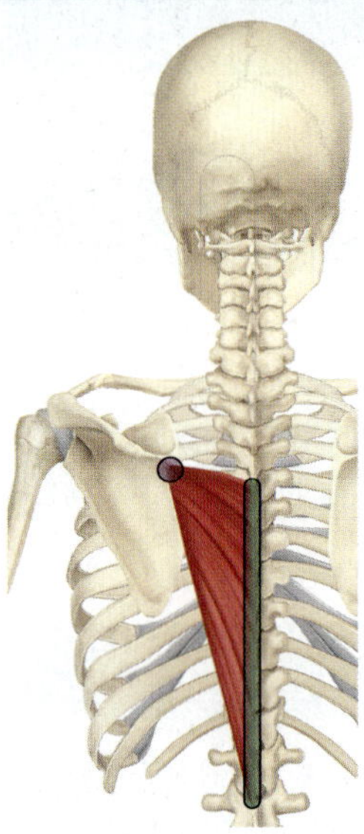

The Shoulder Girdle Retractors

Three muscles primarily retract the shoulder girdle. They are the *trapezius (middle portion)*, the *rhomboideus minor*, and the *rhomboideus major* muscles. The rhomboideus minor and major muscles are commonly referred to as the *rhomboids*, but they are actually two separate muscles. The shoulder girdle retractors will be summarized in **TABLE 14.2** at the end of this section.

The **middle trapezius** runs transversely from the spinous processes of the last cervical vertebra (C7) and the first three thoracic vertebrae (T1–T3) (**FIGURE 14.10**). It then inserts on the spine of the scapula and the acromion (acromial process).

The **rhomboideus** (rom-BOY-dee-us) **minor** passes obliquely in the upper back originating from the spinous processes of the last cervical vertebra (C7) and the first thoracic vertebra (T1) and inserts along the medial border of the scapula from the superior angle to the base of the scapular spine (**FIGURE 14.11**).

The **rhomboideus** (rom-BOY-dee-us) **major** lies obliquely just inferior to the rhomboideus minor muscle originating from the spinous processes of the second through fifth thoracic vertebrae (T2–T5). It inserts on the medial border from the base of the scapular spine to the inferior angle (**FIGURE 14.12**).

FIGURE 14.10 The Middle Section of the Trapezius Muscle.

ORIGINS	spinous processes of C7–T3
INSERTIONS	scapular spine; acromion
OTHER MOTIONS	none
NERVE	Accessory

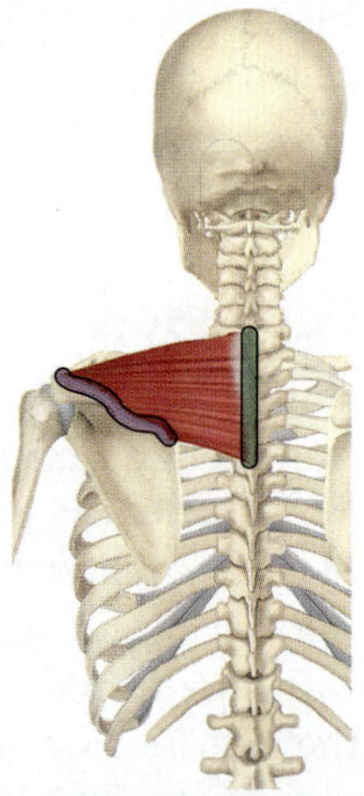

FIGURE 14.11 The Rhomboideus Minor Muscle.

ORIGINS

spinous processes of C7–T1

INSERTION

vertebral border of scapula from superior angle to base of scapular spine

OTHER MOTIONS

scapular downward rotation

shoulder girdle elevation

NERVE

Dorsal Scapular

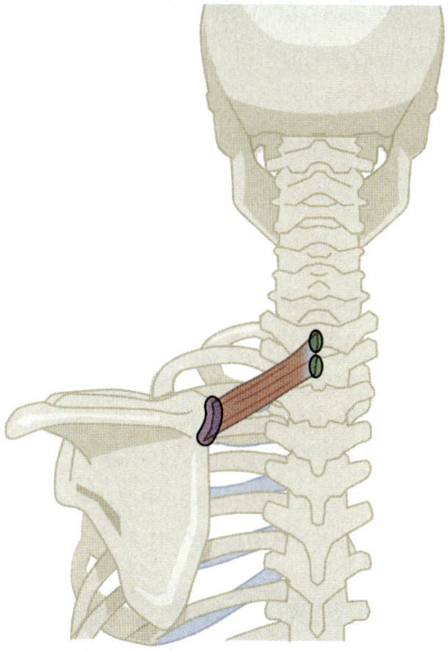

FIGURE 14.12 The Rhomboideus Major Muscle.

ORIGINS

spinous processes of T2–T5

INSERTION

vertebral border of scapula from the base of scapular spine to the inferior angle

OTHER MOTIONS

scapular downward rotation

shoulder girdle elevation

NERVE

Dorsal Scapular

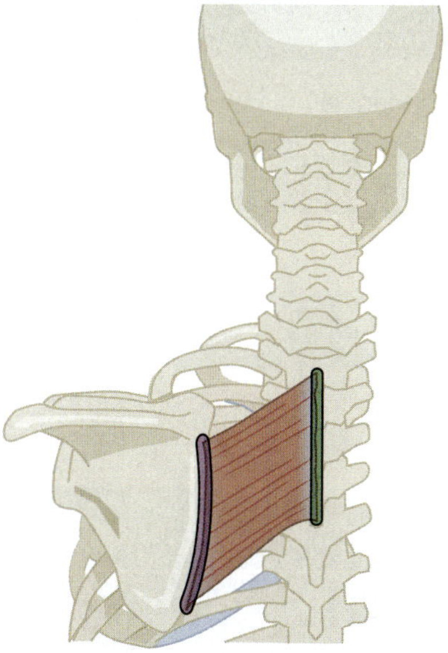

TABLE 14.2 Summary of the Shoulder Girdle Retractors

Muscle	Origins	Insertions	Nerves
Middle Trapezius	spinous processes of C7–T3	scapular spine & acromion	Accessory
Rhomboideus Minor	spinous processes of C7–T1	vertebral border of scapula from superior angle to base of scapular spine	Dorsal Scapular
Rhomboideus Major	spinous processes of T2–T5	vertebral border of scapula from the base of the spine to the inferior angle	

PAUSE TO CHECK FOR UNDERSTANDING

Table 14.2 should serve as a good review for much of the material, but here are a few other checks for understanding

1. How many muscles are in the shoulder girdle retractor group? List them.
2. What motion occurs if these muscles act eccentrically?
3. Where is the general location for the muscle group?

The Shoulder Girdle Protractors

Two muscles primarily protract the shoulder girdle. They are the *pectoralis minor* and the *serratus anterior*. The shoulder girdle protractors will be summarized in **TABLE 14.3** at the end of this section.

The **pectoralis** (PECK-tore-AL-liss) **minor** muscle lies deep in the anterolateral pectoral region (**FIGURE 14.13**). It originates from the anterior side of the third through fifth ribs and inserts on the coracoid process of the scapula.

The **serratus** (sear-RAY-tus) **anterior** muscle passes from the lateral rib cage beneath the anterior scapula (**FIGURE 14.14**). More specifically, it originates from the lateral side of the first eight ribs and inserts on the anterior side of the vertebral (medial) border of the scapula.

FIGURE 14.13 The Pectoralis Minor Muscle.

ORIGINS	anterior side of ribs 3–5
INSERTION	coracoid process of scapula
OTHER MOTIONS	none
NERVE	Medial Pectoral

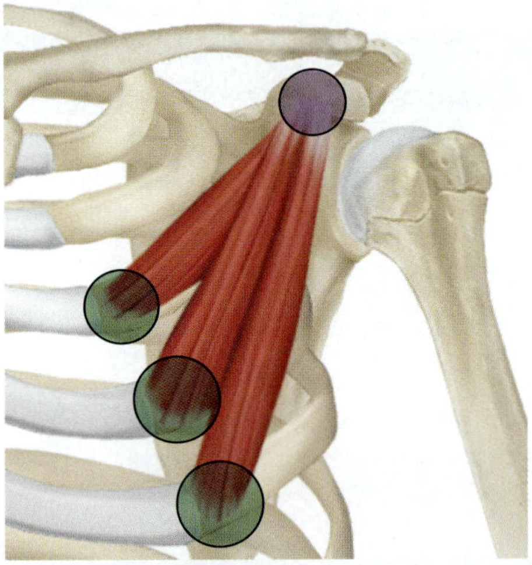

FIGURE 14.14 The Serratus Anterior Muscle.

ORIGINS	lateral side of ribs 1–8
INSERTION	anterior side of vertebral border of scapula
OTHER MOTIONS	scapular upward rotation
NERVE	Long Thoracic

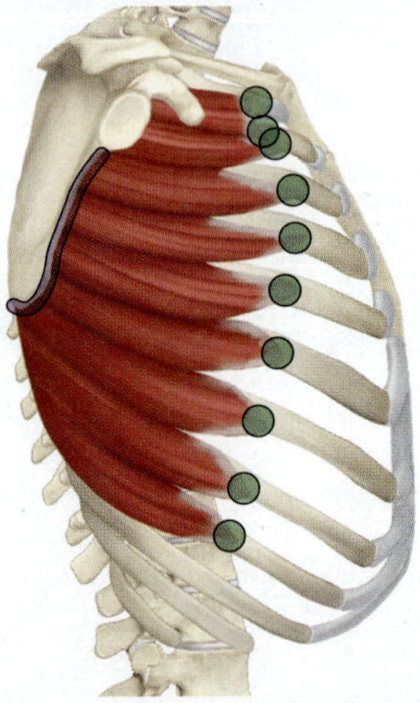

TABLE 14.3 Summary of the Shoulder Girdle Protractors

Muscle	Origins	Insertions	Nerves
Pectoralis Minor	anterior side of ribs 3–5	coracoid process of scapula	Medial Pectoral
Serratus Anterior	lateral side of ribs 1–8	anterior side of vertebral border of scapula	Long Thoracic

PAUSE TO CHECK FOR UNDERSTANDING

Table 14.3 should serve as a good review for much of the material, but here are a few other checks for understanding:

1. How many muscles are in the shoulder girdle protractor group? List them.
2. What motion occurs if these muscles act eccentrically?
3. Where is the general location for the muscle group?

© BraunS/E+/Getty Images

SECTION 5

The Upper Extremities

© BraunS/E+/Getty Images

CHAPTER 15

The Shoulder

CHAPTER OBJECTIVES

After completing this chapter, the student will be able to:

1. summarize the joints of the shoulder;
2. comprehend the shoulder motions;
3. identify the names and locations of the shoulder muscles;
4. recall the attachment sites and nerves of the shoulder muscles; and
5. apply the motions produced by the actions of the shoulder muscles.

The shoulder is a region of the body where the arm intersects the trunk (**FIGURE 15.1**). The shoulder has the greatest mobility of any part of the body and has been classified as a ball and socket joint. Its bony make-up along with the muscles of the shoulder allow for eight different motions.

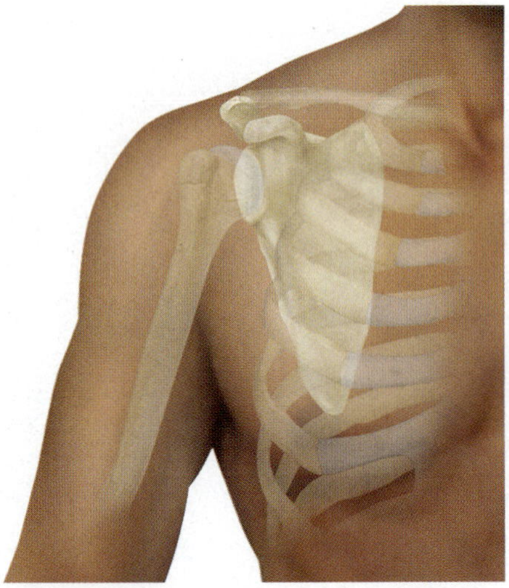

FIGURE 15.1 The Shoulder.

The Joints of the Shoulder

The main joint of the shoulder is the *glenohumeral joint* (**FIGURE 15.2**). Technically, the glenohumeral joint does move in isolation to the joints of the shoulder girdle discussed in Chapter 14. The others

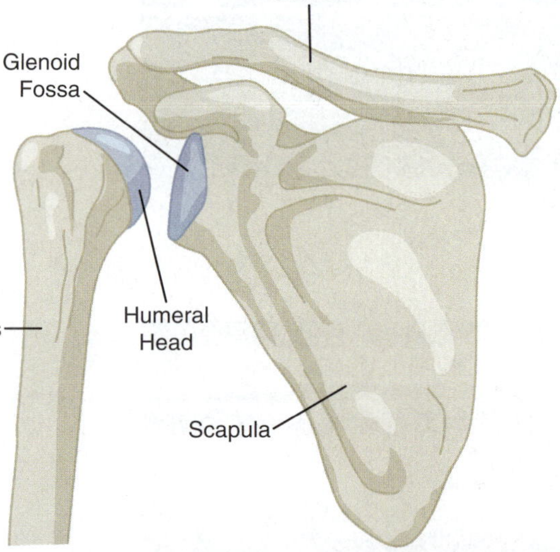

FIGURE 15.2 The Glenohumeral Joint.

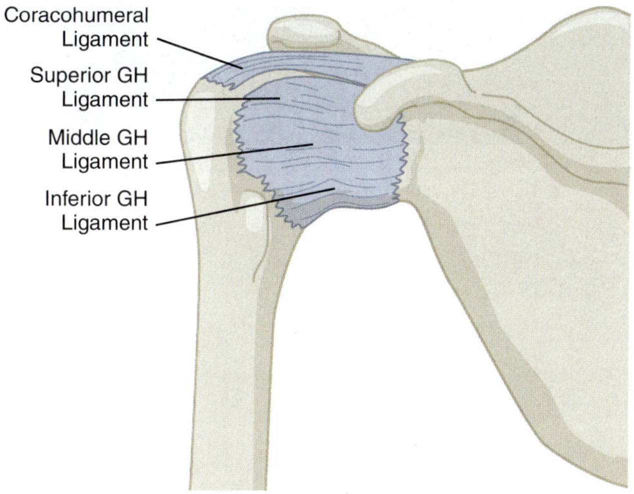

FIGURE 15.3 Glenohumeral Ligaments.

are the scapulothoracic articulation (technically not a joint), the acromioclavicular (AC) joint, and the sternoclavicular (SC) joint. Since these "other" three joints have already been described in Chapter 14, they will not be covered again in this chapter.

The **glenohumeral (GH) joint** is the articulation between the humerus and scapula (Figure 15.2). Its name is derived from the more specific articulation between the glenoid fossa (socket) of the scapula and the head of the humerus (ball). Like the hip, fibrocartilage called the **labrum** deepens the glenoid fossa for the head of the humerus.

A joint capsule encloses and stabilizes the GH joint. The posterior side is stabilized by the posterior joint capsule, but the anterior portion of the capsule is separated into four, distinct main ligaments (**FIGURE 15.3**). The most superior is the **coracohumeral ligament** which connects the coracoid process of the scapula and the humerus. The next three are the **superior**, **middle**, and **inferior glenohumeral ligaments**. They stabilize the rest of the anterior joint by connecting the glenoid fossa and the humeral head. One more ligament of note is the **transverse ligament**, sometimes called the **bicipital ligament**. This ligament, unlike the others, does not connect two bones. Rather, it attaches on both sides of the *intertubercular sulcus* (*bicipital groove*). The purpose of this ligament is to stabilize the tendon of the long head of the biceps brachii muscle as it passes through the groove and under the ligament.

The Motions of the Shoulder

Eight motions occur at the GH joint and is by far the most mobile joint in the average person (**FIGURE 15.4**). These motions are *extension, flexion, abduction, adduction, horizontal abduction, horizontal adduction, external rotation,* and *internal rotation*. **Extension** and **flexion**, opposing motions in the sagittal plane, occur about the frontal axis. Abduction ("take away") and adduction ("add to") rotate around the sagittal axis in the frontal plane. **Horizontal abduction** and **horizontal adduction** have the same root meaning as abduction and adduction, but move in the transverse plane about the longitudinal axis. **Internal** (medial) and **external** (lateral) **rotation** of the GH joint move in the same plane and rotate about the same axis as horizontal abduction and adduction.

The average ranges of motions of the GH joint are summarized in **TABLE 15.1**. Extension and flexion averages about 230 degrees. Abduction and adduction is about 180 degrees while horizontal abduction and

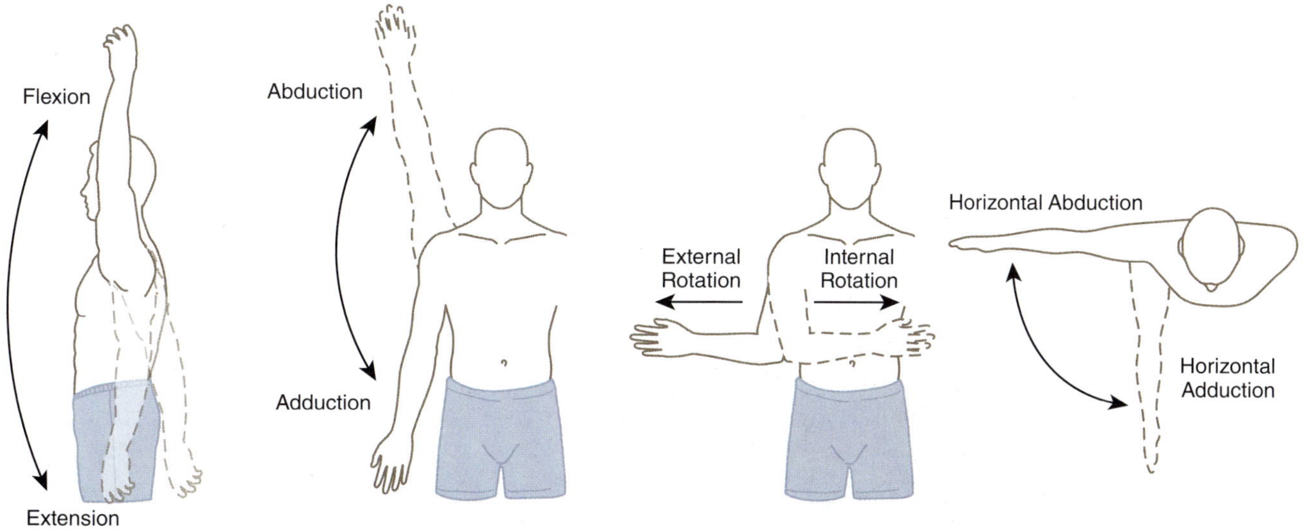

FIGURE 15.4 Motions of the Glenohumeral Joint.

TABLE 15.1 Average Glenohumeral Joint Ranges of Motions

Motion	Average Degrees of Motion
Extension	50–60°
Flexion	170–180°
Combined	220–240°
Adduction	0°
Abduction	170–180°
Combined	170–180°
Horizontal Abduction	45°
Horizontal Adduction	120°
Combined	165°
Internal Rotation	80–90°
External Rotation	80–90°
Combined	160–180°

adduction is about 165 degrees. Internal and external rotation average about 90 degrees apiece for a total of about 180 degrees.

Additionally, there are two accessory motions of the scapula during abduction and adduction of the GH joint (**FIGURE 15.5**). They are *upward* and *downward scapular rotation*. **Upward scapular rotation** is a turning of the scapula in a counterclockwise direction. **Downward scapular rotation** is a motion in a clockwise direction. During the first 30 degrees of GH abduction, the scapula does not move. However, during about 30–90° of GH abduction, the scapula rotates upward about one degree for every two degrees of abduction. Then, from about 90 degrees of GH abduction to the end of this motion, the scapula rotates upward one1 degree for every degree of GH abduction. During GH adduction, the scapula rotates downward (clockwise) with the same ratios at the same points of GH motion.

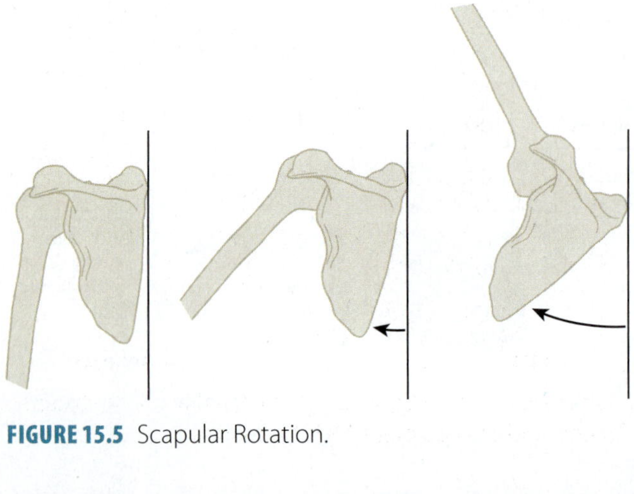

FIGURE 15.5 Scapular Rotation.

Circumduction is sometimes listed as a motion (**FIGURE 15.6**). This text considers circumduction to be a complex joint motion rather than a simple joint motion. As defined before, a simple motion is displacement in one direction within a single plane rotating about a single axis of rotation. Circumduction definitely breaks this definition. However, it is worth mentioning because of its commonality and to illustrate how great the ability of the shoulder is to move. Average GH flexion is limited to about 180 degrees; however, most people can continue to rotate the GH joint for a complete revolution until the arm returns to its starting position. This is circumduction and it is only possible if the person internally rotates the GH joint in addition to flexion to complete the revolution which is the use of two motions to perform the *movement* of circumduction.

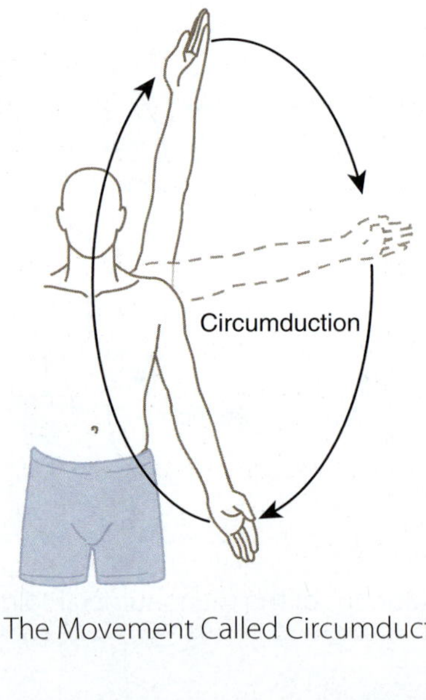

FIGURE 15.6 The Movement Called Circumduction.

PAUSE TO CHECK FOR UNDERSTANDING

1. List the joints of the shoulder.
2. Then, tell which bones form each joint.
3. What are the eight motions of the shoulder?

The Muscles That Move the Glenohumeral Joint

The rest of this chapter is dedicated to learning the muscles that cause the motions of the glenohumeral joint; their attachment sites; the *primary*, main motions they produce; and the nerves that innervate them. Muscles work in groups and so each muscle within the group supplies a certain percentage of the force needed to move a body segment and that amount (percentage) is not the same for each muscle. Furthermore, the force each muscle contributes varies depending on the starting position of the body segment and changes throughout the range of motion. This can be confusing to the beginning kinesiology student. So, this text has made certain choices in an attempt to simplify the muscle groups based on initial motion from anatomical position. The exception to this is horizontal abduction and horizontal adduction which begins from a starting position of 90 degrees of glenohumeral abduction.

The muscle groups are based on simple motion caused by their *concentric* action (i.e., The shoulder flexors flex the shoulder when they concentrically act). However, it should not be forgotten that all muscles can and will act *eccentrically* and *isometrically*. The reader should also notice that the table for each individual muscle has a section called "Other Motions." *Other Motions* means that this muscle aides, to some degree, in other motion(s) besides the group it has been placed in. Some muscles may appear in more than one group because their role is too great to ignore.

Before describing the individual muscles within each muscle group, four muscles should be divided into sections like the trapezius muscle learned in Chapter 14. This is because each section of each muscle primarily causes a different motion than the other sections. These muscles are the *deltoid*, the *triceps brachii*, the *pectoralis major*, and the *biceps brachii* (**FIGURES 15.7** and **15.8**).

The **deltoid** muscle is commonly divided into three sections: anterior, middle, and posterior. The **triceps brachii** is also divided into three sections called the long head, the medial head, and the lateral head. The **pectoralis major** has a clavicular section and a sternal section. As well, the **biceps brachii** has two sections called the long head and short head.

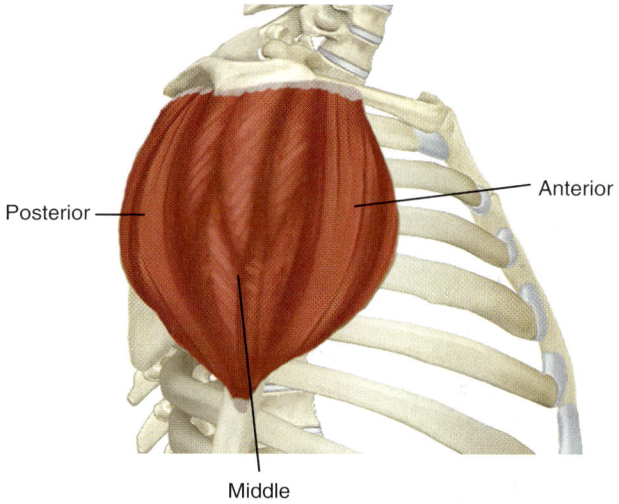

FIGURE 15.7 The Sections of the Deltoid Muscles.

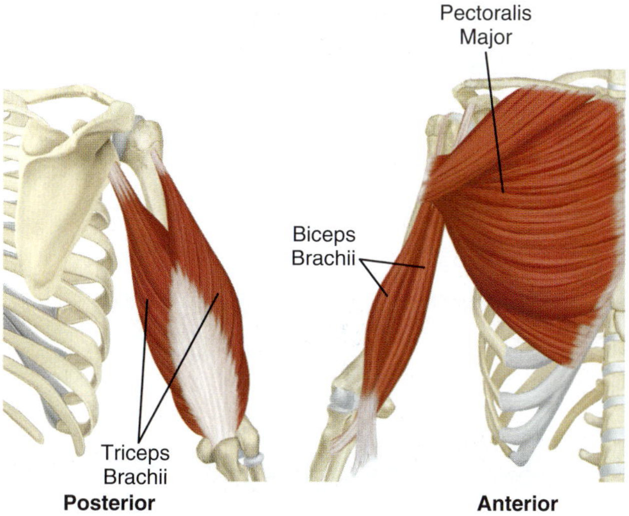

FIGURE 15.8 The Sections of the Triceps Brachii, the Biceps Brachii, and the Pectoralis Major Muscles.

Since there are eight motions, there are eight muscle groups. The eight groups are the *shoulder flexors, shoulder extensors, shoulder abductors, shoulder adductors, shoulder horizontal adductors, shoulder horizontal abductors, shoulder external rotators*, and *shoulder internal rotators*.

The Shoulder Flexors

Four muscles are mainly responsible for flexing the shoulder (**FIGURE 15.9**). They are the *deltoid* (*anterior section*), the *pectoralis major* (*clavicular section*), the *coracobrachialis*, and the *biceps brachii* (*short head*). The shoulder flexors are summarized in **TABLE 15.2** at the end of this section.

The **anterior deltoid** covers the anterior portion of the shoulder (**FIGURE 15.10**). It originates from the distal one-third of the clavicle and inserts on the deltoid tuberosity with the rest of the sections of the deltoid muscle.

The **clavicular section** of the **pectoralis** (PECK-tore-AL-liss) **major** resides superficially in the superior, lateral pectoral (chest) region of the thorax (**FIGURE 15.11**). It runs obliquely originating from the proximal one-third of the clavicle and inserts on the lateral lip of the intertubercular (bicipital) groove.

The **coracobrachialis** (CORE-ah-co-BRAY-key-al-is) is a relatively slender muscle that passes through the medial side of the brachial region (**FIGURE 15.12**). It originates from the coracoid process of the scapula and inserts on the medial side of the diaphysis of the humerus.

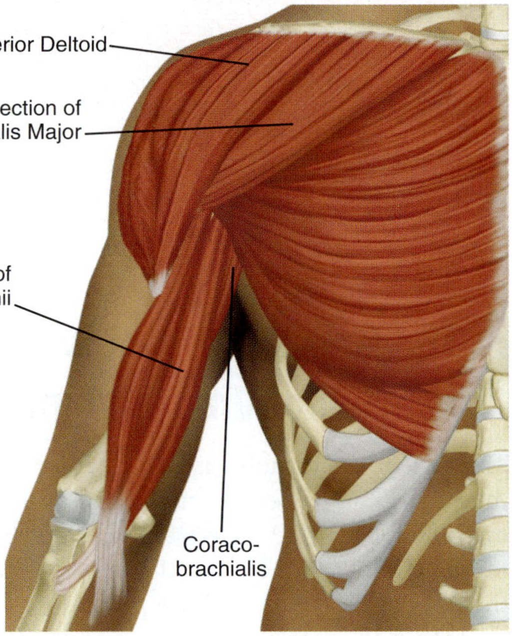

FIGURE 15.9 The Glenohumeral Flexors.

The **biceps brachii** (BRAY-key-eye) is a two-jointed muscle running superficially across the shoulder, through the anterior brachial region, and over the elbow (**FIGURE 15.13**). It has two heads (biceps) or sections, but only the **short head** is significantly involved in flexion. It originates from the coracoid process of the scapula, merges with the long head, and inserts on the radial tuberosity which is on the anterior side of the radius just below the proximal epiphysis.

FIGURE 15.10 The Anterior Section of the Deltoid Muscle.

ORIGIN	
distal third of clavicle	
INSERTION	
deltoid tuberosity	
OTHER MOTIONS	
GH horizontal adduction	
NERVE	
Axillary	

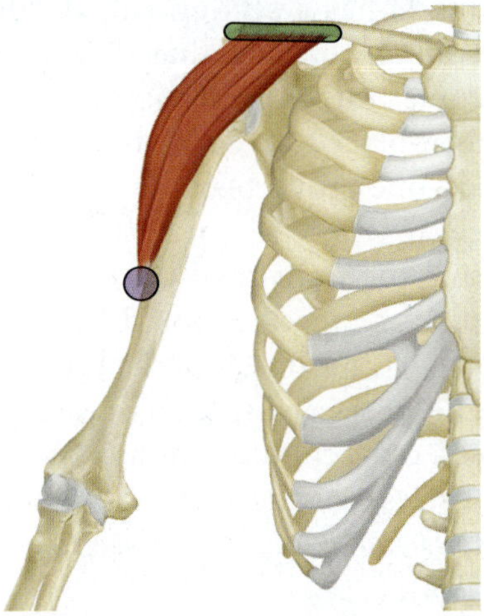

FIGURE 15.11 The Clavicular Section of the Pectoralis Major Muscle.

ORIGIN	
proximal third of clavicle	
INSERTION	
lateral lip of intertubercular groove	
OTHER MOTIONS	
GH horizontal adduction	
GH internal rotation	
NERVE	
Pectoral	

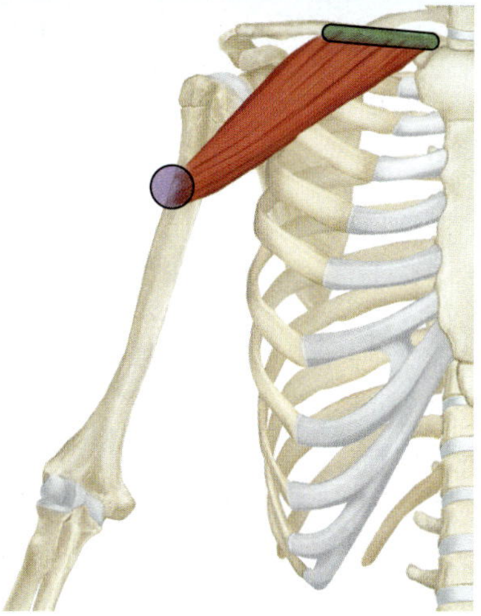

FIGURE 15.12 The Coracobrachialis Muscle.

ORIGIN	
coracoid process	
INSERTION	
medial side of humeral diaphysis	
OTHER MOTIONS	
GH horizontal adduction	
NERVE	
Musculocutaneous	

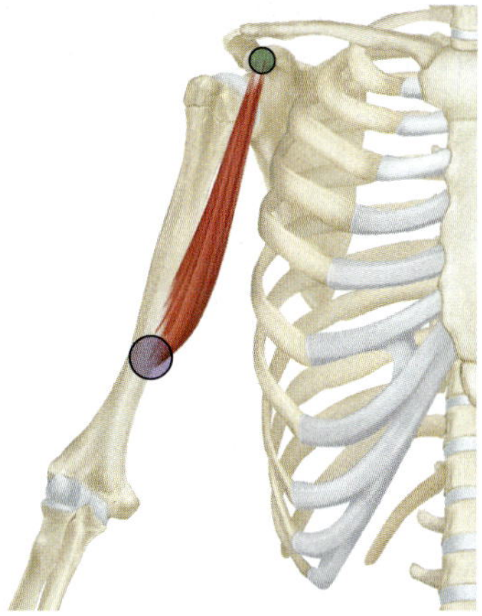

FIGURE 15.13 The Short Head of the Biceps Brachii Muscle.

ORIGIN	
coracoid process	
INSERTION	
radial tuberosity	
OTHER MOTIONS	
elbow flexion	
NERVE	
Musculocutaneous	

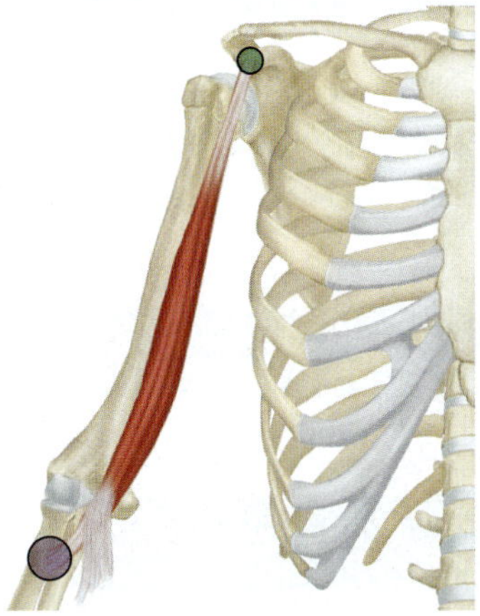

TABLE 15.2 Summary of the Shoulder Flexors

Muscle	Origins	Insertions	Nerve
Deltoid (Anterior)	distal 1/3 of clavicle	deltoid tuberosity	Axillary
Pectoralis Major (Clavicular)	proximal 1/3 of clavicle	lateral lip of intertubercular groove	Pectoral
Coracobrachialis	coracoid process	medial side of humeral diaphysis	Musculocutaneous
Biceps Brachii (SH)	coracoid process	radial tuberosity	

PAUSE TO CHECK FOR UNDERSTANDING

Table 15.2 should serve as a good review for much of the material, but here are a few other checks for understanding:

1. How many muscles are in the shoulder flexor group? List them.
2. What motion occurs if these muscles act eccentrically?
3. Where is the general location for the muscle group?

The Shoulder Extensors

Two muscles are mainly responsible for extending the shoulder (**FIGURE 15.14**). They are the *deltoid* (*posterior section*) and the *triceps brachii* (*long head*). The shoulder flexors are summarized in **TABLE 15.3** at the end of this section.

The **posterior deltoid** covers the posterior portion of the shoulder (**FIGURE 15.15**). It originates

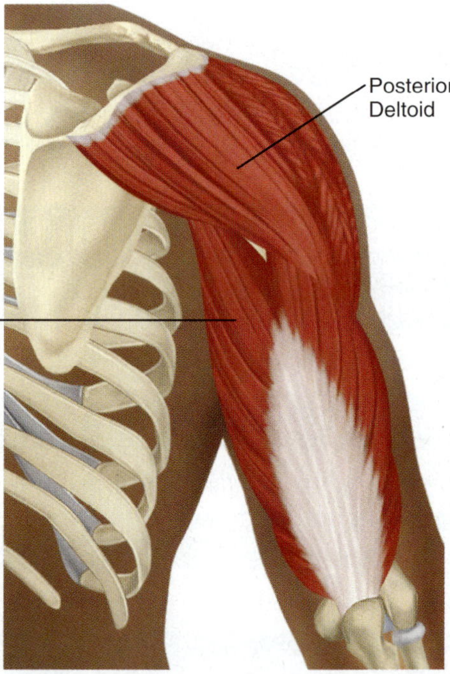

FIGURE 15.14 The Shoulder Extensors.

FIGURE 15.15 The Posterior Section of the Deltoid Muscle.

ORIGIN	
spine of scapula	
INSERTION	
deltoid tuberosity	
OTHER MOTIONS	
GH horizontal abduction	
NERVE	
Axillary	

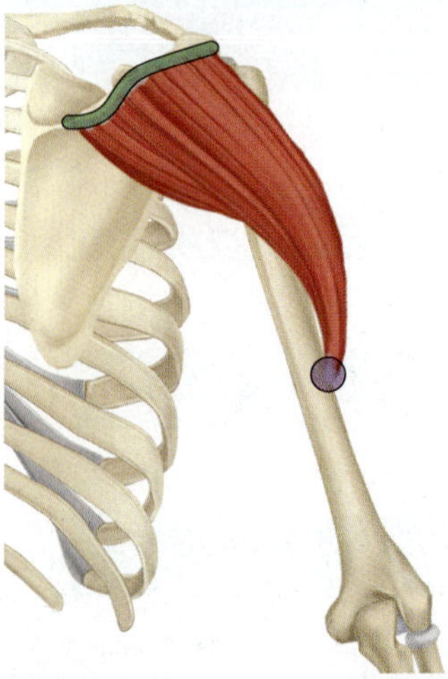

from the scapular spine and inserts on the deltoid tuberosity with the rest of the portions of the deltoid muscle.

The **triceps brachii** (BRAY-key-eye) lies in the posterior brachial region crossing both the shoulder and the elbow (**FIGURE 15.16**). The **long head** originates from the inferior side of the glenoid fossa of the scapula, crosses the shoulder, merges with the other two heads, and inserts onto the olecranon process of the ulna.

FIGURE 15.16 The Long Head Section of the Triceps Brachii Muscle.

ORIGIN
infraglenoid fossa
INSERTION
olecranon process
OTHER MOTIONS
elbow extension
NERVE
Radial

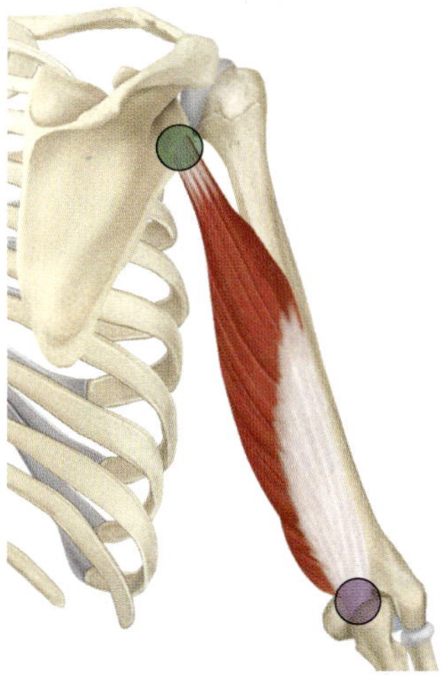

TABLE 15.3 Summary of the Shoulder (GH Joint) Extensors

Muscle	Origins	Insertions	Nerve
Posterior Deltoid	acromion	deltoid tuberosity	Axillary
Triceps Brachii (LH)	infraglenoid fossa	olecranon	Radial

PAUSE TO CHECK FOR UNDERSTANDING

Table 15.3 should serve as a good review for much of the material, but here are a few other checks for understanding:

1. How many muscles are in the shoulder extensor group? List them.
2. What motion occurs if these muscles act eccentrically?
3. Where is the general location for the muscle group?

The Shoulder Abductors

Two main muscles abduct the shoulder (**FIGURE 15.17**). They are the *supraspinatus* and the *deltoid* (*middle section*). The shoulder abductors are summarized in **TABLE 15.4** at the end of this section.

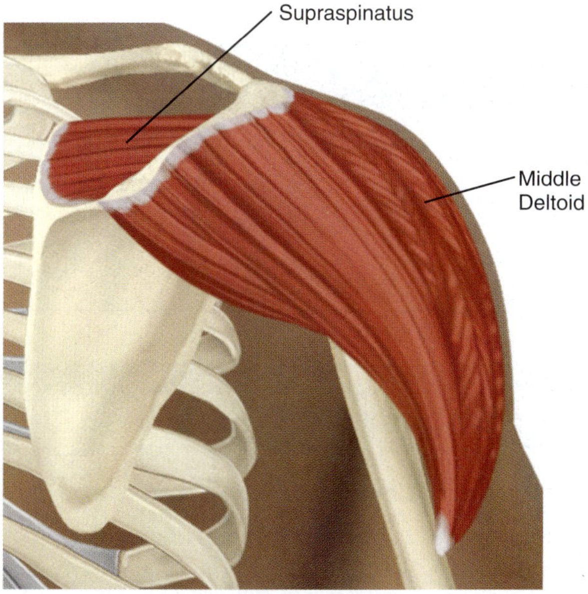

FIGURE 15.17 The Glenohumeral Abductors.

The **supraspinatus** (SUE-prah-spah-NAH-tus) originates from the supraspinous fossa of the scapula (**FIGURE 15.18**). From there, its belly passes horizontally across the area superior to the scapular spine and then under the acromioclavicular joint. It inserts on the greater tuberosity of the humerus.

The **middle deltoid** covers the lateral shoulder and is in between the anterior and posterior sections of the muscle (**FIGURE 15.19**). It originates from the acromion and has a common insertion with the rest of the deltoid muscle on the deltoid tuberosity of the humerus.

FIGURE 15.18 The Supraspinatus Muscle.

ORIGIN
supraspinous fossa
INSERTION
greater tuberosity of the humerus
OTHER MOTIONS
none
NERVE
Suprascapular

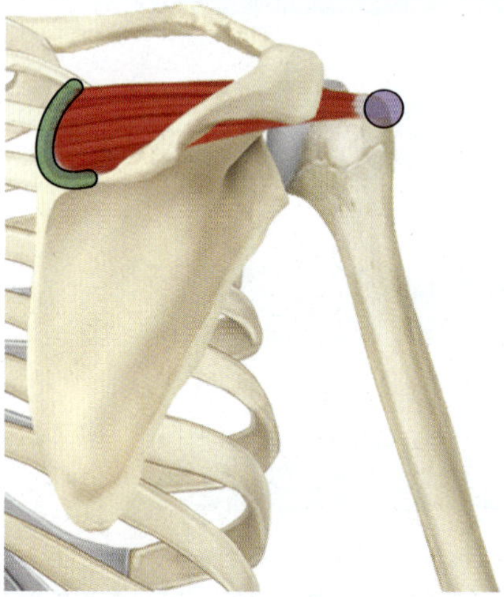

FIGURE 15.19 The Middle Section of the Deltoid Muscle.

ORIGIN
acromion
INSERTION
deltoid tuberosity
OTHER MOTIONS
none
NERVE
Axillary

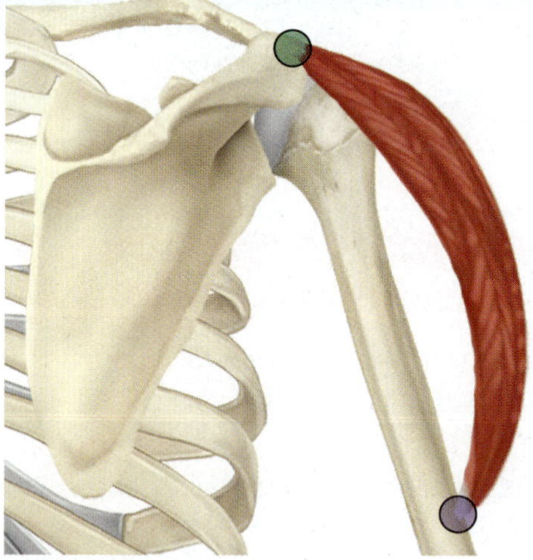

TABLE 15.4 Summary of the Shoulder (GH Joint) Abductors

Muscle	Origins	Insertions	Nerve
Supraspinatus	supraspinous fossa	humeral greater tuberosity	Suprascapular
Posterior Deltoid	acromion	deltoid tuberosity	Axillary

PAUSE TO CHECK FOR UNDERSTANDING

Table 15.4 should serve as a good review for much of the material, but here are a few other checks for understanding:

1. How many muscles are in the shoulder abductor group? List them.
2. What motion occurs if these muscles act eccentrically?
3. Where is the general location for the muscle group?

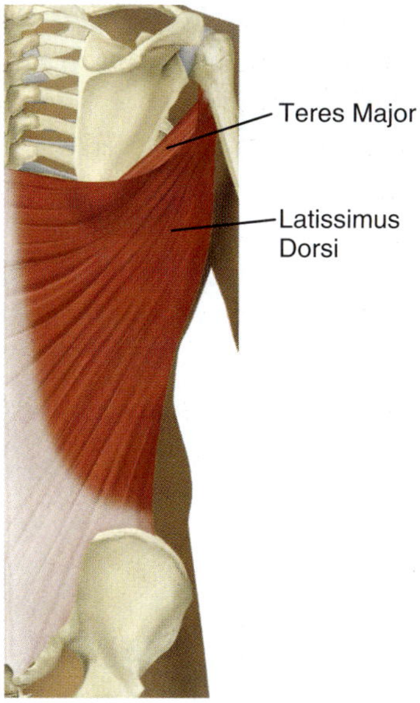

FIGURE 15.20 The Glenohumeral Adductors.

The Shoulder Adductors

Two muscles are mainly responsible for shoulder adduction (**FIGURE 15.20**). They are the *latissimus dorsi* and the *teres major* muscles. These muscles are summarized in **TABLE 15.5** at the end of this section.

The **latissimus** (lah-TISS- ah-mus) **dorsi** (DOOR-sigh) lies on each side of the mid to lower back (**FIGURE 15.21**). It originates from four areas: 1) the spinous processes of T7–L5 vertebrae, 2) the posterior side of sacrum, 3) the iliac spine, and 4) the posterior side of ribs 10–12. From these origins, it passes underneath the arm forming the posterior border of the axilla and inserts on the medial side of the humerus just inferior to the intertubercular groove.

The **teres** (TEAR-ease) **major** resides lateral to the scapula and runs through the axilla (**FIGURE 15.22**). It originates from the inferior angle of the scapula, passes beneath the arm forming the posterior border of the axilla with the latissimus dorsi, and inserts on the medial lip of the intertubercular (bicipital) groove.

FIGURE 15.21 The Latissimus Dorsi Muscle.

ORIGINS	
spinous process of T7–L5	
posterior sacrum	
iliac spine	
posterior side of ribs 10–12	
INSERTIONS	
anteromedial side of humerus just inferior to intertubercular groove	
OTHER MOTIONS	
GH extension	
GH internal rotation	
NERVE	
Thoracodorsal	

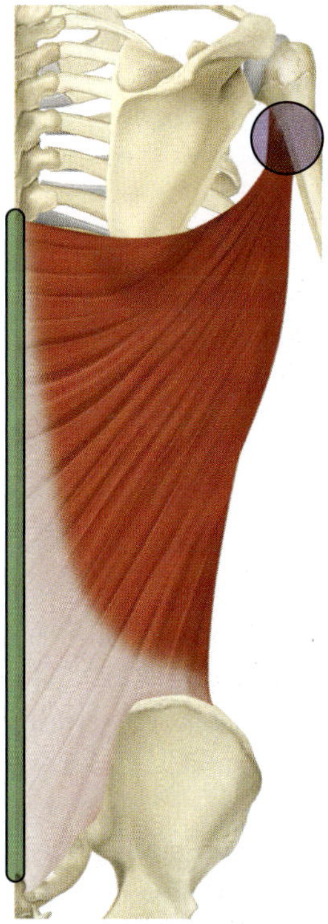

FIGURE 15.22 The Teres Major Muscle.

ORIGIN

inferior angle of scapula

INSERTION

medial lip of intertubercular groove

OTHER MOTIONS

GH extension

GH internal rotation

NERVE

Subscapular

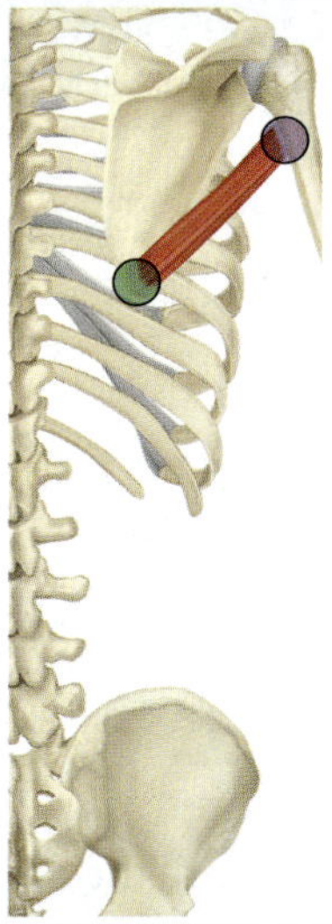

TABLE 15.5 Summary of the Shoulder (GH Joint) Adductors

Muscle	Origins	Insertions	Nerve
Latissimus Dorsi	spinous processes of T7–L5, posterior sacrum, iliac spine, & posterior side of ribs 10–12	anteromedial side of humerus just inferior to intertubercular groove	Thoracodorsal
Teres Major	inferior angle of scapula	medial lip of intertubercular groove	Subscapular

PAUSE TO CHECK FOR UNDERSTANDING

Table 15.5 should serve as a good review for much of the material, but here are a few other checks for understanding

1. How many muscles are in the shoulder adductor group? List them.
2. What motion occurs if these muscles act eccentrically?
3. Where is the general location for the muscle group?

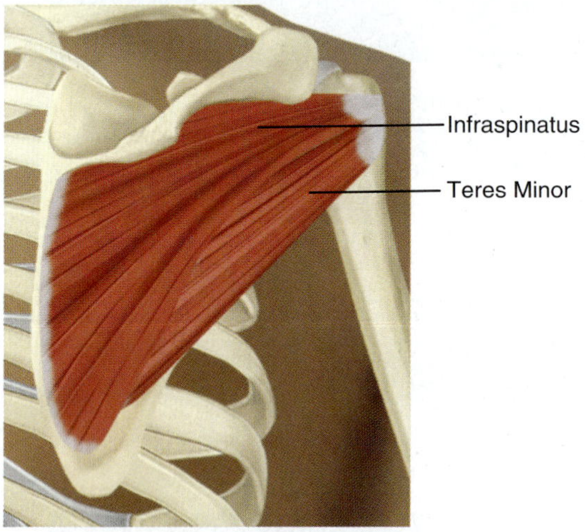

FIGURE 15.23 The Glenohumeral External Rotators.

The Shoulder External Rotators

Two main muscles are mainly responsible for external rotation of the shoulder (**FIGURE 15.23**). They are the *infraspinatus* and the *teres minor*. The shoulder external rotators are summarized in **TABLE 15.6** at the end of this section.

The **infraspinatus** (IN-frah-spah-NAH-tus) originates from the infraspinous fossa of the scapula (**FIGURE 15.24**). From there, its belly runs laterally and crosses the posterior side of the GH joint and humeral head. It inserts on the greater tuberosity of the humerus.

The **teres** (TEAR-ease) **minor** originates from the lateral border of the scapula just superior to the origin site of the teres major muscle (**FIGURE 15.25**). Its belly runs just inferior to the infraspinatus before inserting on the greater tuberosity of the humeral head.

FIGURE 15.24 The Infraspinatus Muscle.

ORIGIN

infraspinous fossa

INSERTION

greater tuberosity of humerus

OTHER MOTIONS

GH horizontal abduction

NERVE

Suprascapular

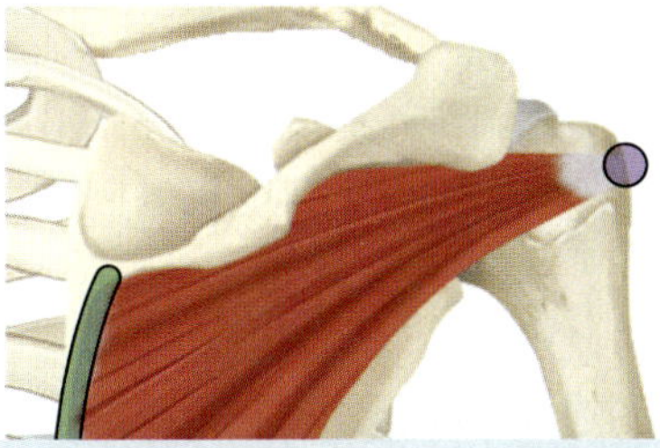

FIGURE 15.25 The Teres Minor Muscle.

ORIGIN

lateral border of scapula just superior to teres major origin

INSERTION

greater tuberosity of humerus

OTHER MOTIONS

GH horizontal abduction

NERVE

Axillary

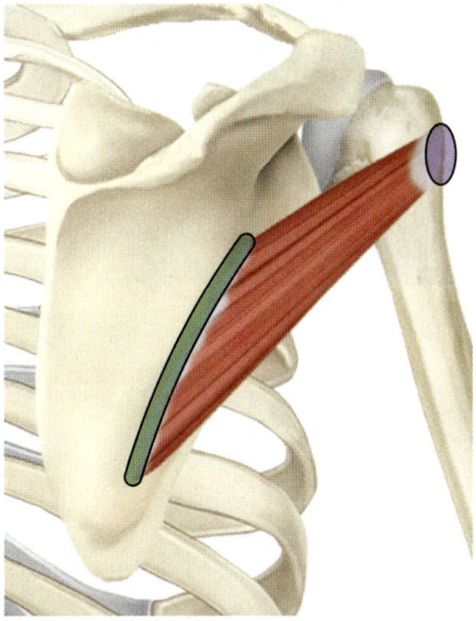

TABLE 15.6 Summary of the Shoulder (GH Joint) External Rotators

Muscle	Origins	Insertions	Nerve
Infraspinatus	infraspinous fossa	greater tuberosity of humerus	Suprascapular
Teres Minor	lateral border of scapula	greater tuberosity of humerus	Axillary

PAUSE TO CHECK FOR UNDERSTANDING

Table 15.6 should serve as a good review for much of the material, but here are a few other checks for understanding

1. How many muscles are in the shoulder external rotator group? List them.
2. What motion occurs if these muscles act eccentrically?
3. Where is the general location for the muscle group?

The Shoulder Internal Rotators

Two muscles are primarily responsible for shoulder internal rotation. They are the *subscapularis* and the *teres major* (**FIGURE 15.26**). The description of the teres major will not be repeated here (see Shoulder Adductors). The shoulder horizontal adductors are summarized in **TABLE 15.7** at the end of this section.

The **subscapularis** originates from the subscapular fossa on the anterior side of the scapula (**FIGURE 15.27**). Its belly runs laterally across the scapula through the axilla and anterior to the humeral head. It then inserts on the lesser tuberosity of the humeral head.

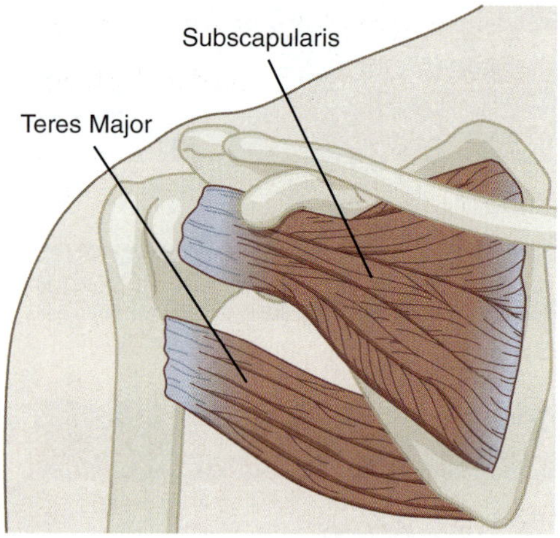

FIGURE 15.26 The GH Internal Rotators.

FIGURE 15.27 The Subscapularis Muscle.

ORIGIN	subscapular fossa
INSERTION	lesser tuberosity of humerus
OTHER MOTIONS	none
NERVE	Subscapular

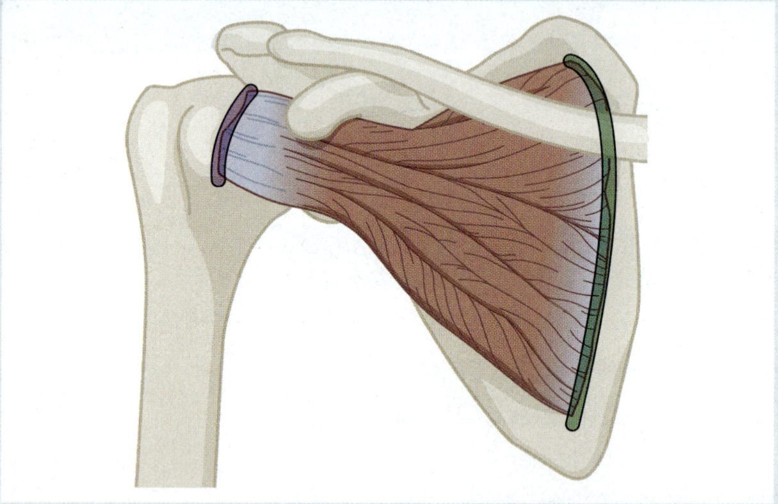

TABLE 15.7 Summary of the Shoulder (GH Joint) Internal Rotators

Muscle	Origins	Insertions	Nerve
Subscapularis	subscapular fossa	lesser tuberosity of the humerus	Subscapular
Teres Major	inferior angle of scapula	medial lip of intertubercular groove	

PAUSE TO CHECK FOR UNDERSTANDING

Table 15.7 should serve as a good review for much of the material, but here are a few other checks for understanding

1. How many muscles are in the shoulder flexor group? List them.
2. What motion occurs if these muscles act eccentrically?
3. Where is the general location for the muscle group?

The Shoulder Horizontal Abductors

Three muscles are mainly involved in horizontal abduction (**FIGURE 15.28**). They are the *posterior section of the deltoid*, the *infraspinatus*, and the *teres minor*. The posterior deltoid has already been described in the Shoulder Extensors. And, the infraspinatus and the teres minor have already been described in the Shoulder External Rotators. Therefore, they will not be repeated here.

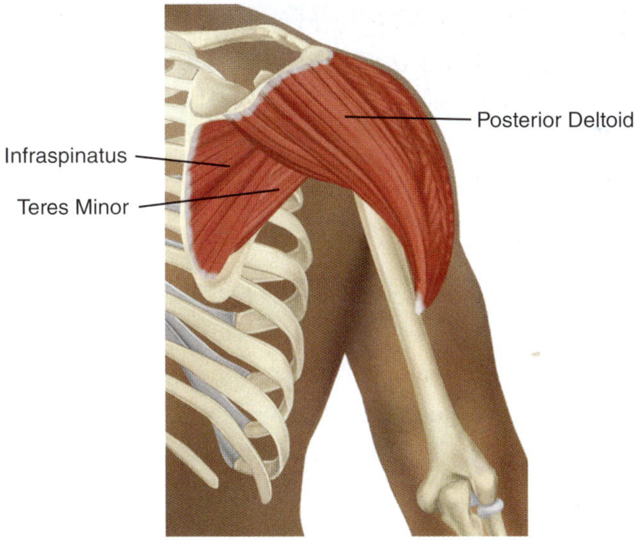

FIGURE 15.28 The GH Horizontal Abductors.

PAUSE TO CHECK FOR UNDERSTANDING

1. How many muscles are in the shoulder horizontal abductor group? List them.
2. What motion occurs if these muscles act eccentrically?
3. Where is the general location for the muscle group?

The Shoulder Horizontal Adductors

Three muscles are mainly responsible for abducting the shoulder (**FIGURE 15.29**). They are the *pectoralis major* (both sections), the *deltoid (anterior section)*, and the *coracobrachialis*. The anterior deltoid and the coracobrachialis have already been described in the section on the Shoulder Flexors, but the sternal section of the pectoralis major has not. The shoulder horizontal adductors are summarized in **TABLE 15.8** at the end of this section.

The **pectoralis** (PECK-tore-AL-liss) **major** resides superficially in the pectoral (chest) region (**FIGURE 15.30**). Its fibers run much more horizontal than the oblique clavicular section. It originates from the lateral side of the sternum as well as the costal cartilage of ribs 1–6, merges with the clavicular section, and finally inserts on the lateral lip of the intertubercular (bicipital) groove.

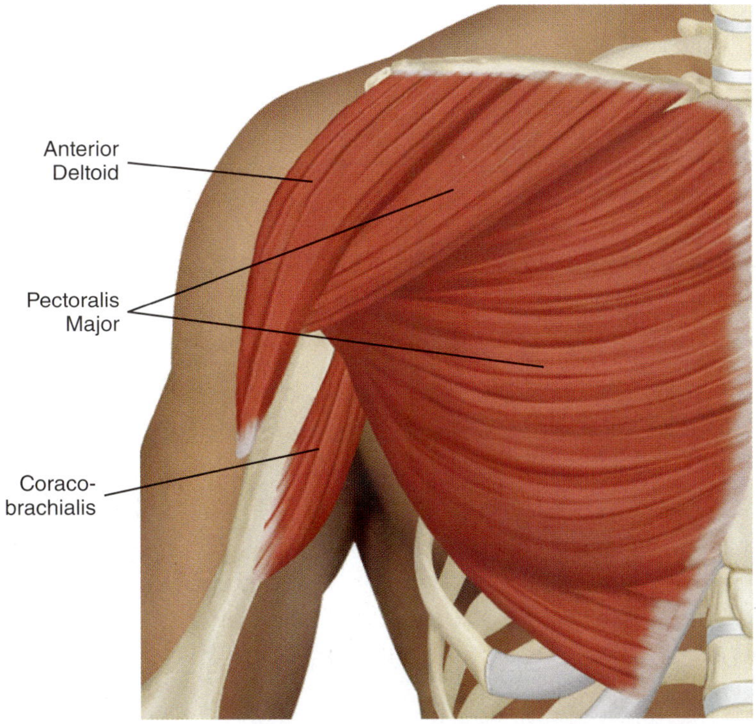

FIGURE 15.29 The Glenohumeral Horizontal Adductors.

FIGURE 15.30 The Sternal Section of the Pectoralis Major Muscle.

ORIGINS
lateral side of sternum
costal cartilage of ribs 1–6
INSERTION
lateral lip of intertubercular groove
OTHER MOTIONS
GH adduction
GH internal rotation
NERVE
Pectoral

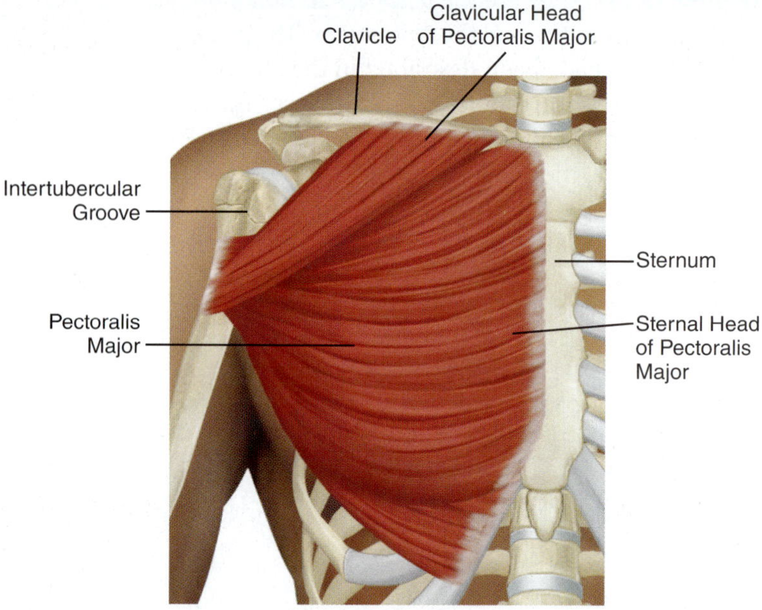

TABLE 15.8 Summary of the Horizontal Shoulder (GH Joint) Adductors

Muscle	Origins	Insertions	Nerve
Pectoralis Major	proximal 1/3 of clavicle, lateral side of sternum & costal cartilage of ribs 1–6	lateral lip of intertubercular groove	Pectoral
Anterior Deltoid	distal 1/3 of clavicle	deltoid tuberosity	Axillary
Coracobrachialis	coracoid process	medial side of humeral diaphysis	Musculocutaneous

PAUSE TO CHECK FOR UNDERSTANDING

Table 15.8 should serve as a good review for much of the material, but here are a few other checks for understanding

1. How many muscles are in the shoulder horizontal adductor group? List them.
2. What motion occurs if these muscles act eccentrically?
3. Where is the general location for the muscle group?

CHAPTER 16

The Elbow

CHAPTER OBJECTIVES

After completing this chapter, the student will be able to:

1. summarize the joints of the elbow region;
2. comprehend the elbow motions;
3. identify the names and locations of the elbow muscles;
4. recall the attachment sites and nerves of the elbow muscles; and
5. apply the motions produced by the actions of the elbow muscles.

The elbow is a region of the body where the (upper) arm intersects the forearm (lower arm) (**FIGURE 16.1**). Despite its traditional classification as a hinge joint, its anatomical make-up and the actions of its muscles produce four different motions. Two of these motions are supination and pronation. Traditionally, these two motions have been credited to the forearm. It is true that the forearm moves during these motions, but the root of the motion takes place at the elbow. This is no different than when the elbow extends, the forearm moves. There is some displacement at the wrist during these motions, but the vast majority of the motion occurs at the elbow.

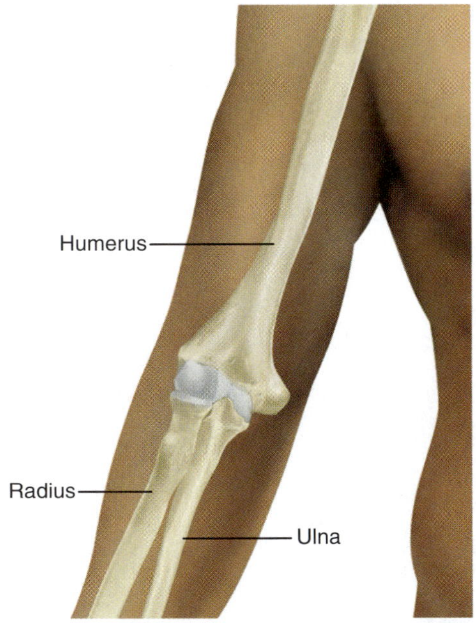

FIGURE 16.1 The Elbow.

The Joints of the Elbow

The elbow is comprised of three joints (**FIGURE 16.2**). They are the *humeroradial joint, humeroulnar joint*, and the *proximal radioulnar joint.*

The **humeroradial joint** is the articulation between the capitulum (lateral condyle) of the humerus and the radial head. The **humeroulnar joint** is the articulation between the trochlea (lateral condyle) of the humerus and the trochlear notch of the proximal ulna. The **proximal radioulnar joint** is the dynamic relationship between the radial head and the radial notch of the ulna. Three

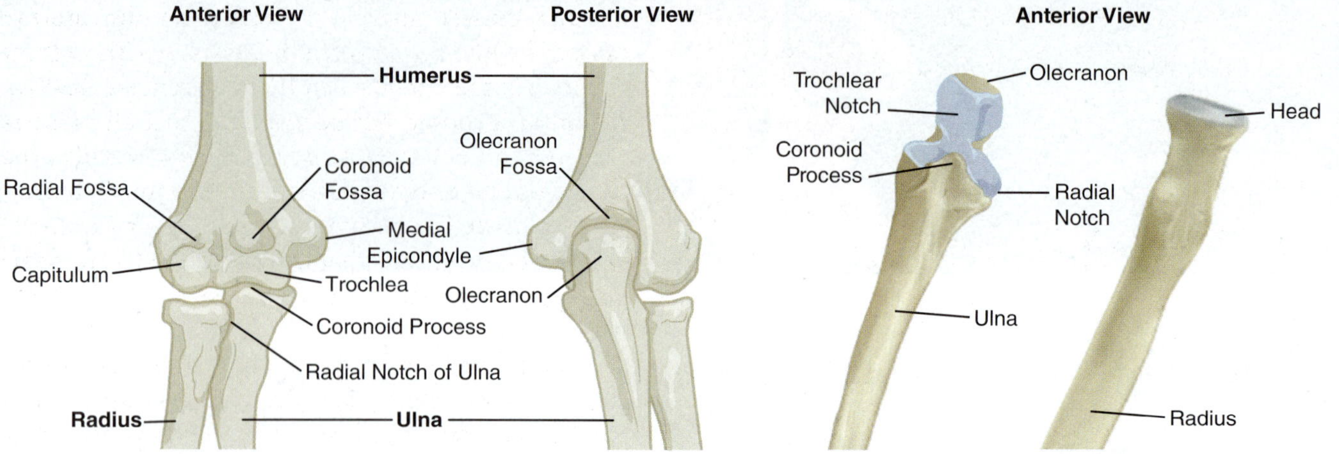

FIGURE 16.2 The Joints of the Elbow.

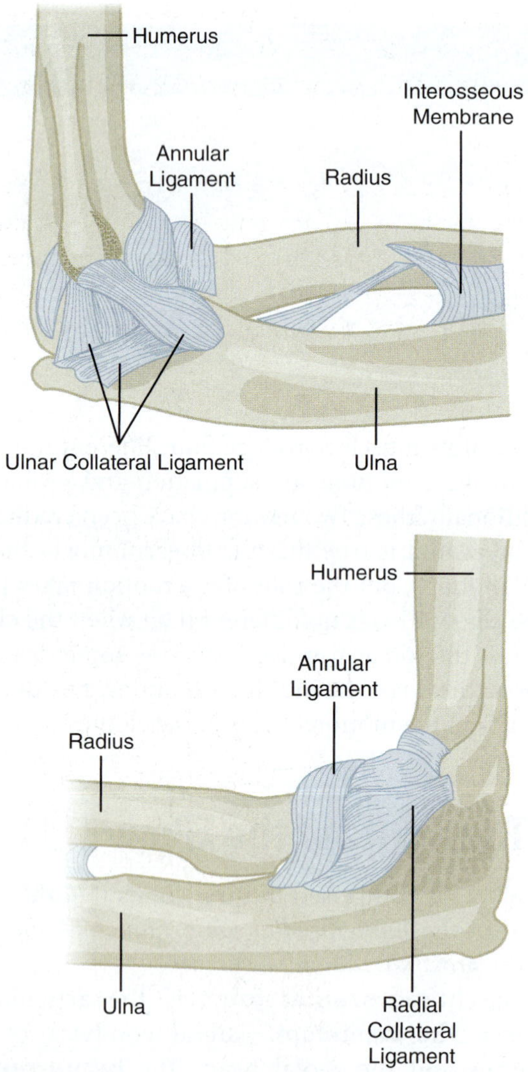

FIGURE 16.3 The Ligaments of the Elbow.

main ligaments stabilize the bones of the elbow (**FIGURE 16.3**). The **annular ligament** stabilizes the radius (radial head) in the radial notch of the ulna. It wraps around the radial head attaching on either side of it on the ulna. The **radial collateral ligament**, also called *lateral collateral ligament*, stabilizes the lateral side of the elbow. Despite its name, it actually does not attach to the radius. This is so the radius is free to rotate around the ulna. It attaches from the lateral epicondyle of the humerus to the annular ligament and the lateral side of the ulna. The **ulnar collateral ligament**, also called the *medial collateral ligament*, stabilizes the medial side of the elbow. It attaches from medial epicondyle of the humerus and fans out to attach on medial parts of the ulna.

The Motions of the Elbow

Four motions occur at the elbow (**FIGURE 16.4**). These motions are *flexion*, *extension*, *pronation*, and *supination*. **Elbow extension** (straightening) and **flexion**

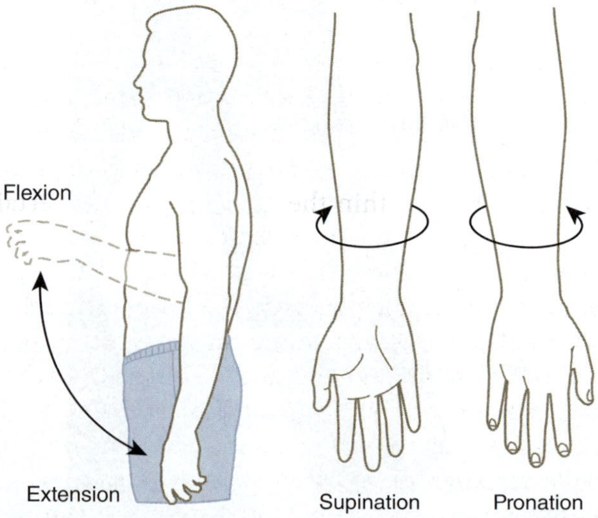

FIGURE 16.4 Motions of the Elbow.

TABLE 16.1 Average Elbow Range of Motions

Motion	Average Degrees of Motion
Extension	0°
Flexion	145–155°
Combined	145–155°
Supination	90°
Pronation	90°
Combined	180°

(bending), opposing motions, occur in the sagittal plane rotating about the frontal axis. **Elbow pronation**, turning of forearm toward the midline, and **supination**, turning away from the midline, occur in the transverse plane about the transverse (vertical) axis.

TABLE 16.1 summarizes the average ranges of motions of the elbow. Average elbow extension is 0 degrees. Average elbow flexion is about 150 degrees. Both supination and pronation average 90 degrees.

PAUSE TO CHECK FOR UNDERSTANDING

1. List the joints of the elbow.
2. Then, tell which bones form each joint.
3. What are the four motions of the elbow?

The Muscles That Move the Elbow

The rest of this chapter is dedicated to learning the elbow muscles; their attachment sites; the *primary*, main motions they produce; and the nerves that innervate them. Muscles work in groups and so each muscle within the group supplies a certain percentage of the force needed to move a body segment and that amount (percentage) is not the same for each muscle. Furthermore, the force each muscle contributes varies depending on the starting position of the body segment and changes throughout the range of motion. This can be confusing to the beginning kinesiology student. So, this text has made certain choices in an attempt to simplify the muscle groups based on initial motion from anatomical position.

The muscle groups are based on the motion caused by their *concentric* action (i.e., The elbow flexors flex the elbow when they act concentrically.). However, it should not be forgotten that all muscles can and will act *eccentrically* and *isometrically*. The reader should also notice that the table for each individual muscle has a section called "Other Motions." *Other Motions* means that this muscle aides, to some degree, in other motion(s) besides the group it has been placed in.

The elbow muscles are grouped into four categories corresponding to the four elbow motions. They are the *elbow flexors*, *elbow extensors*, *elbow pronators*, and *elbow supinators*.

The Elbow Flexors

Three muscles are mainly responsible for elbow flexion (**FIGURE 16.5**). They are the *biceps brachii*, the *brachialis*, and the *brachioradialis*. The elbow flexors are summarized in **TABLE 16.2** at the end of this section.

The **biceps brachii** (**FIGURE 16.6**) is a two-jointed muscle running superficially across the shoulder, through the anterior brachial region, and over the elbow. It has two heads (biceps). The long head originates on the superior side of the glenoid fossa

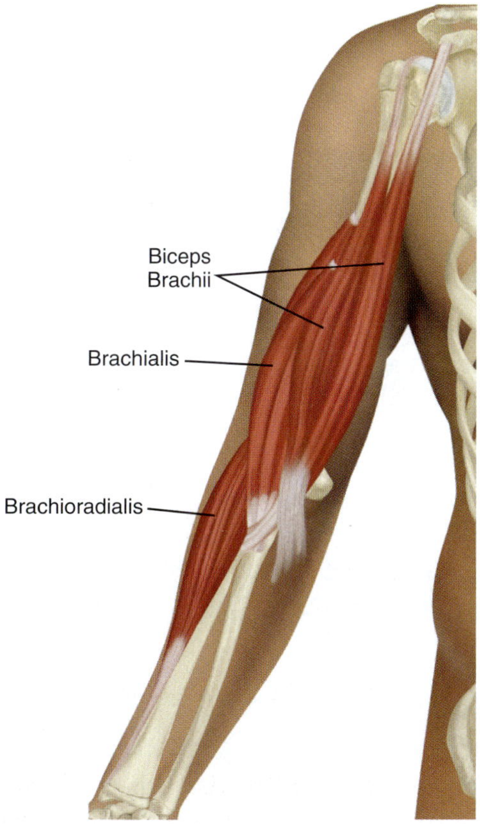

FIGURE 16.5 The Elbow Flexors.

and the short head from the coracoid process of the scapula. It inserts on the radial tuberosity of the anterior radius just below the proximal epiphysis.

The **brachialis** muscle (**FIGURE 16.7**) resides in the distal end of the anterior side of the brachial region. It originates from the anterior side of the distal one-half of the humerus. It inserts on the coronoid process and the ulnar tuberosity.

The brachioradialis muscle (**FIGURE 16.8**) lies on the lateral side of the forearm. It originates from the lateral supracondylar ridge and inserts on the styloid process of the radius.

FIGURE 16.6 The Biceps Brachii Muscle.

ORIGINS
LH: supraglenoid fossa
SH: coracoid process
INSERTION
radial tuberosity
OTHER MOTIONS
SH: shoulder (GH) flexion
elbow supination
NERVE
Musculocutaneous

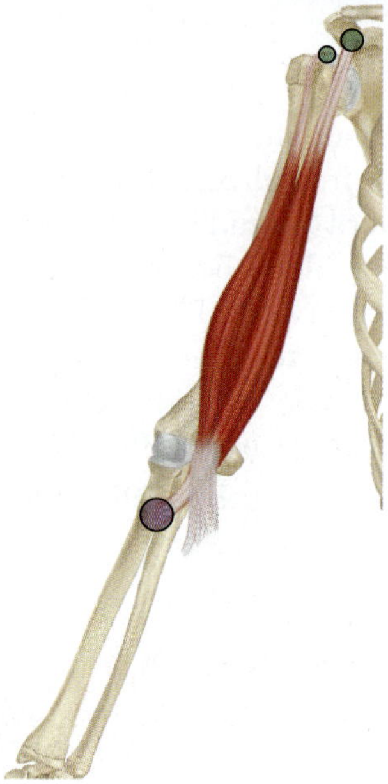

FIGURE 16.7 The Brachialis Muscle.

ORIGIN
posterior side of distal half of humerus
INSERTIONS
coronoid process
ulnar tuberosity
OTHER MOTIONS
none
NERVE
Musculocutaneous

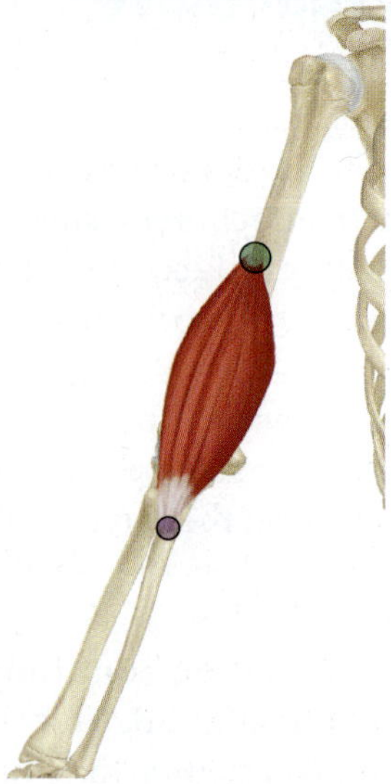

FIGURE 16.8 The Brachioradialis Muscle.

ORIGIN

lateral supracondylar ridge of humerus

INSERTION

styloid process of radius

OTHER MOTIONS

none

NERVE

Radial

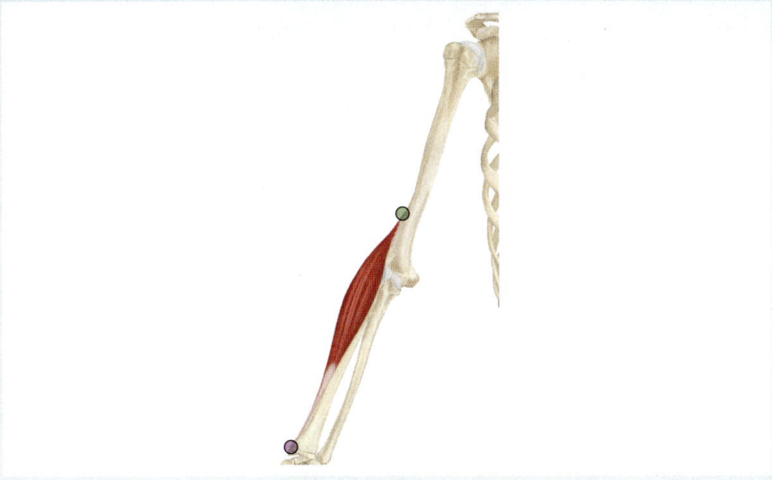

TABLE 16.2 Summary of the Elbow Flexors

Muscle	Origins	Insertions	Nerves
Biceps Brachii	supraglenoid fossa (LH) & coracoid process (SH)	radial tuberosity	Musculocutaneous
Brachialis	posterior side of distal half of humerus	coronoid process & ulnar tuberosity	Musculocutaneous
Brachioradialis	lateral supracondylar ridge of humerus	styloid process of radius	Radial

PAUSE TO CHECK FOR UNDERSTANDING

Table 16.2 should serve as a good review for much of the material, but here are a few other checks for understanding

1. How many muscles are in the elbow flexor group? List them.
2. What motion occurs if these muscles act eccentrically?
3. Where is the general location for the muscle group?

The Elbow Extensors

The **triceps brachii** is the only muscle primarily responsible for extending the elbow (**FIGURE 16.9**). It lies in the posterior brachial region crossing both the shoulder and the elbow (two-jointed muscle). Triceps means three heads and this muscle has three origins. The **long head** is the most medial and originates on the inferior side of the glenoid fossa of the scapula. The head in between the other two, called the **medial head**, originates from the entire posterior side of the humeral diaphysis. The **lateral head** originates from the posterolateral humerus just inferior to the greater tuberosity. The three heads merge and insert into the olecranon process of the ulna.

PAUSE TO CHECK FOR UNDERSTANDING

1. How many muscles are in the elbow extensor group? List them.
2. What motion occurs if these muscles act eccentrically?
3. Where is the general location for the muscle group?

FIGURE 16.9 The Triceps Brachii Muscle.

ORIGINS	
LH: infraglenoid fossa	
Med H: entire posterior humeral diaphysis	
Lat H: posterolateral humerus just inferior to greater tuberosity	
INSERTION	
olecranon process	
OTHER MOTIONS	
LH: shoulder (GH) extension	
NERVE	
Radial	

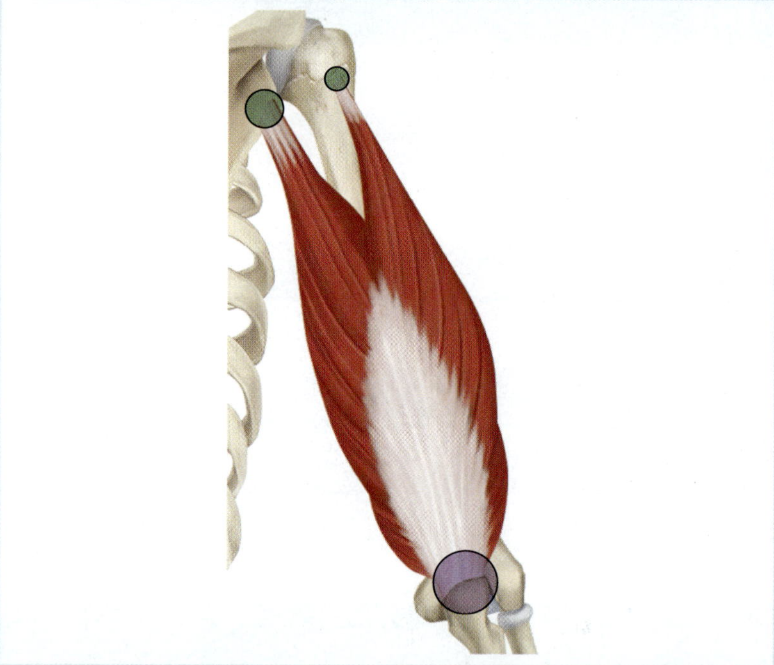

The Elbow Pronators

Two main muscles pronate the elbow. They are the *pronator teres* and the *pronator quadratus* muscles (**FIGURE 16.10**). The pronator teres is proximal to the elbow while the pronator quadratus is more proximal to the wrist. The elbow pronators will be summarized in **TABLE 16.3** at the end of this section.

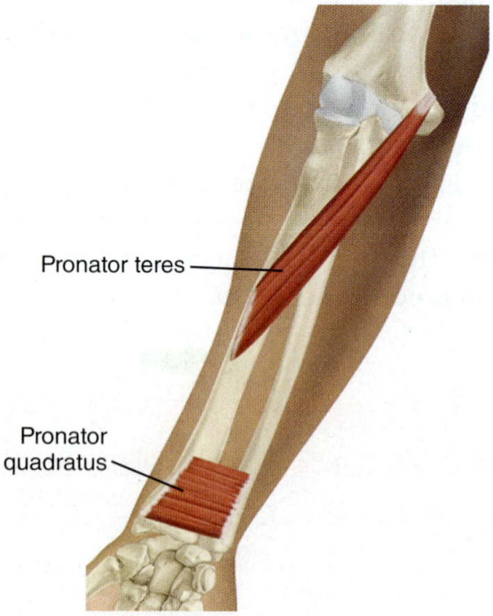

FIGURE 16.10 The Elbow Pronator Muscles.

The **pronator teres** (**FIGURE 16.11**) lies opposite the supinator muscle on the medial elbow and its belly crosses the proximal, anterior forearm. It originates from the medial epicondyle of the humerus as well as the coronoid process of the ulna. It inserts on the lateral side of the radius approximately half way on its diaphysis.

The **pronator quadratus** (**FIGURE 16.12**) lies just proximal from the wrist and is rectangular in shape, hence its name quadratus. It originates from the distal one fourth of the anterior side of the ulna and inserts on the distal one fourth of the anterior side of the radius.

PAUSE TO CHECK FOR UNDERSTANDING

Table 16.3 should serve as a good review for much of the material, but here are a few other checks for understanding

1. How many muscles are in the elbow pronator group? List them.
2. What motion occurs if these muscles act eccentrically?
3. Where is the general location for the muscle group?

FIGURE 16.11 The Pronator Teres Muscle.

ORIGINS

medial epicondyle of humerus

coronoid process

INSERTION

lateral side of mid radial diaphysis

OTHER MOTIONS

none

NERVE

Median

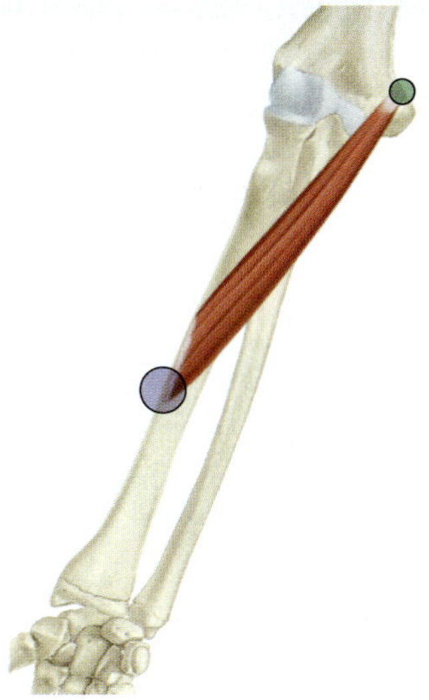

FIGURE 16.12 The Pronator Quadratus Muscle.

ORIGIN

distal 1/4 of anterior ulna

INSERTION

distal 1/4 of anterior radius

OTHER MOTIONS

none

NERVE

Median

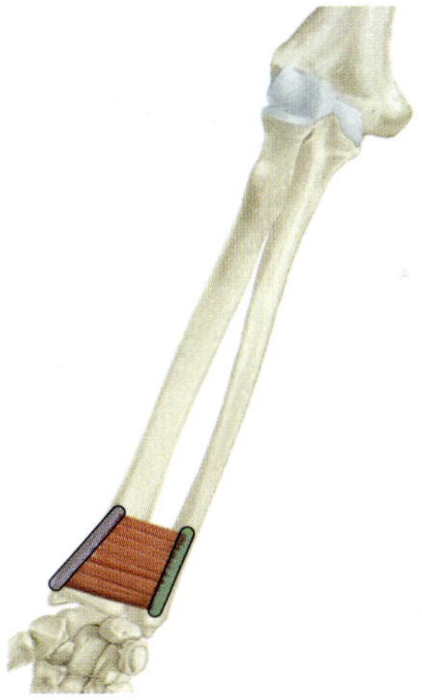

TABLE 16.3 Summary of the Elbow Pronators

Muscle	Origins	Insertions	Nerve
Pronator Teres	medial epicondyle of humerus & coronoid process	lateral side of mid radial diaphysis	Median
Pronator Quadratus	distal 1/4 of anterior ulna	distal 1/4 of anterior radius	

The Elbow Supinators

Only one muscle is primarily responsible for elbow supination. The belly of the **supinator** muscle lies in the proximal, lateral forearm (**FIGURE 16.13**). It originates from the lateral epicondyle of the humerus and from the ulna just inferior to the radial notch. It inserts on the anterolateral side of the proximal one third radius.

PAUSE TO CHECK FOR UNDERSTANDING

1. How many muscles are in the elbow supinator group? List them.
2. What motion occurs if these muscles act eccentrically?
3. Where is the general location for the muscle group?

FIGURE 16.13 The Supinator Muscle.

ORIGIN
lateral epicondyle of humerus
just inferior to radial notch of ulna
INSERTION
anterolateral side of proximal 1/3 radius
OTHER MOTIONS
none
NERVE
Radial

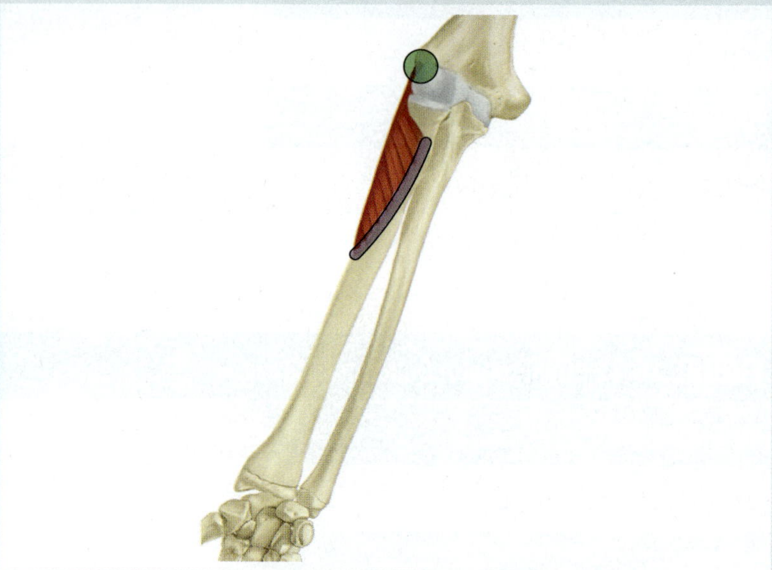

CHAPTER 17

The Wrist

CHAPTER OBJECTIVES

After completing this chapter, the student will be able to:

1. summarize the joints of the wrist region;
2. comprehend the wrist motions;
3. identify the names and locations of the wrist muscles;
4. recall the attachment sites and nerves of the wrist muscles; and
5. apply the motions produced by the actions of the wrist muscles.

The wrist is a region of the body where the forearm and hand intersect (**FIGURE 17.1**). The radius, ulna, and the proximal carpal bones fall within this region. This and the muscles acting on the wrist produce four motions.

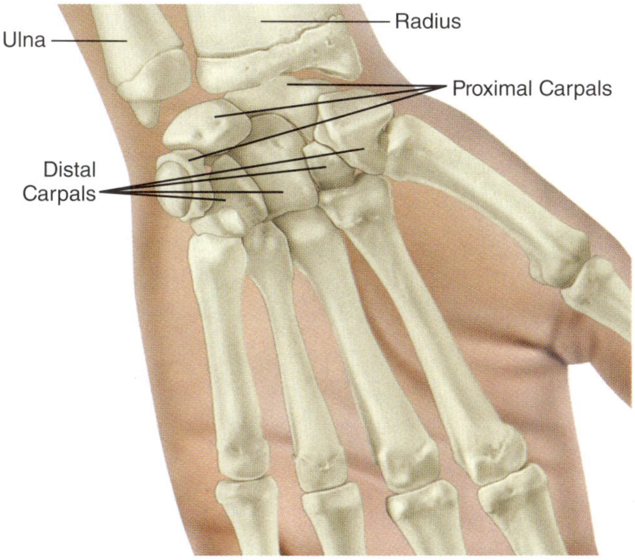

FIGURE 17.1 The Wrist.

The Joints of the Wrist

There are two joints in the wrist (**FIGURE 17.2**). One of the joints of the wrist is the **distal radioulnar joint**. This joint is the articulation between the distal ends of the radius and ulna. Very minimal pronation and supination occurs here; by far, the majority of these

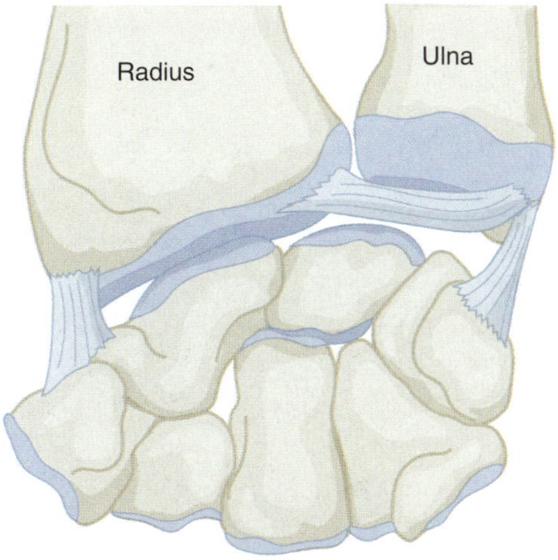

FIGURE 17.2 The Distal Radioulnar and Radiocarpal Joints.

motions occur at the elbow (see Chapter 16). The other joint is commonly called the **radiocarpal joint** which is the articulation between the distal end of the radius and the proximal carpal bones. However, these designations ignore the apparent articulation between the distal end of the ulna and a few of the proximal carpals.

Despite the proximity of the distal ulna, it is widely accepted that it does not articulate with any of the carpal bones because of an articular disk that separates them. This author does not consider this a viable argument. There are several examples of tissues separating the ends of two bones. For example, there is a vertebral disk in between each vertebrae of the spinal column, yet their articulation is not denied. Therefore, it is the opinion of this author that the presence of this articular disk should not exclude the ulna from being considered as an articulation with some of the proximal carpals. Like the talocrural joint, this would then be another exception to the rule that only two bones intersect at a joint. A new term would improve the accuracy and description of the structural anatomy. This text suggests that the radiocarpal joint be renamed as the **radioulnar-carpal joint**.

There are four main ligaments of the wrist (**FIGURE 17.3**). The **distal ulnar collateral ligament** stabilizes the medial wrist attaching at the ulnar styloid process to some carpal bones. Some its fibers connect to the pisiform and the rest goes to the triquetrum. The **distal radial collateral ligament** stabilizes the lateral wrist connecting the radius to the scaphoid bone. The **palmar radiocarpal ligament** stabilizes the palm (anterior) side of the wrist and the **dorsal radiocarpal ligament** stabilizes the posterior side of the wrist. Both attach from the radius to the carpal bones on their respective sides.

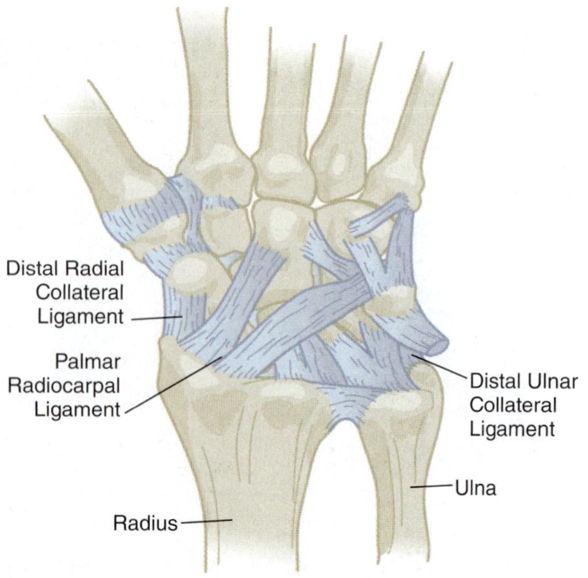

FIGURE 17.3 The Wrist Ligaments.

The Motions of the Wrist

Four motions occur at the wrist (**FIGURE 17.4**). The motions are *flexion*, *extension*, *radial deviation*, and *ulnar deviation*. **Wrist flexion**, bending toward the anterior, and **extension**, straightening toward the posterior, are opposite motions occurring in the sagittal plane rotating about the frontal axis. **Radial deviation** is bending sideways toward the radius (laterally) or away from midline. **Ulnar deviation** is also bending sideways, but toward the ulna (medially) or toward midline. These opposing motions occur in the frontal plane about the sagittal axis. Sometimes they have been called radial and ulnar flexion. Far less commonly, they have also been called medial and lateral wrist flexion as well as wrist abduction and adduction.

TABLE 17.1 summarizes the average ranges of motions of the wrist. Average wrist flexion is up to 90 degrees. Average wrist extension is less at 75–85°, respectively. Average radial deviation is 20 degrees while average ulnar deviation is greater at 35 degrees.

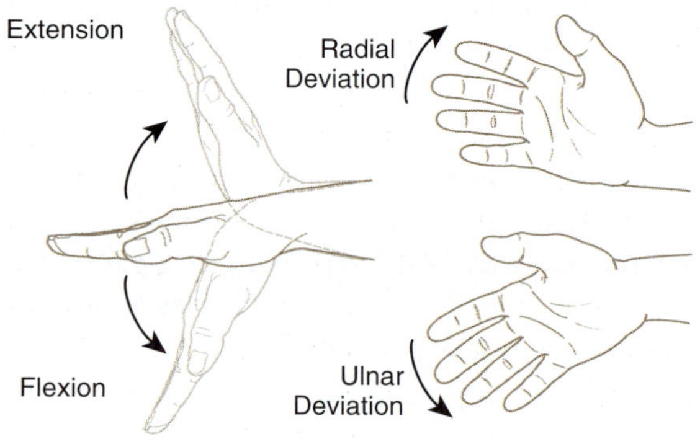

FIGURE 17.4 Wrist Motions.

TABLE 17.1 Average Wrist Range of Motions

Motion	Average Degrees of Motion
Extension	75–85°
Flexion	80–90°
Combined Extension-Flexion	155–175°
Radial Deviation	20°
Ulnar Deviation	35°
Combined Radial-Ulnar Deviation	55°

PAUSE TO CHECK FOR UNDERSTANDING

1. List the joints of the wrist.
2. Then, tell which bones form each joint.
3. What are the four motions of the wrist?

The Muscles That Move the Wrist

The rest of this chapter is dedicated to learning the wrist muscles; their attachment sites; the *primary*, main motions they produce; and the nerves that innervate them. Muscles work in groups and so each muscle within the group supplies a certain percentage of the force needed to move a body segment and that amount (percentage) is not the same for each muscle. Furthermore, the force each muscle contributes varies depending on the starting position of the body segment and changes throughout the range of motion. This can be confusing to the beginning kinesiology student. So, this text has made certain choices in an attempt to simplify the muscle groups based on initial motion from anatomical position.

The muscle groups are based on the motion caused by their *concentric* action (i.e., the wrist flexors flex the wrist when they act concentrically.). However, it should not be forgotten that all muscles can and will act *eccentrically* and *isometrically*. The reader should also notice that the table for each individual muscle has a section called "Other Motions." *Other Motions* means that this muscle aides, to some degree, in other motion(s) besides the group it has been placed in. Some muscles may appear in more than one group because their role is too great to ignore.

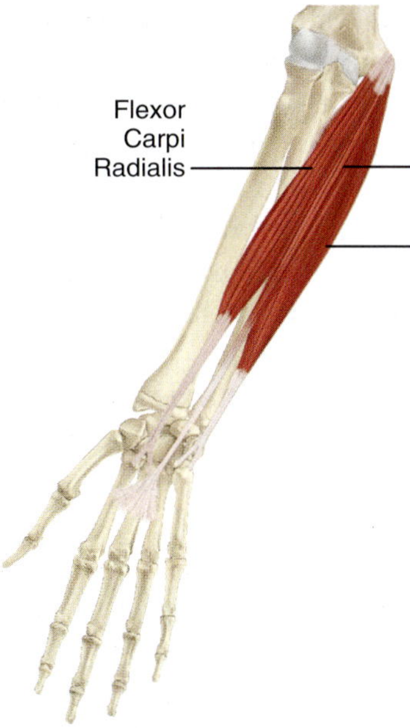

FIGURE 17.5 The Wrist Flexors.

There are four groups which correlate with the four motions. The categories are the *wrist flexors*, *wrist extensors*, *wrist radial deviators*, and *wrist ulnar deviators*.

The Wrist Flexors

Three muscles flex the wrist (**FIGURE 17.5**). They are the *flexor carpi radialis*, the *palmaris longus*, and the *flexor carpi ulnaris* muscles. All of them originate on the medial side of the elbow. Their bellies pass longitudinally through the anterior forearm and attach on various locations on the anterior (palmar) side of the hand bones. The wrist flexors will be summarized in **TABLE 17.2** at the end of this section.

The **flexor carpi radialis** muscle originates from the medial epicondyle of the humerus (**FIGURE 17.6**).

FIGURE 17.6 The Flexor Carpi Radialis Muscle.

ORIGIN	
medial epicondyle of humerus	
INSERTIONS	
anterior side of metacarpal bases II & III	
OTHER MOTIONS	
radial deviation	
NERVE	
Median	

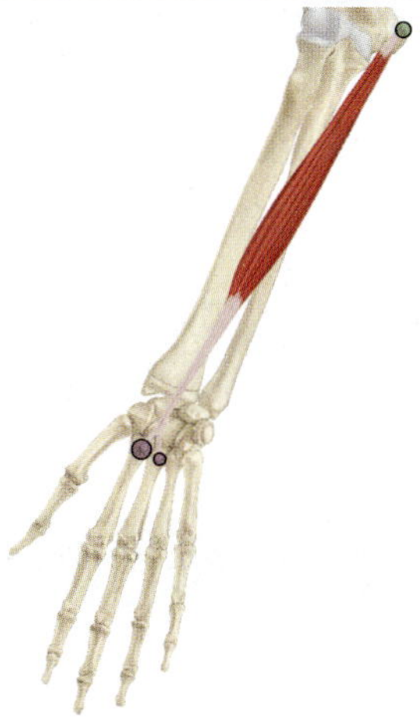

It is the most lateral of this group of muscles passing through the anterior side of the forearm. It then inserts on the anterior side of the base of the second and third metacarpals.

The **palmaris longus** muscle also originates from the medial epicondyle of the humerus (**FIGURE 17.7**). Its belly lies lateral to the flexor carpi ulnaris muscle and inserts on the palmar fascia which is in the anterior side (palmar) of the hand.

Like the other two wrist flexors, the **flexor carpi ulnaris** muscle originates from the medial epicondyle of the humerus (**FIGURE 17.8**). Its belly passes relatively

PAUSE TO CHECK FOR UNDERSTANDING

Table 17.2 should serve as a good review for much of the material, but here are a few other checks for understanding

1. How many muscles are in the wrist flexor group? List them.
2. What motion occurs if these muscles act eccentrically?
3. Where is the general location for the muscle group?

FIGURE 17.7 The Palmaris Longus Muscle.

ORIGIN
medial epicondyle of humerus
INSERTION
palmar fascia
OTHER MOTIONS
none
NERVE
Median

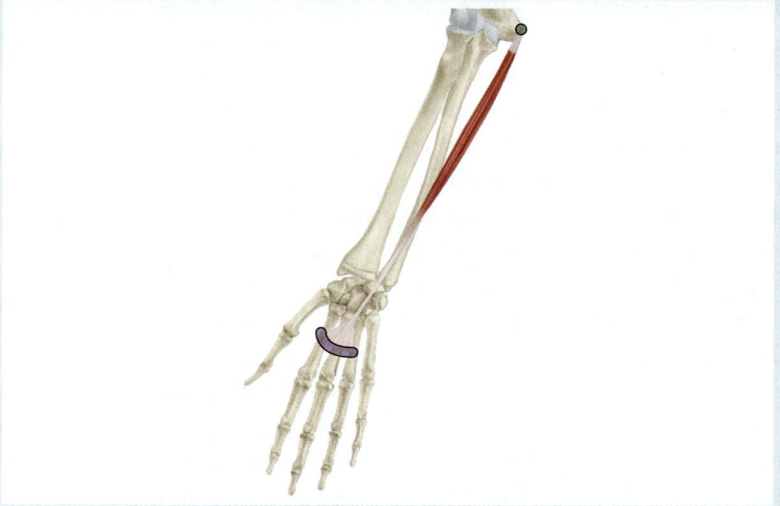

FIGURE 17.8 The Flexor Carpi Ulnaris Muscle.

ORIGIN
medial epicondyle of humerus
INSERTIONS
anterior side of base of metacarpal V
pisiform
OTHER MOTIONS
ulnar deviation
NERVE
Ulnar

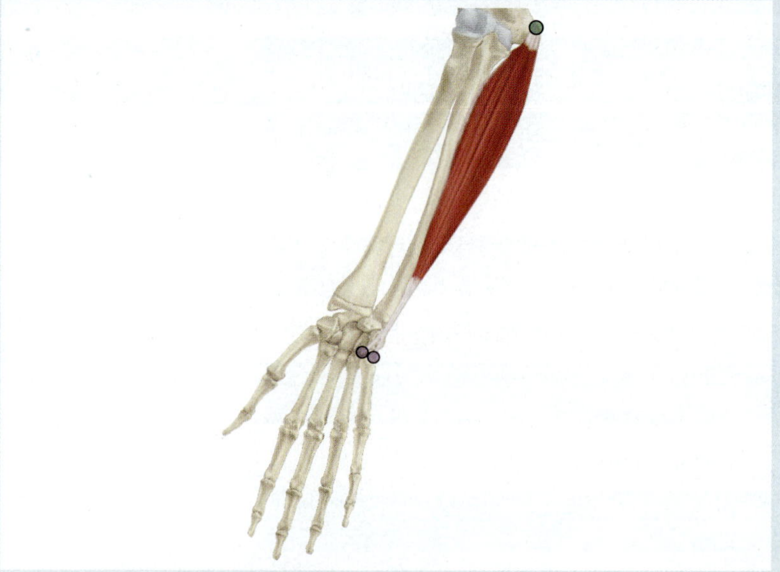

TABLE 17.2 Summary of the Wrist Flexors

Muscle	Origins	Insertions	Nerve
Flexor Carpi Radialis		anterior side of metacarpal bases II & III	Median
Palmaris Longus	medial epicondyle of humerus	palmar fascia	
Flexor Carpi Ulnaris		anterior side of base of metacarpal V & pisiform	Ulnar

straight from there along the medial side of the anterior forearm and inserts on the anterior sides of the base of the fifth metacarpal and the pisiform (a carpal bone).

The Wrist Extensors

Three muscles extend the wrist (**FIGURE 17.9**). They are the *extensor carpi radialis longus*, the *extensor radialis brevis*, and the *extensor carpi ulnaris* muscles. All of them originate on the lateral side of the elbow. Their bellies pass longitudinally through the posterior forearm and attach on various locations on the posterior (dorsal) side of the hand bones. The wrist extensors will be summarized in **TABLE 17.3** at the end of this section.

The **extensor carpi radialis longus** muscle originates from the distal end of the lateral supracondylar ridge of the humerus (**FIGURE 17.10**). Its belly "runs" along the lateral side of the posterior forearm and inserts on the posterior side of the base of the second metacarpal.

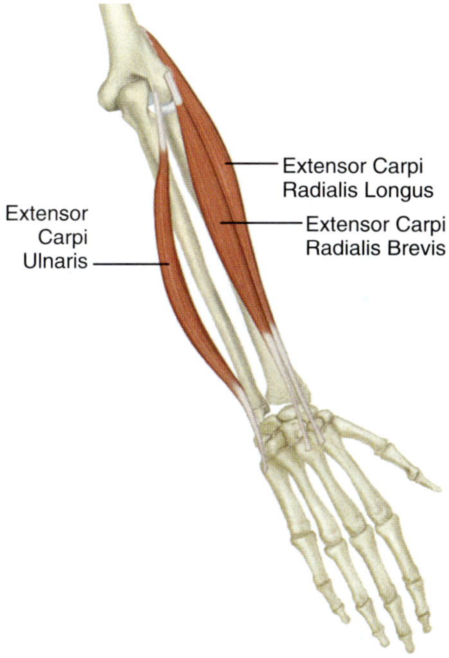

FIGURE 17.9 The Wrist Extensors.

FIGURE 17.10 The Extensor Carpi Radialis Longus Muscle.

ORIGIN

distal end of lateral supracondylar ridge of humerus

INSERTION

posterior side of base of metacarpal II

OTHER MOTIONS

radial deviation

NERVE

Radial

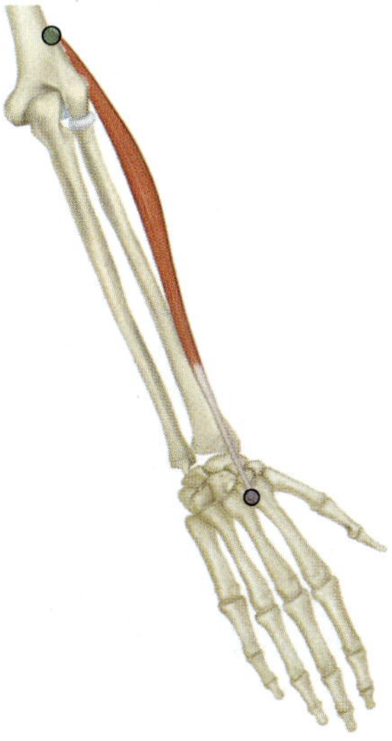

FIGURE 17.11 The Extensor Carpi Radialis Brevis Muscle.

ORIGIN	lateral epicondyle of humerus
INSERTION	posterior side of base of metacarpal III
OTHER MOTIONS	none
NERVE	Radial

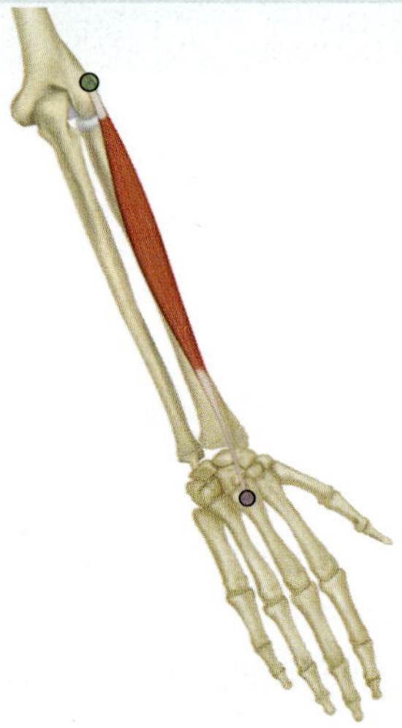

FIGURE 17.12 The Extensor Carpi Ulnaris Muscle.

ORIGIN	lateral epicondyle of humerus
INSERTION	posterior side of base of metacarpal V
OTHER MOTIONS	ulnar deviation
NERVE	Radial

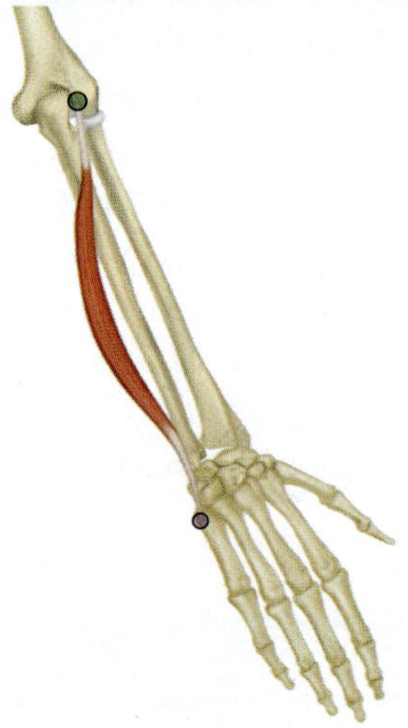

The **extensor carpi radialis brevis** muscle originates from the lateral epicondyle of the humerus (**FIGURE 17.11**). Its belly is more medial than the extensor radialis longus muscle passing relatively through the middle of the posterior forearm. It inserts on the posterior side of the base of the third metacarpal.

The **extensor carpi ulnaris** muscle also originates from the lateral epicondyle of the humerus (**FIGURE 17.12**). Its belly runs obliquely to and then through the medial side of the posterior forearm. It then inserts on the posterior side of the base of the fifth metacarpal.

PAUSE TO CHECK FOR UNDERSTANDING

Table 17.3 should serve as a good review for much of the material, but here are a few other checks for understanding

1. How many muscles are in the wrist extensor group? List them.
2. What motion occurs if these muscles act eccentrically?
3. Where is the general location for the muscle group?

TABLE 17.3 Summary of the Wrist Extensors

Muscle	Origins	Insertions	Nerve
Extensor Carpi Radialis Longus	distal end of lateral supracondylar ridge of humerus	posterior side of base of metacarpal II	
Extensor Carpi Radialis Brevis	lateral epicondyle of humerus	posterior side of base of metacarpal III	Radial
Extensor Carpi Ulnaris		posterior side of metacarpal V	

The Wrist Radial Deviators

When acting together, two muscles, already discussed, deviate (or flex) the wrist toward the radius (**FIGURE 17.13**). They are the **extensor carpi radialis longus** and the **flexor carpi radialis** muscles. For their specific origins, insertions, and nerves, refer to their respective sections related to the wrist extensors and flexors. Knowing that both muscles attach to the lateral side of the proximal hand with one from the anterior side and one for the posterior side should give an understanding of how, when acting together, they radially deviate the wrist.

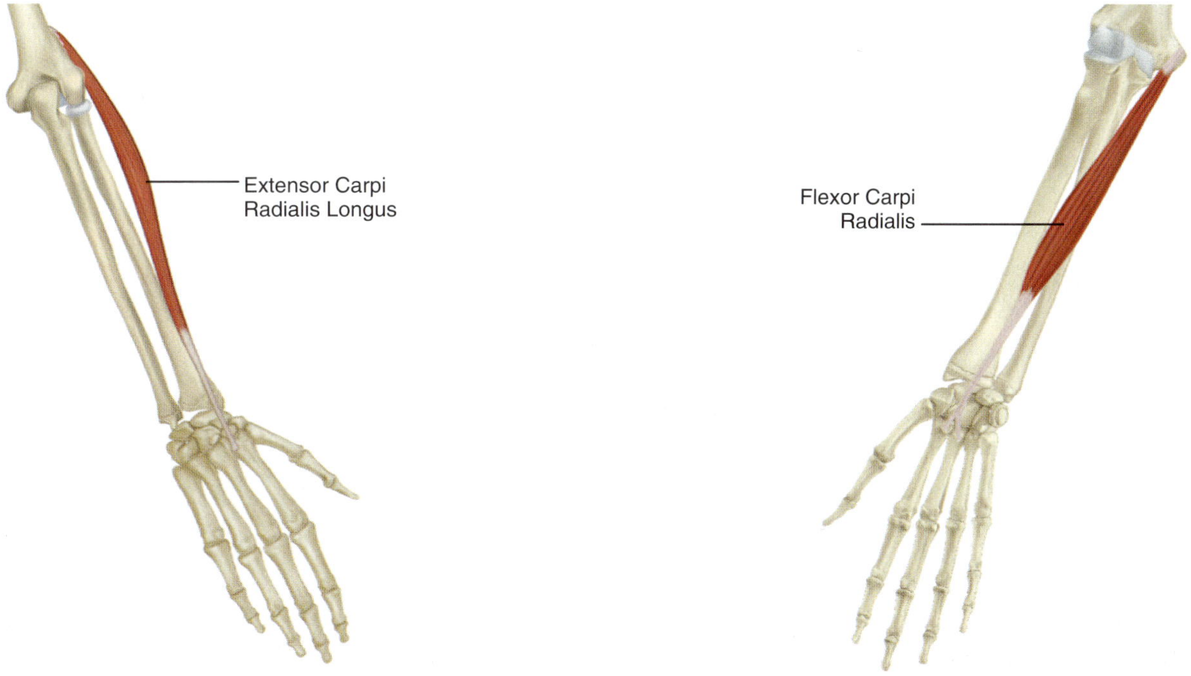

FIGURE 17.13 The Wrist Radial Deviators.

PAUSE TO CHECK FOR UNDERSTANDING

1. How many muscles are in the radial deviator group? List them.
2. What motion occurs if these muscles act eccentrically?
3. Where is the general location for the muscle group?

The Wrist Ulnar Deviators

Like the wrist radial deviators, two muscles on opposite sides of the forearm deviate (or flex) the wrist toward the ulna (**FIGURE 17.14**). They are the **extensor carpi ulnaris** and the **flexor carpi ulnaris** muscles. For their specific origins, insertions, and nerves, refer to their respective sections related to the wrist extensors and flexors. Like the radial wrist deviators except on the medial side of the wrist, both muscles insert on the proximal hand. With this knowledge, the reader may imagine that when they act, they move the wrist toward the ulna.

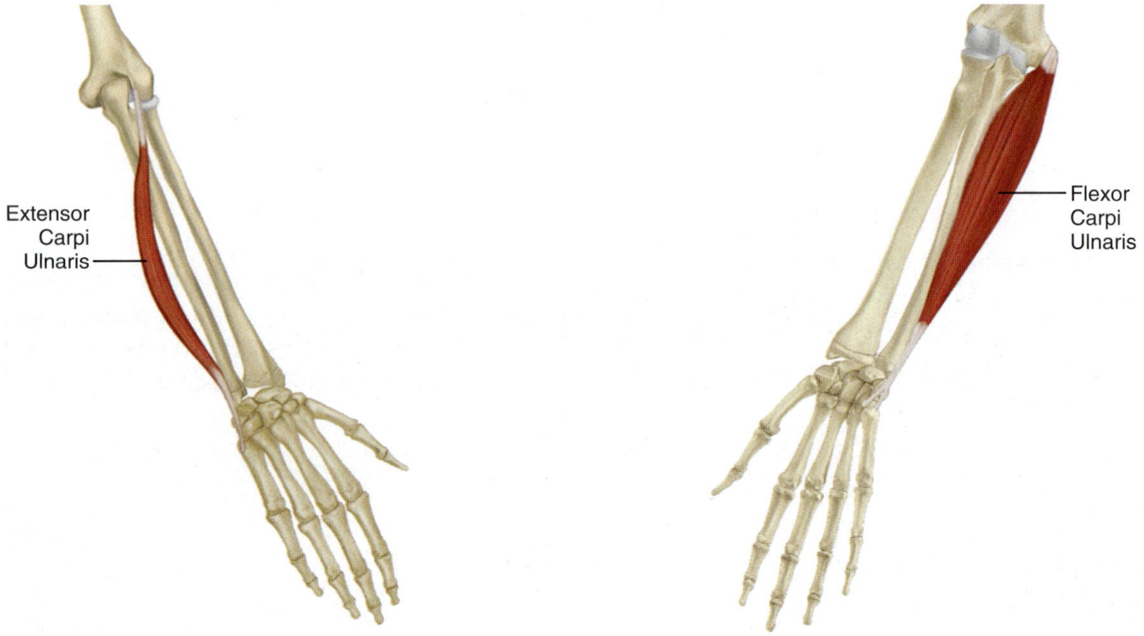

FIGURE 17.14 The Wrist Ulnar Deviators.

PAUSE TO CHECK FOR UNDERSTANDING

1. How many muscles are in the ulnar deviator group? List them.
2. What motion occurs if these muscles act eccentrically?
3. Where is the general location for the muscle group?

CHAPTER 18
The Hand

CHAPTER OBJECTIVES

After completing this chapter, the student will be able to:

1. summarize the joints of the hand region;
2. comprehend the hand motions;
3. identify the names and locations of the hand muscles;
4. recall the attachment sites and nerves of the hand muscles; and
5. apply the motions produced by the actions of the hand muscles.

The hand (manus) is the region of the body distal to the wrist (**FIGURE 18.1**). The hand can be divided into two parts. The part just distal to the wrist contains the five metacarpal bones.

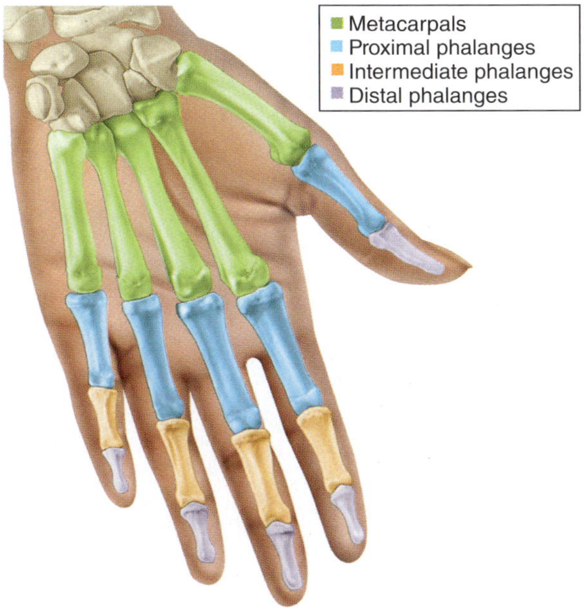

FIGURE 18.1 The Hand.

The anterior side is called the palm and the posterior is the dorsum. Beyond that are the five fingers (digits) that contain the phalanges. There are five digits numbered I–V from the most lateral to the most medial. Each digit also has a name; digit I (thumb), digit II (index), digit III (middle), digit IV (ring), and digit V (little or pinky). The arrangement of the bones and actions of the muscles produce six motions of the digits.

The Joints of the Hand

There are numerous joints between all the bones of the hand and fingers (**FIGURE 18.2**). Collectively, the **carpometacarpal joints** are all the joints between the distal carpal bones and the five metacarpal bones. The**metacarpophalangeal (MP) joints** are between the five metacarpal bones and the five proximal phalanges of the fingers. They are sequentially numbered from the lateral side (MP joint I) to the medial side (MP joint V) of the hand. In the same fashion, the **interphalangeal (IP) joints** are sequentially numbered. The first digit (the thumb) only contains two phalanges; therefore, there is only one IP joint (IP joint I) in this digit. However, the rest of the digits (II–V) contain three phalanges apiece and therefore have two IP joints per digit. The **proximal interphalangeal (PIP) joints** are between the proximal phalanges and the middle phalanges. The **distal interphalangeal (DIP) joints** are between the middle phalanges and the distal phalanges.

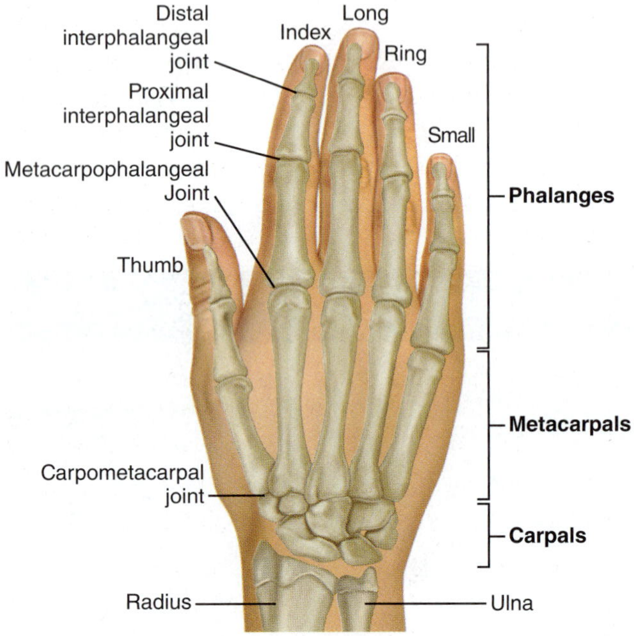

FIGURE 18.2 The Joints of the Hand.

The Motions of the Hand

There is only slight motion of the carpometacarpal joints and therefore will not be considered. However, there are certainly significant motions of the MP joints and the IP joints (**FIGURE 18.3**). The average ranges of motions of the MP and IP joints are summarized in **TABLE 18.1**. The second through fifth MP joints are capable of *extension, flexion, abduction*, and *adduction*. In addition to these same motions, the first MP joint of the thumb is also capable of *opposition* and *reposition*. The IP joints are only capable of *extension* and *flexion*.

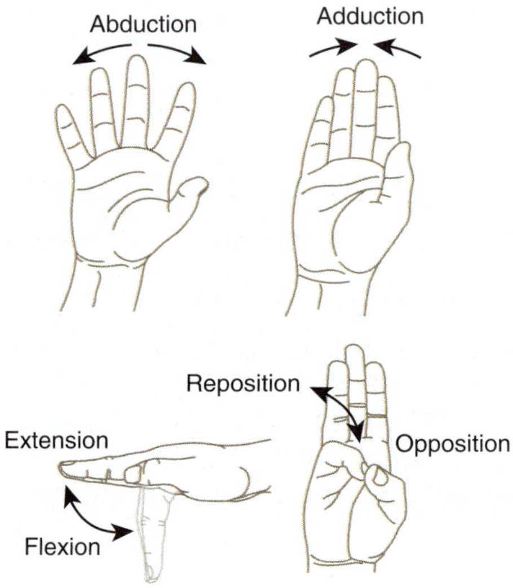

FIGURE 18.3 The Motions of the Fingers.

TABLE 18.1 Average Finger Ranges of Motions

Joint	Motion	Degrees
MP I – V	Extension Flexion Combined	20–30° 85–105° 105–135°
MP I – V	Adduction Abduction Combined	0° 20–25° 20–25°
IP I	Extension Flexion Combined	0° 80–90° 80–90°
PIP II – V	Extension Flexion Combined	0° 110–120° 110–120°
DIP II – V	Extension Flexion Combined	0° 80–90° 80–90°

PAUSE TO CHECK FOR UNDERSTANDING

1. What bones are involved in an MP joint?
2. Is the first digit the thumb or little finger?
3. What motions are the fingers capable of?

The Muscles That Move the Fingers

Numerous muscles move the fingers. Like the toe muscles, the finger muscles are categorized as either *intrinsic* or *extrinsic*. **Intrinsic muscles** are muscles whose origin, belly, and insertion are located within the region of the joint(s) it acts upon. **Extrinsic muscles** are distinguished because their origin and belly are outside the region of the joint(s) it acts upon. The origins and bellies of the extrinsic finger muscles reside in the forearm whereas the intrinsic finger muscles completely reside within the hand. **TABLE 18.2** summarizes the muscles that act upon the fingers including whether they are intrinsic or extrinsic as well as the nerves that innervate them. Of note, the finger repositioners are not listed because those muscles are already listed in the table under their other actions of finger extension and abduction.

TABLE 18.2 The Finger Muscles

Muscle	Location	Nerve
Finger Flexors		
Flexor Pollicis Brevis	Intrinsic	Median
Flexor Pollicis Longus	Extrinsic	Median
Flexor Digitorum Superficialis	Extrinsic	Median
Flexor Digitorum Profundus	Extrinsic	Median & Ulnar
Lumbricals	Intrinsic	Median & Ulnar
Flexor Digiti Minimi	Intrinsic	Ulnar
Finger Extensors		
Extensor Pollicis Brevis	Intrinsic	Radial
Extensor Pollicis Longus	Extrinsic	
Extensor Indicis	Extrinsic	
Extensor Digitorum	Extrinsic	
Extensor Digiti Minimi	Extrinsic	
Finger Abductors		
Abductor Pollicis Brevis	Intrinsic	Median
Abductor Pollicis Longus	Extrinsic	Radial
Dorsal Interossei	Intrinsic	Ulnar
Abductor Digiti Minimi	Intrinsic	Ulnar
Finger Adductors		
Adductor Pollicis	Intrinsic	Ulnar
Palmar Interossei		
Thumb Oppositioners		
Opponens Pollicis	Intrinsic	Median
Opponens Digiti Minimi		Ulnar

Additionally, the lumbricals, the dorsal interossei, and the palmar interossei are actually three muscle groups of four (4) muscles per group.

The rest of this chapter is dedicated to learning the finger muscles; their attachment sites; the *primary*, main motions they produce; and the nerves that innervate them. Each of the following sections is a muscle group and they are divided based on the main motion produced by their collective concentric action. However, it should not be forgotten that all muscles also can and will act and *eccentrically* and *isometrically*.

Since there are six motions, there are six groups. The groups are the *finger flexors, finger extensors, finger abductors, finger adductors, finger oppositioners*, and *finger repositioners*. Furthermore, each group of finger muscles will be summarized at the end of each section including the specific joints they act upon, their other motions, and the nerves that innervate them. It should be noted that each of these summary tables includes a column for "Other Motions." *Other Motions* means that this muscle also plays some lesser, *secondary* role in some other motion(s).

PAUSE TO CHECK FOR UNDERSTANDING

1. What is the difference between an intrinsic and extrinsic finger muscle?
2. Create a three-column chart. a) In the first column, list all the toe muscles. b) In the second column from memory, write what motion occurs during a concentric action. Double check your correct answers for accuracy. c) In the third column, record what motion occurs when the muscles act eccentrically.

The Finger Flexors

There are nine muscles that flex the fingers. Five of them are the *flexor pollicis brevis, flexor pollicis longus, flexor digitorum superficialis, flexor digitorum profundus*, and *flexor digiti minimi* muscles. The other four of them are collectively called the *lumbricals*.

The **flexor pollicis brevis** acts upon the thumb (digit I) (**FIGURE 18.4**). It lies on the anteromedial side of the thumb originating from the anterior side of the trapezium (a carpal bone) and the flexor retinaculum. It inserts on the palmar side of the base of the first proximal phalange.

The **flexor pollicis longus** also acts upon the thumb (**FIGURE 18.5**). It originates from about the proximal one-half of the anterior side of the radial diaphysis. It passes through the lateral side of the anterior forearm and inserts on the palmar side of the base of the first distal phalange.

The **flexor digitorum superficialis** acts on the second through fifth digits (**FIGURE 18.6**). It originates from the medial epicondyle of the humerus. From there, its fibers pass through the anterior forearm before converging into a tendon that inserts on the bases of the second through fifth middle phalanges.

The **flexor digitorum profundus** also acts on the second through fifth digits (**FIGURE 18.7**). It lies underneath (deep to) the flexor digitorum in the medial side of the anterior forearm. It originates from

FIGURE 18.4 The Flexor Pollicis Brevis Muscle.

ORIGINS	
trapezium	
flexor retinaculum	
INSERTION	
palmar side of base of proximal phalange I	
OTHER MOTIONS	
none	
NERVE	
Median	

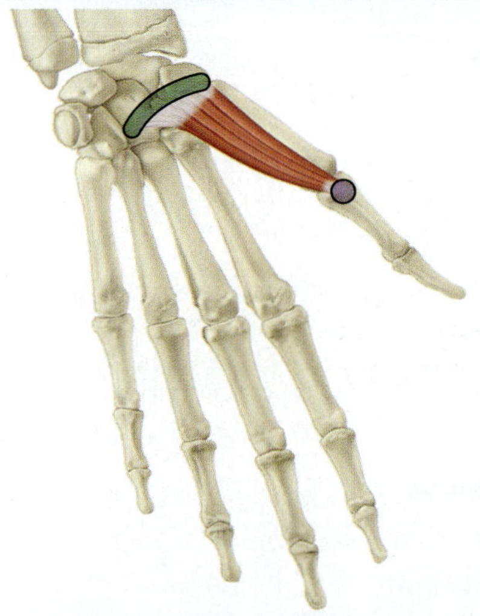

FIGURE 18.5 The Flexor Pollicis Longus Muscle.

ORIGIN	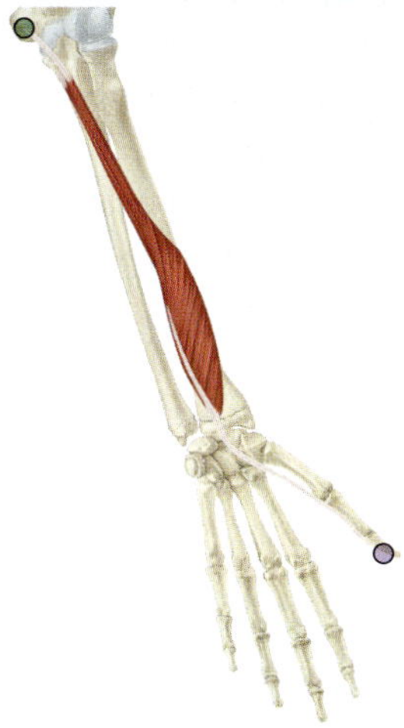
anterior side of proximal 1/2 radius	
INSERTION	
palmar base of distal phalange I	
OTHER MOTIONS	
none	
NERVE	
Median	

FIGURE 18.6 The Flexor Digitorum Superficialis Muscle.

ORIGIN	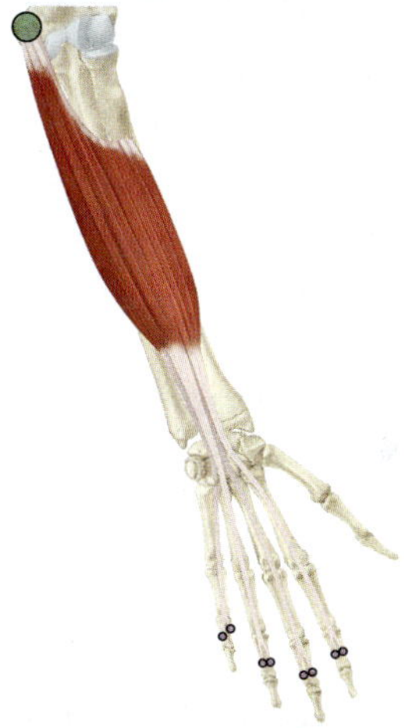
medial epicondyle of humerus	
INSERTIONS	
palmar bases of middle phalanges II–V	
OTHER MOTIONS	
none	
NERVE	
Median	

FIGURE 18.7 The Flexor Digitorum Profundus Muscle.

ORIGIN	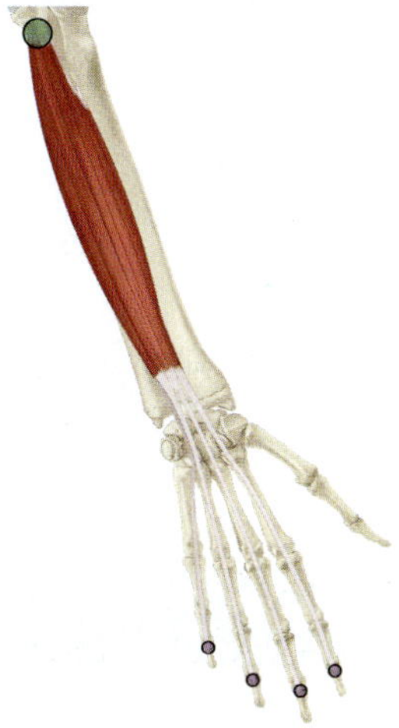
anterior side of the proximal 3/4 of ulna	
INSERTIONS	
palmar bases of distal phalanges II–V	
OTHER MOTIONS	
none	
NERVES	
Median & Ulnar	

the proximal three quarters of the anterior diaphysis of the ulna and inserts on the bases of the second through fifth distal phalanges.

The **lumbricals** are actually a group of four separate muscles that reside in between the metacarpals (**FIGURE 18.8**). They can be indicated sequentially as the first through fourth lumbrical muscles from the lateral side of the hand to the medial side. The first lumbrical acts on the second digit (index finger), the second lumbrical on the third digit (middle finger), the third lumbrical on the fourth digit (ring finger), and the fourth muscle on the fifth digit (little). The four muscles originate from the tendons of the flexor digitorum profundus muscle and insert on the lateral sides of the diaphyses of the second through fifth proximal phalanges.

The **flexor digiti minimi** only acts on the fifth (little finger) digit (**FIGURE 18.9**). It passes longitudinally over the palmar side of the fifth metacarpal. It is sometimes called the flexor digiti minimi brevis despite there is no flexor digiti longus. It originates from the hamate bone and the flexor retinaculum and inserts on the medial side of the base of the fifth proximal phalange.

FIGURE 18.8 The Four Lumbrical Muscles.

ORIGIN
tendons of flexor digitorum profundus
INSERTIONS
lateral side of the diaphyses of proximal phalanges II–V
OTHER MOTIONS
none
NERVES
Median (first 2) & Ulnar (second 2)

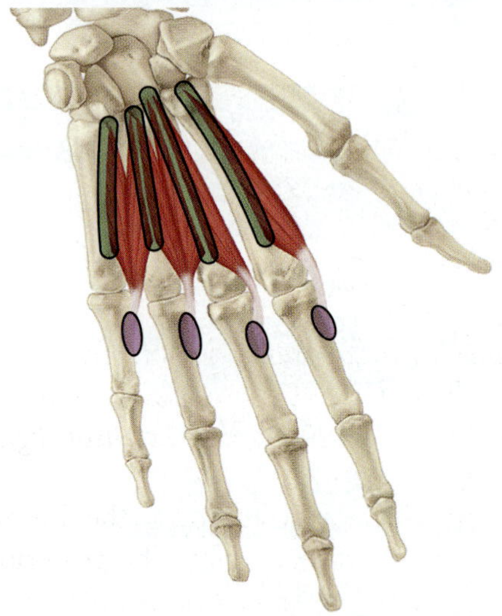

FIGURE 18.9 The Flexor Digiti Minimi Muscle.

ORIGINS
hamate
flexor retinaculum
INSERTION
medial base of proximal phalange V
OTHER MOTIONS
none
NERVE
Ulnar

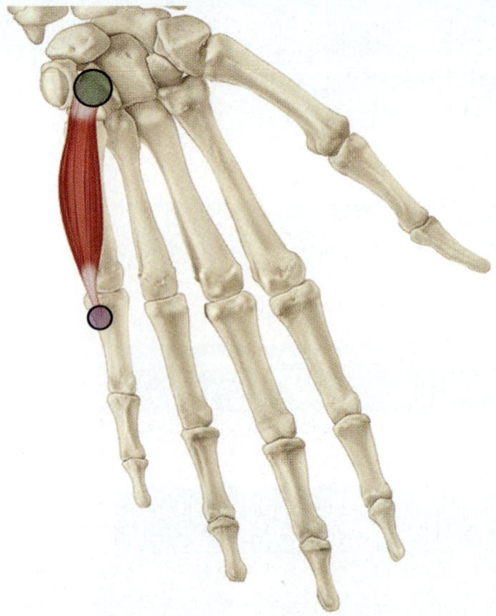

TABLE 18.3 The Finger Flexors

Muscle	Digit(s) Acted On	Joint(s) Acted On	Nerve
Flexor Pollicis Brevis	Digit I	MP	
Flexor Pollicis Longus	Digit I	MP & IP	Median
Flexor Digitorum Superficialis	Digits II–V	MP & PIP	
Flexor Digitorum Profundus	Digits II–V	MP, PIP, & DIP	Median & Ulnar
Lumbricals	Digits II–V	PIP & DIP	Median (first 2) Ulnar (second 2)
Flexor Digiti Minimi	Digit V	MP	Ulnar

PAUSE TO CHECK FOR UNDERSTANDING

TABLE 18.3 should serve as a good review for much of the material, but you should also review the origins and insertions.

The Finger Extensors

There are five muscles that extend the fingers. They are the *extensor pollicis brevis*, *extensor pollicis longus*, *extensor indicis*, *extensor digitorum*, and *extensor digiti minimi* muscles.

The **extensor pollicis brevis** acts upon the thumb (digit I) (**FIGURE 18.10**). It originates from the posterior side of the distal radius. From there, it passes obliquely across the radius, crosses the wrist, and then inserts on the lateral side of the base of first proximal phalange.

The **extensor pollicis longus** also acts upon the thumb (digit I) (**FIGURE 18.11**). Its belly lies obliquely across the distal forearm from the ulna. It originates from the posterior side of the mid ulna, crosses the wrist, and inserts on the dorsal side of the base of the first distal phalange.

The **extensor indicis** muscle acts upon the second digit (index finger) (**FIGURE 18.12**). It originates from the posterior side of the distal ulna. Its belly runs longitudinally down the distal mid forearm and across the wrist inserting on the dorsal side of the base of the second distal phalange.

The **extensor digitorum** muscle resides in the middle of the posterior forearm acting on the second through fifth digit (**FIGURE 18.13**). It originates from the lateral epicondyle of the humerus and inserts on

FIGURE 18.10 The Extensor Pollicis Brevis Muscle.

ORIGIN

posterior side of distal radius

INSERTION

lateral base of proximal phalange I

OTHER MOTIONS

reposition of thumb

NERVE

Radial

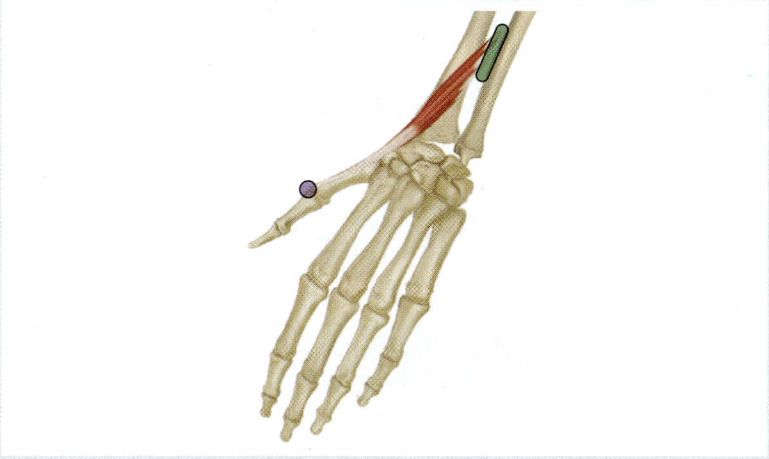

FIGURE 18.11 The Extensor Pollicis Longus Muscle.

ORIGIN
posterior side of mid ulna
INSERTION
dorsal side of base of distal phalange I
OTHER MOTIONS
reposition of thumb
NERVE
Radial

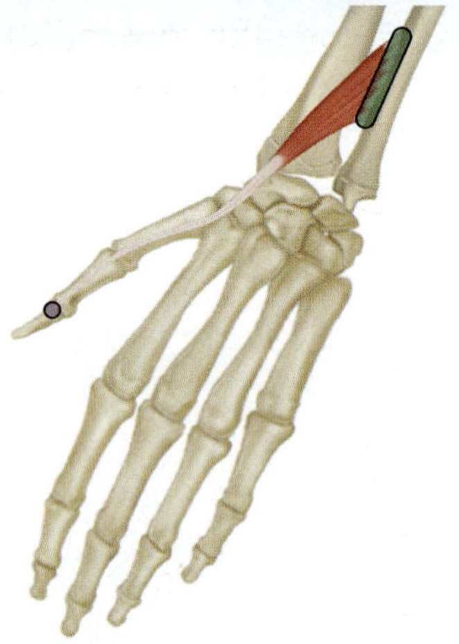

FIGURE 18.12 The Extensor Indicis Muscle.

ORIGIN
posterior side of distal ulna
INSERTION
dorsal side of base of distal phalange II
OTHER MOTIONS
none
NERVE
Radial

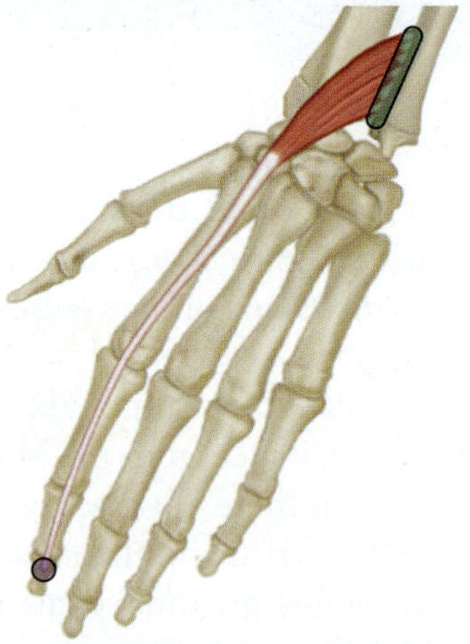

FIGURE 18.13 The Extensor Digitorum Muscle.

ORIGIN
lateral epicondyle of humerus
INSERTIONS
dorsal bases of the distal phalanges II–V
OTHER MOTIONS
none
NERVE
Radial

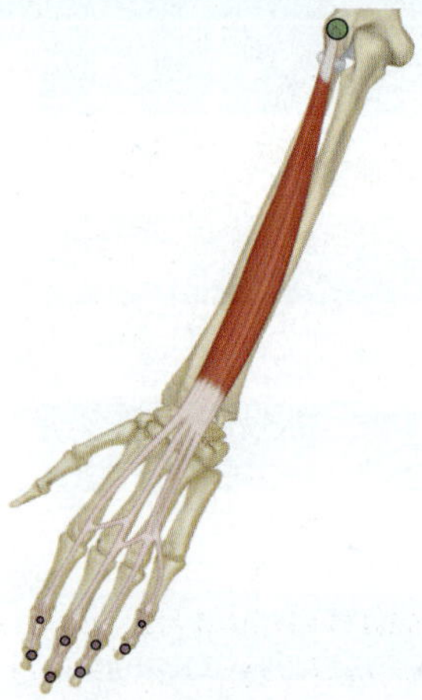

the dorsal side of the bases of the second through fifth distal phalanges.

The **extensor digiti minimi** muscle, as the name implies, acts on the fifth digit (little finger) (**FIGURE 18.14**). It originates from the lateral epicondyle of the humerus. It passes longitudinally over the ulna in the medial side of the posterior forearm. After crossing the wrist, it then inserts on the dorsal side of the base of the fifth distal phalange.

FIGURE 18.14 The Extensor Digiti Minimi Muscle.

ORIGIN

lateral epicondyle of humerus

INSERTION

dorsal side of base of distal phalange V

OTHER MOTIONS

reposition of finger V

NERVE

Radial

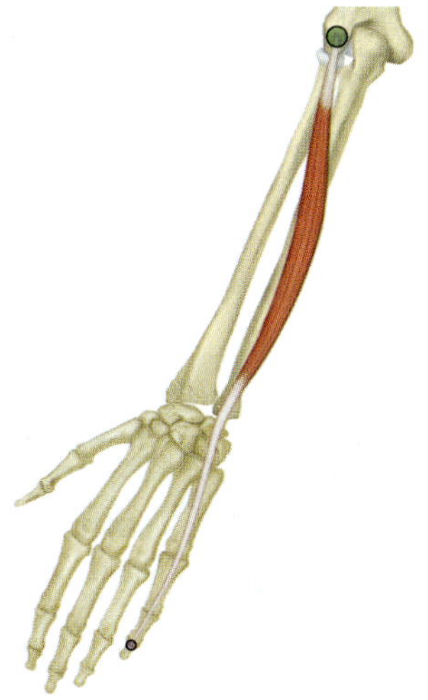

TABLE 18.4 The Muscles that Extend the Fingers

Muscle	Digit(s) Acted On	Joint(s) Acted On	Nerve
Extensor Pollicis Brevis	Digit I	MP	
Extensor Pollicis Longus	Digit I	Carpometacarpal, MP, & IP	
Extensor Indicis	Digit II	MP, PIP, & DIP	Radial
Extensor Digitorum	Digits II–V	MP, PIP, & DIP	
Extensor Digiti Minimi	Digit V	MP, PIP, & DIP	

PAUSE TO CHECK FOR UNDERSTANDING

TABLE 18.4 should serve as a good review for much of the material, but you should also review the origins and insertions.

The Finger Abductors

There are seven muscles that abduct the fingers. Three of them are the *abductor pollicis brevis, abductor pollicis longus, and abductor digiti minimi*. The other four are typically expressed as a group called the *dorsal interossei*.

The **abductor pollicis brevis** is an intrinsic muscle that acts on the first digit (thumb) (**FIGURE 18.15**). Its belly lies just lateral to that of the flexor pollicis brevis muscle which is over the palmar side of the first metacarpal. It originates from the scaphoid and trapezium (carpal bones) and inserts on the lateral side of the base of the first proximal phalange.

The **abductor pollicis longus** is an extrinsic muscle that also acts on the thumb (digit I) (**FIGURE 18.16**). It passes obliquely through the posterior side of the forearm originating from the posterior side of the ulna about hallway down the diaphysis. It inserts on the base of the first metacarpal.

The **dorsal interossei** is another group of four muscles that reside in between the metacarpals (**FIGURE 18.17**). They can be indicated sequentially from the lateral side of the hand to the medial side. The first dorsal interossei muscle acts on the second digit, the second and third muscles on the third digit, and the fourth muscle on the fourth digit. The four muscles originate from the medial and lateral sides of the diaphyses of the metacarpals I–V. The first muscle inserts on the lateral side of the base of the proximal phalange II. The second and third muscles insert on the medial and lateral sides of the base of the proximal phalange III, respectively. The fourth muscle inserts on the medial side of the base of the proximal phalange IV.

The **abductor digiti minimi** only acts on the fifth digit (little finger) (**FIGURE 18.18**). It resides on the anteromedial side of the hand adjacent to the flexor digiti minimi muscle. It originates from the pisiform (carpal bone) and inserts on the medial side of the base of the fifth proximal phalange.

FIGURE 18.15 The Abductor Pollicis Brevis Muscle.

ORIGINS
scaphoid
trapezium
INSERTION
lateral base of proximal phalange I
OTHER MOTIONS
reposition of thumb
NERVE
Median

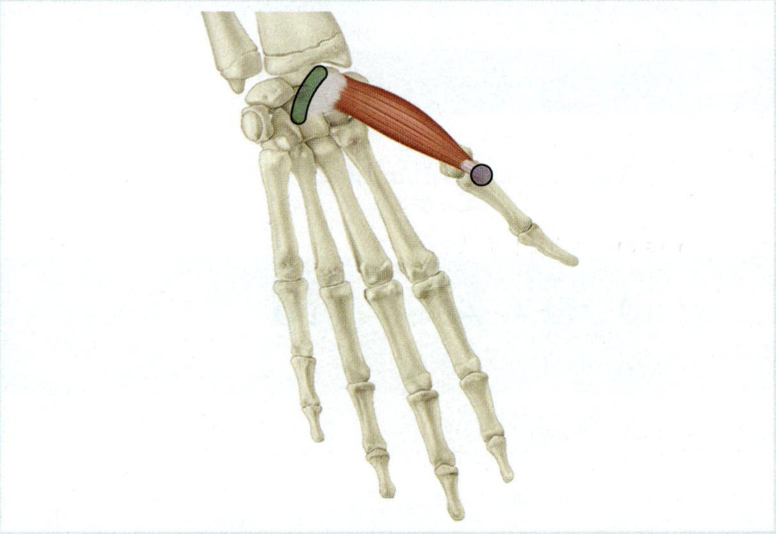

FIGURE 18.16 The Abductor Pollicis Longus Muscle.

ORIGIN
posterior side of mid ulna
INSERTION
dorsal base of metacarpal I
OTHER MOTIONS
reposition of thumb
NERVE
Radial

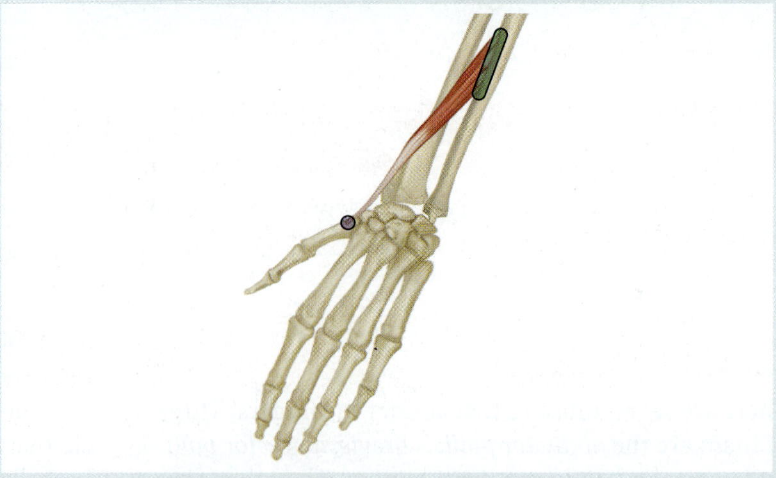

FIGURE 18.17 The Four Dorsal Interossei Muscles.

ORIGINS

sides of the diaphyses of metacarpals I–V

INSERTIONS

sides of the bases of proximal phalanges II–IV

OTHER MOTIONS

none

NERVE

Ulnar

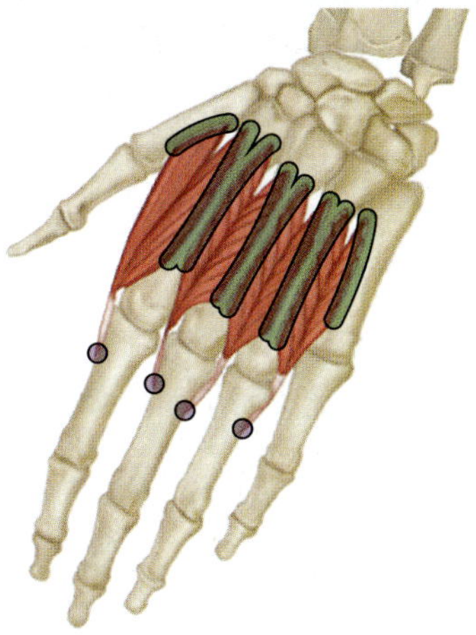

FIGURE 18.18 The Abductor Digiti Minimi Muscle.

ORIGIN

pisiform

INSERTION

medial base of proximal phalange V

OTHER MOTIONS

reposition of finger V

NERVE

Ulnar

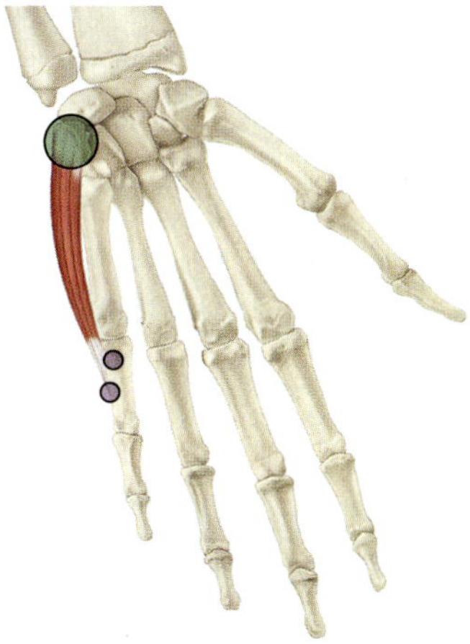

TABLE 18.5 The Finger Abductors

Muscle	Digit(s) Acted On	Joint(s) Acted On	Nerve
Abductor Pollicis Brevis	Digit I	MP	Median
Abductor Pollicis Longus	Digit I	Carpometacarpal	Radial
Dorsal Interrossei	Digits II–IV	MP	Ulnar
Abductor Digiti Minimi	Digits V	MP	

PAUSE TO CHECK FOR UNDERSTANDING

TABLE 18.5 should serve as a good review for much of the material, but you should also review the origins and insertions.

The Finger Adductors

There are five muscles that adduct the fingers. One of them is the *adductor pollicis*. The other four is a group commonly called the *palmar interossei*.

The **adductor pollicis** muscle, like the other pollicis muscles, acts on the thumb (**FIGURE 18.19**). Its fibers pass transversely across the palmar hand from the middle of the hand to the thumb. Specifically, it originates from the palmar side of the bases of the second and third metacarpals, the diaphysis of the third metacarpal, and the capitate carpal bone. It inserts on the medial side of the base of the first proximal phalange.

The **palmar interrossei** is a third group of four muscles that reside in between the metacarpals (**FIGURE 18.20**). They act on the first and second digits

FIGURE 18.19 The Adductor Pollicis Muscle.

ORIGINS
palmar bases of metacarpal II & III
diaphysis of metacarpal III
capitate
INSERTION
medial base of proximal phalange I
OTHER MOTIONS
none
NERVE
Median

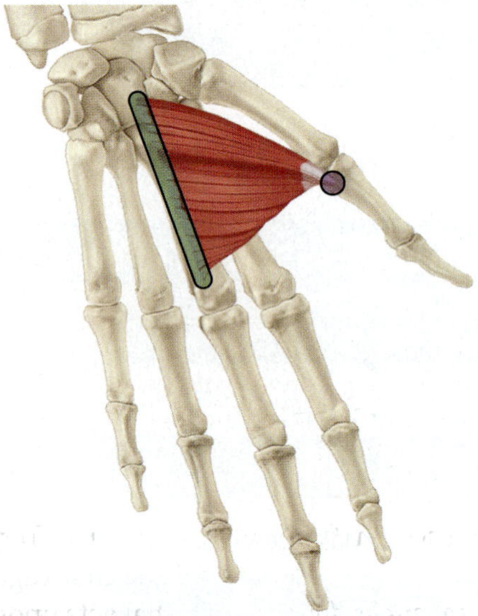

FIGURE 18.20 The Four Palmar Interossei Muscles.

ORIGINS
medial sides of diaphysis of metacarpals I, II, IV, & V
lateral sides of diaphysis of metacarpals I, II, IV, & V
INSERTIONS
medial sides of bases of proximal phalanges I, II, IV, & V
lateral sides of bases of proximal phalanges I, II, IV, & V
OTHER MOTIONS
none
NERVE
Ulnar

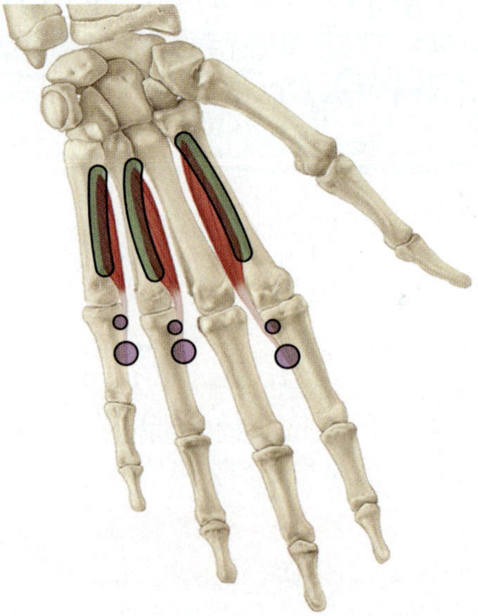

as well as the fourth and fifth digits. The first palmar interossei muscle acts on the thumb, the second on the second digit, the third on the fourth digit, and the fourth on the fifth digit. None of the palmar interossei muscles act on the third digit (middle finger). The first two muscles originate from the medial side of the diaphyses of the first and second metacarpals and insert on the medial side of the bases of the first and second proximal phalanges. The last two muscles originate from the lateral side of the diaphyses of the fourth and fifth metacarpals and insert on the lateral side of the bases of the fourth and fifth proximal phalanges.

TABLE 18.6 The Finger Adductors

Muscle	Digit(s) Acted On	Joint(s) Acted On	Nerve
Adductor Pollicis	Digit I	MP	Median
Palmar Interossei	Digit II, IV, & V (not I or III)	MP	Ulnar

PAUSE TO CHECK FOR UNDERSTANDING

TABLE 18.6 should serve as a good review for much of the material, but you should also review the origins and insertions.

The Finger Oppositioners

There are two muscles that oppose the fingers. They are the *opponens pollicis* which acts on the first digit (thumb) and the *opponens digiti minimi* that acts upon the fifth digit (little finger).

The **opponens pollicis** passes obliquely through the proximal end of the palmar hand over the fifth metacarpal (**FIGURE 18.21**). It originates from the trapezium carpal bone and the flexor retinaculum. It inserts on the lateral side of the diaphysis of the first metacarpal.

The **opponens digiti minimi** originates from the flexor retinaculum and the hamate carpal bone (**FIGURE 18.22**). From there, it passes up the palmar side of the fifth digit and inserts on the diaphysis of the fifth metacarpal.

FIGURE 18.21 The Opponens Pollicis Muscle.

ORIGINS
flexor retinaculum
trapezium
INSERTION
lateral side of diaphysis of metacarpal I
OTHER MOTIONS
none
NERVE
Median

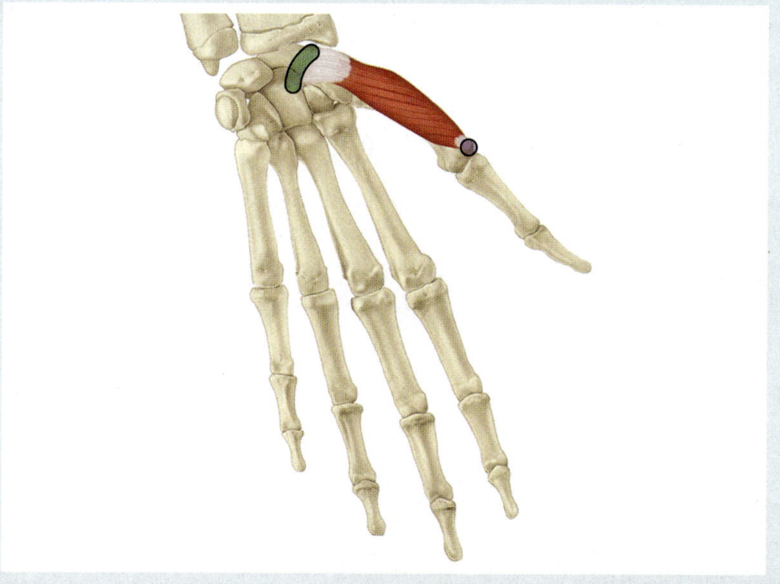

FIGURE 18.22 The Opponens Digiti Minimi Muscle.

ORIGINS	
flexor retinaculum	
hamate	
INSERTION	
palmar diaphysis of metacarpal V	
OTHER MOTIONS	
none	
NERVE	
Ulnar	

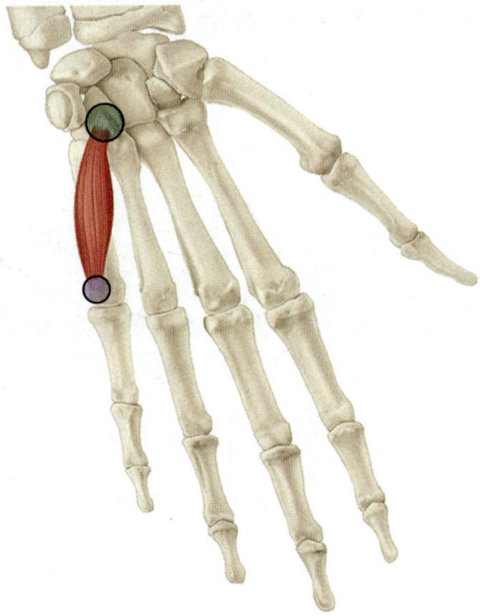

TABLE 18.7 The Finger Oppositioners

Muscle	Digit(s) Acted On	Joint(s) Acted On	Nerve
Opponens Pollicis	Digit I	Carpometacarpal	Median
Opponens Digiti Minimi	Digit V	Carpometacarpal	Ulnar

PAUSE TO CHECK FOR UNDERSTANDING

TABLE 18.7 should serve as a good review for much of the material, but you should also review the origins and insertions.

The Finger Repositioners

Six muscles act as finger repositioners. Four act on the first digit (thumb) and two on the fifth digit (little finger). Working together, the abductor pollicis brevis, the abductor pollicis longus, the extensor pollicis brevis, and the extensor pollicis longus reposition the thumb (digit I). Similarly, the abductor digiti minimi and the extensor digiti minimi act together to reposition the little finger (digit V).

The attachments, nerve innervations, and other actions of all these muscles have already been described and will not be repeated. To review them, see the relevant section labeled "The Finger Extensors" or "The Finger Abductors."

Workbook 1

Name ______________________ Section ________ Date ____________

Fundamentals of Anatomy

Using the word banks in each item, record the anatomical term in the blank that corresponds with what is being pointed at in the image(s) of each item.

1. Body positions.

supine anatomical fundamental prone

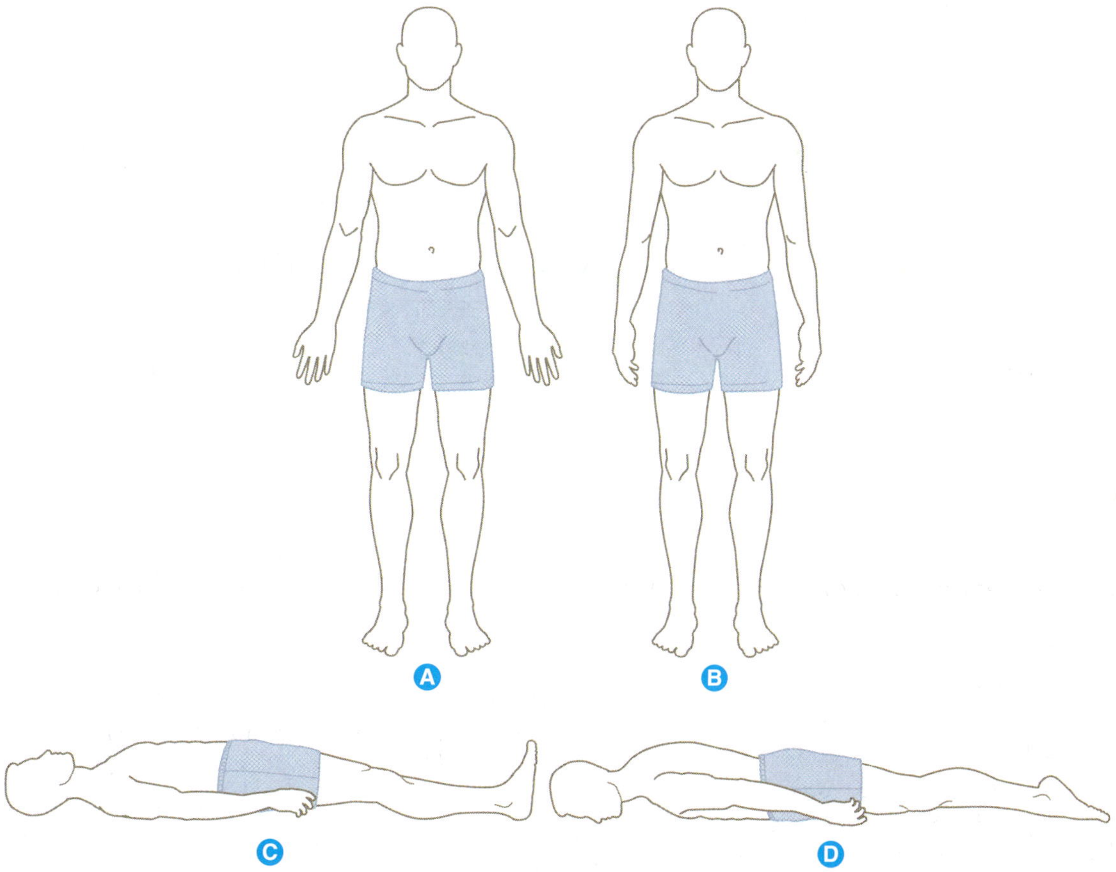

A. anatomical B. Fundmental

C. Supine D. Prone

2. Directional terms.

proximal ~~posterior~~ ~~medial~~ ~~superior~~ ~~distal~~ ~~inferior~~ lateral ~~anterior~~

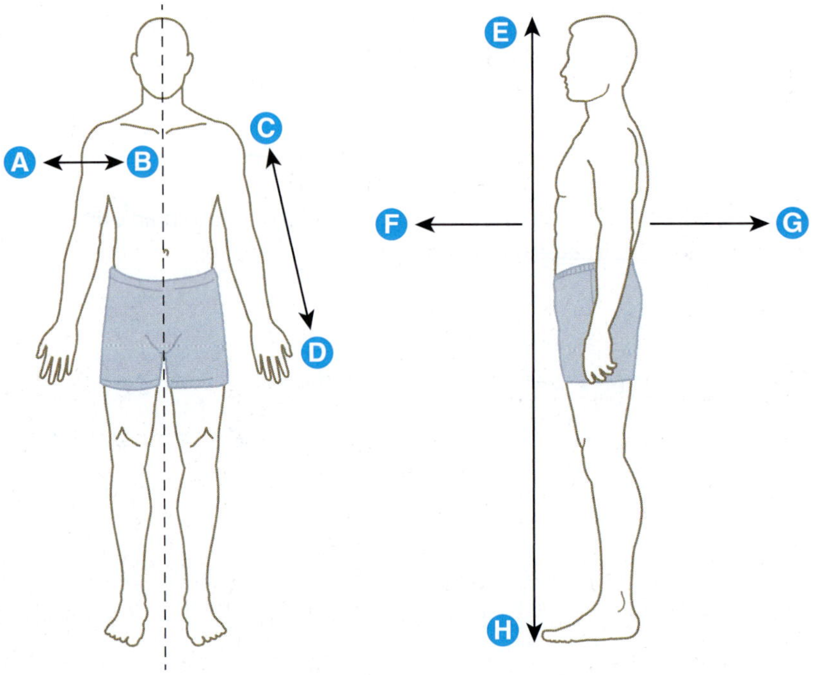

A. Lateral
B. medial
C. Proximal
D. Distal
E. Superior
F. Antrior
G. Posterior
H. inferior

3. Sub-regions of axial region.

cephalic gluteal pelvic thoracic inguinal
scapular abdominal cervical vertebral lumbar

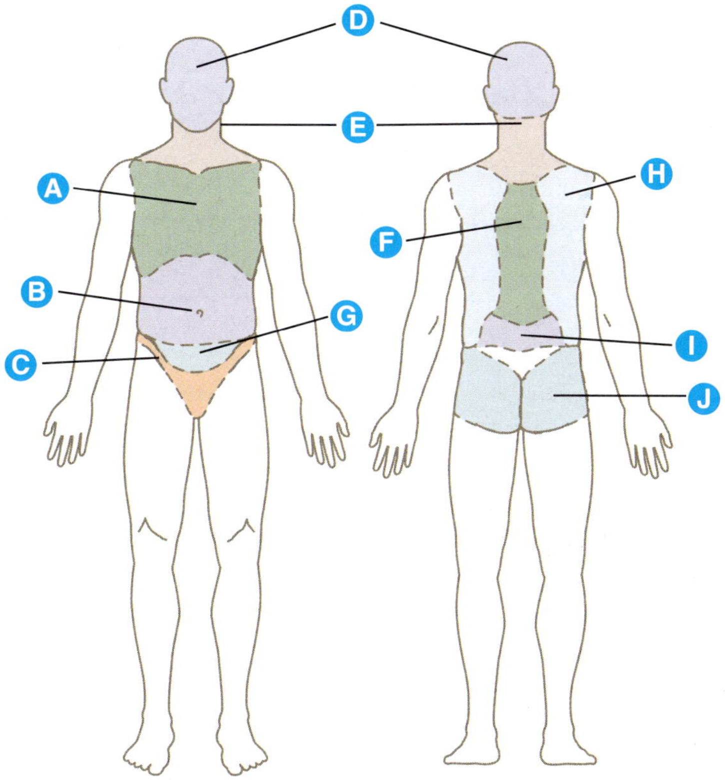

A. Thoracic

B. Abdominal

C. inguinal

D. Cephalic

E. Cervical

F. Vertebral

G. Pelvic

H. Scapular

I. Lumbar

J. Gluteal

4. Sub-regions of the upper extremities.

dorsum acromial cubital manus brachial
carpal axillary antebrachial digits palmar

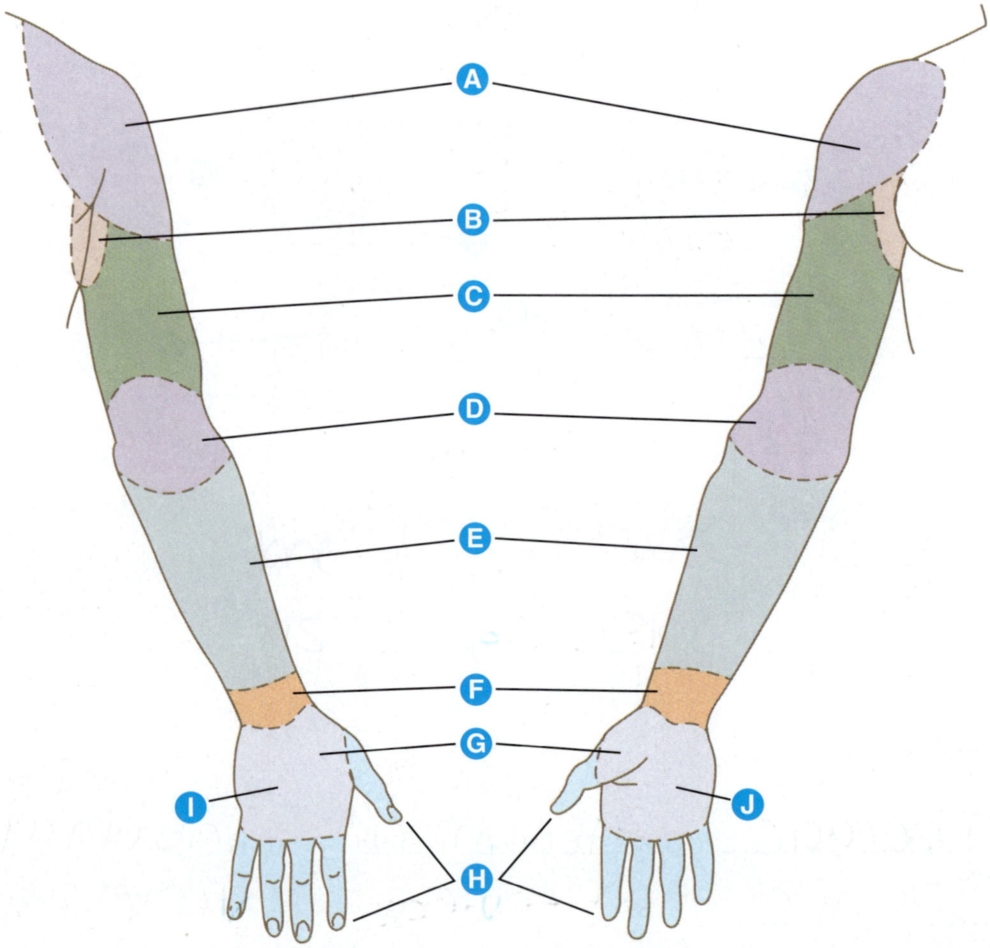

A. acromial B. axillary C. Brachial

D. Cubital E. antebracnial F. Carpal

G. manus H. Digits I. Dorsal

J. Plamer

5. Sub-regions of the lower extremities.

crural plantar pedal femoral popliteal digit
tarsal coxal calcaneal patellar

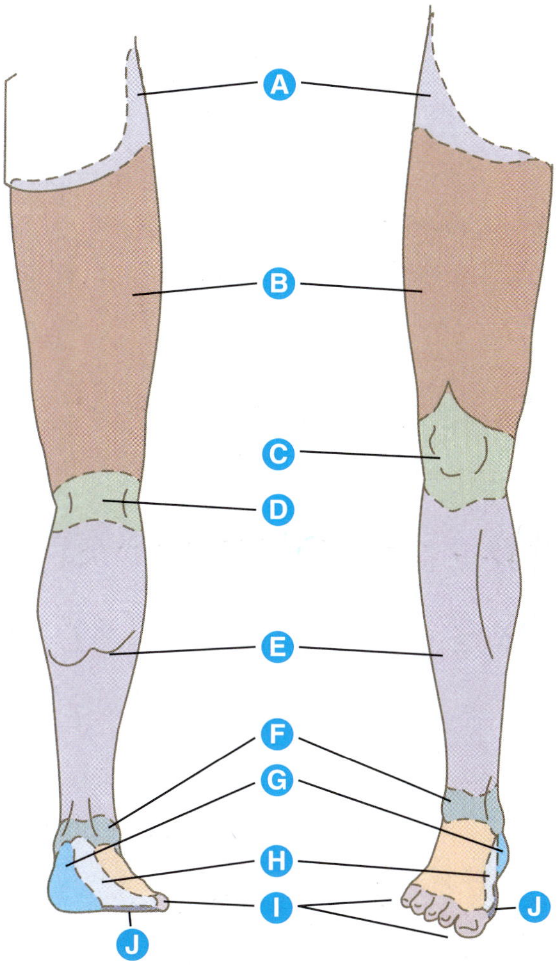

A. Coxal
B. femoral
C. Patellar
D. Pop pateral
E. Crural
F. Tarsal
G. Calcneal
H. Pedal
I. Digits
J. Plantar

6. Body planes and axes of rotation.

transverse plane   sagittal plane   sagittal axis   frontal axis   frontal plane   vertical axis

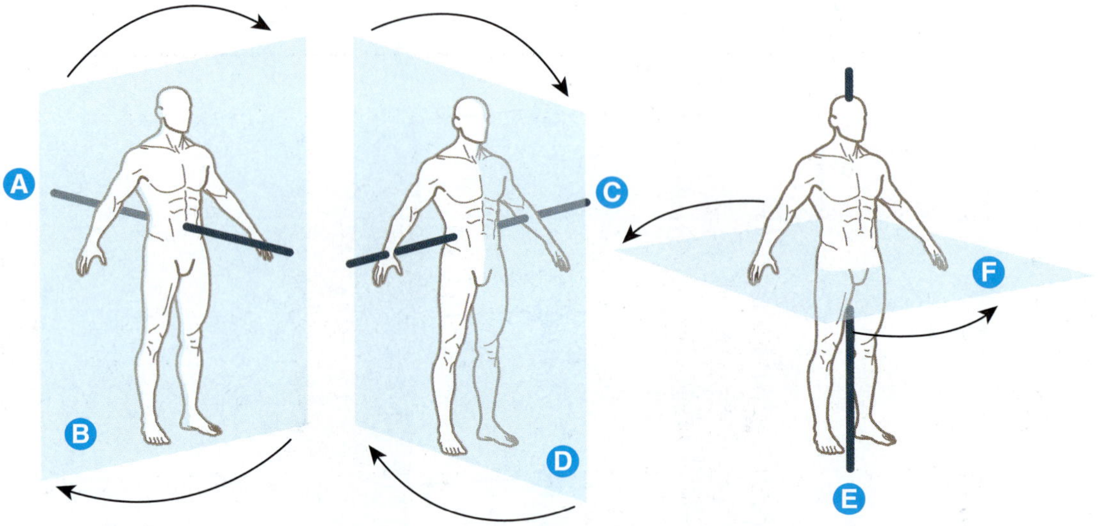

A. Frontal Plane B. Sagittal axis C. Sagital Plane

D. Frontal axis E. Transverse Plane F. verticle axis

Workbook 2

Name ____________________ Section ________ Date ____________

The Skeletal System

Using the word banks in each item, record the anatomical term in the blank that corresponds with what is being pointed at in the image(s) of each item.

1. The tissues of a bone.

marrow cartilage nervous osseous vascular

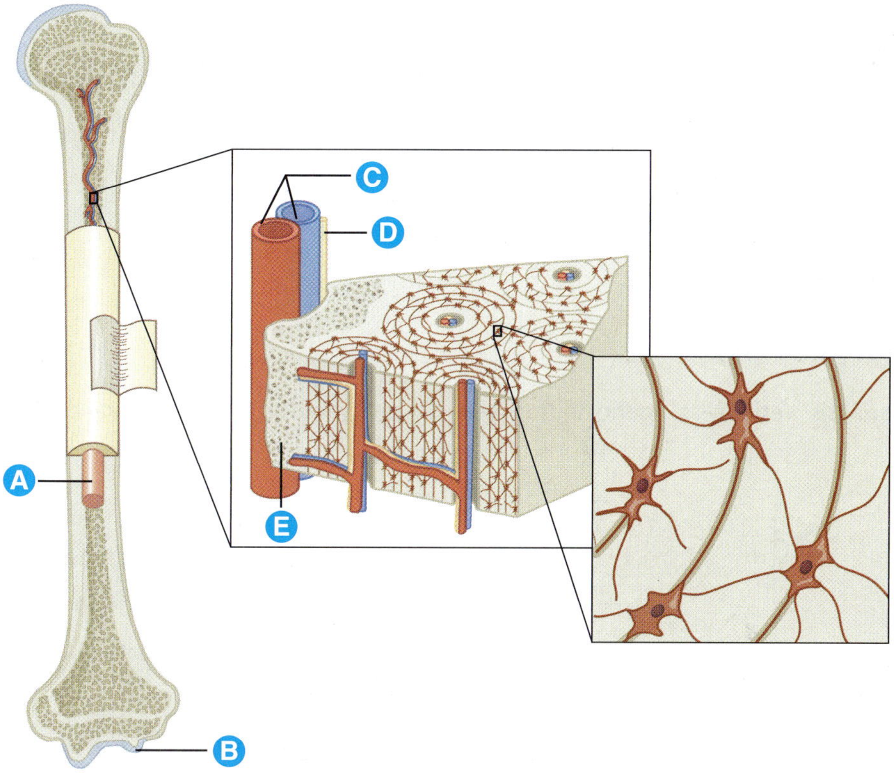

A. Marrow

B. Cartilage

C. Vasular

D. Nervous

E. Osseous

2. Types of bones.

flat short irregular long

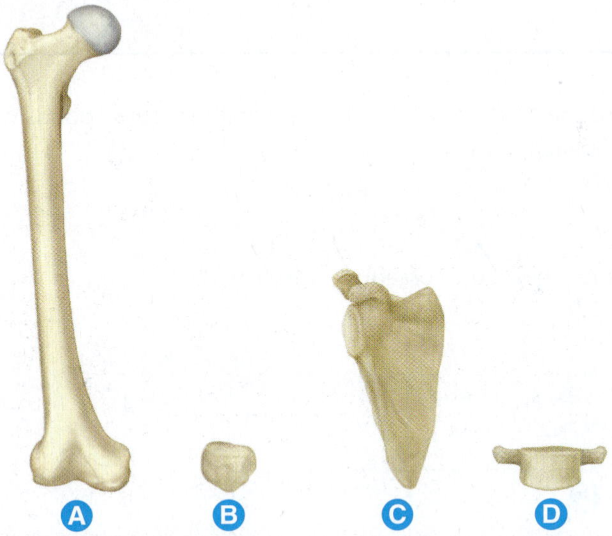

A. ____________________

B. ____________________

C. ____________________

D. ____________________

3. Long bone anatomy.

diaphysis epiphysis periosteum marrow cavity endosteum articular cartilage

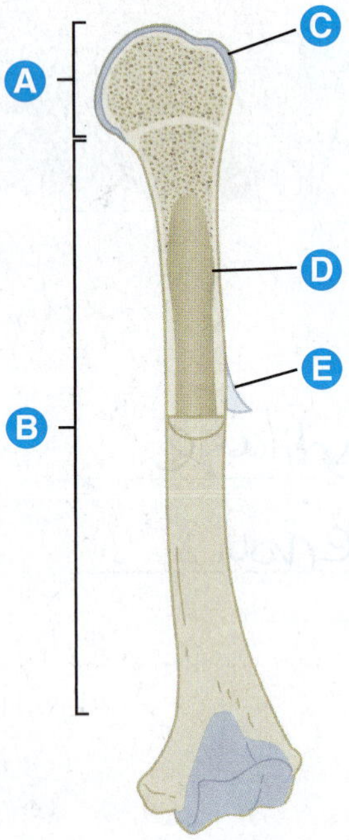

A. ____________________

B. ____________________

C. ____________________

D. ____________________

E. ____________________

4. Structural categories of joints.

cartilaginous bony synovial fibrous

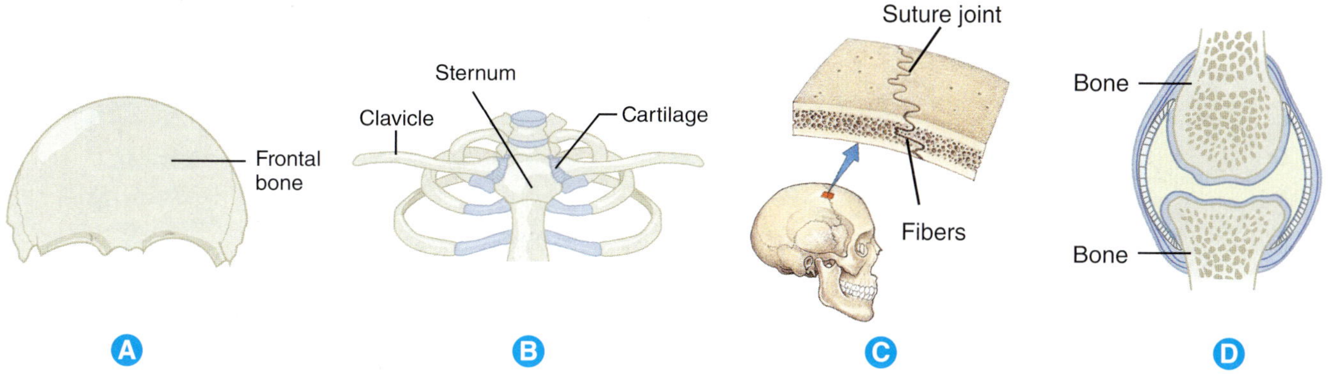

A. Bony B. Cartilaginous

C. Fibrous D. Synovial

5. Synovial joint anatomy.

fibrocartilage synovial membrane articular cartilage ligament joint cavity

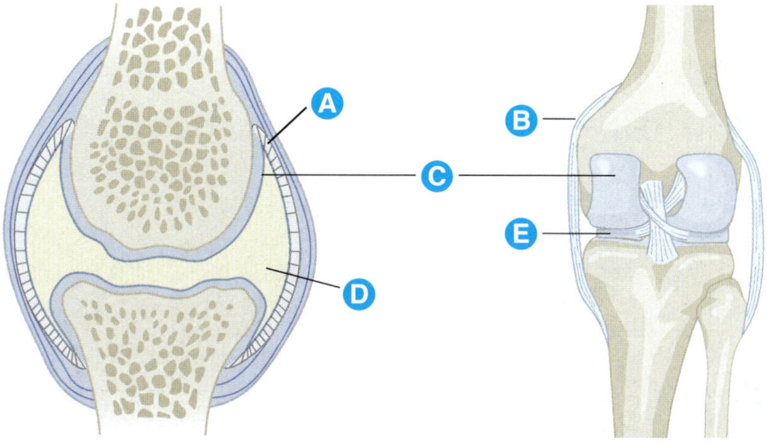

A. Synovial membrane B. Ligament

C. Articular cartilage D. Joint Cavity

E. Fibrocartilage

6. Synovial joint classifications.

ball and socket   plane   pivot   saddle   condylar   hinge

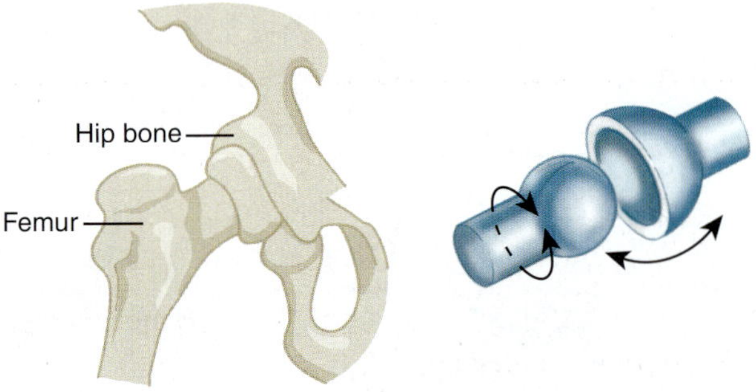

A

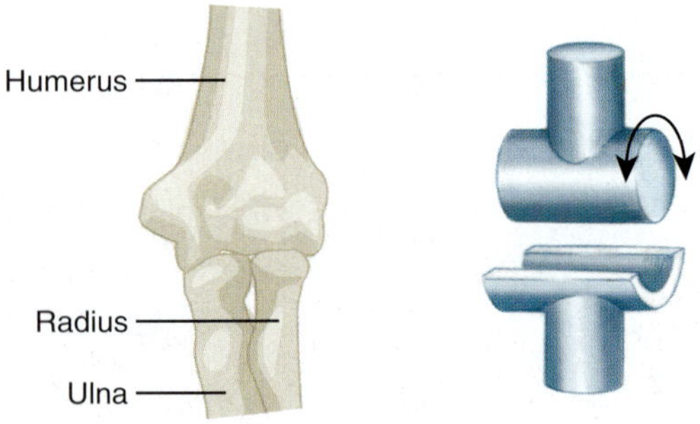

B

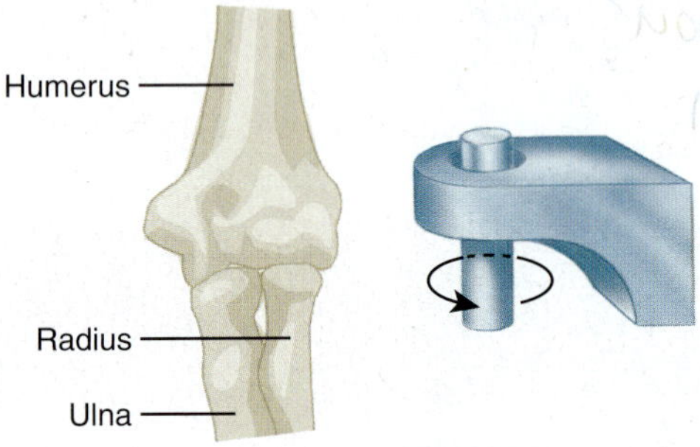

C

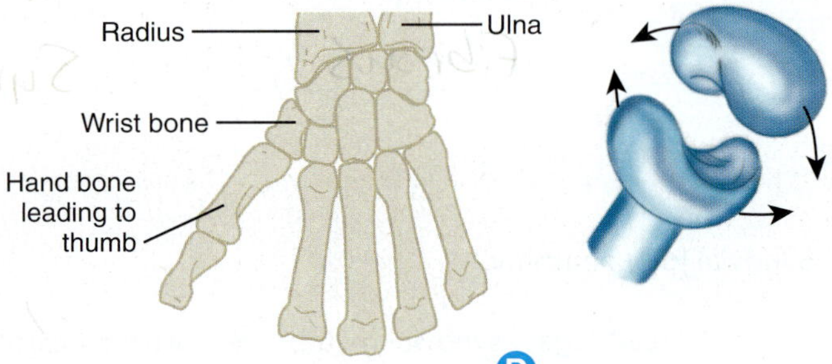

D

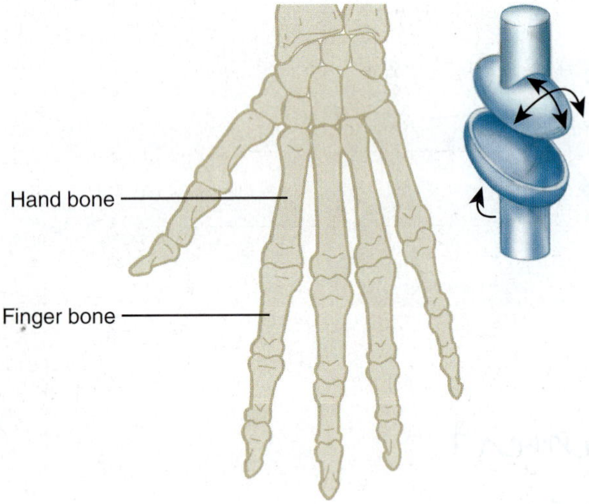

E

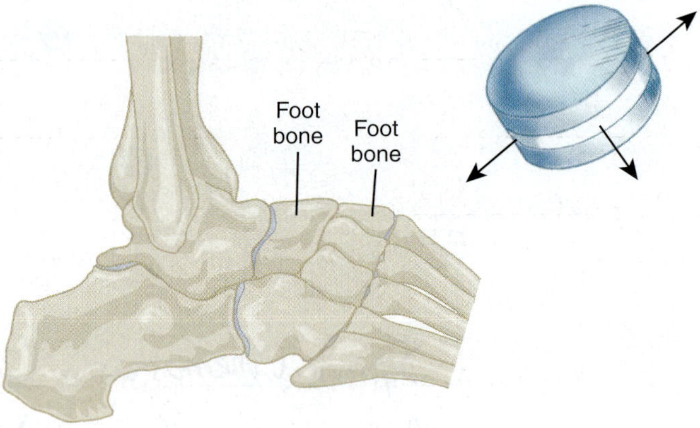

F

A. ______________________ B. ______________________

C. ______________________ D. ______________________

E. ______________________ F. ______________________

Workbook 3

Name ______________ Section ________ Date ______________

The Muscular System

Using the word banks in each item, record the anatomical term in the blank that corresponds with what is being pointed at in the image(s) of each item.

1. The tissues of a muscle.

tendon fibers perimysium fascicles endomysium epimysium fascia

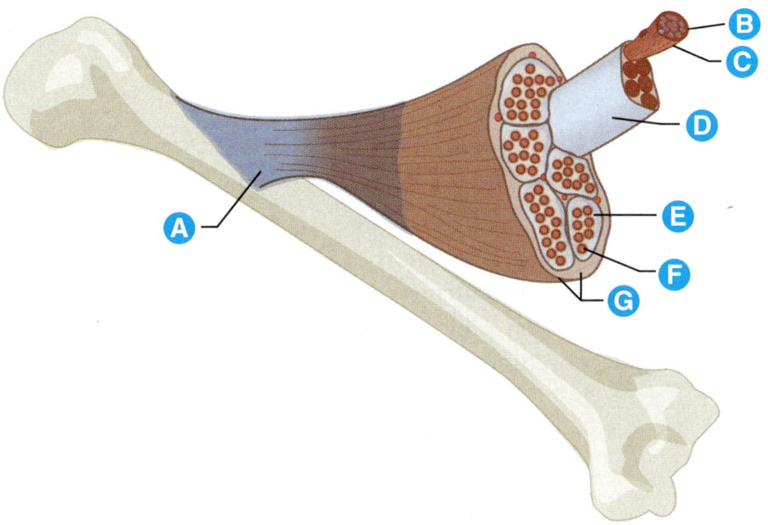

Pg 21

A. Tendon
B. Endomysium
C. Fibers
D. Fascicles
E. ~~Fascia~~ Endomysium
F. Perimysium
G. Epimysium

2. Muscle classes by shape.

parallel fusiform bipennate multipennate convergent unipennate

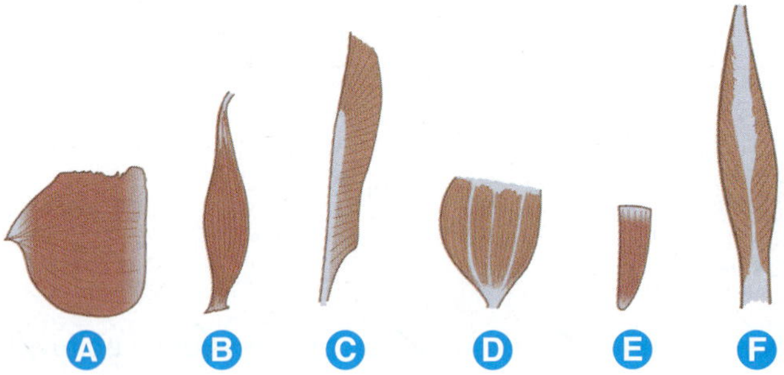

A. ____________ B. ____________ C. ____________

D. ____________ E. ____________ F. ____________

3. Simple motions.

adduction lateral rotation rotation elevation flexion plantar flexion radial deviation horizontal abduction supination inversion medial rotation opposition ulnar deviation dorsiflexion horizontal adduction protraction eversion retraction depression abduction pronation extension reposition

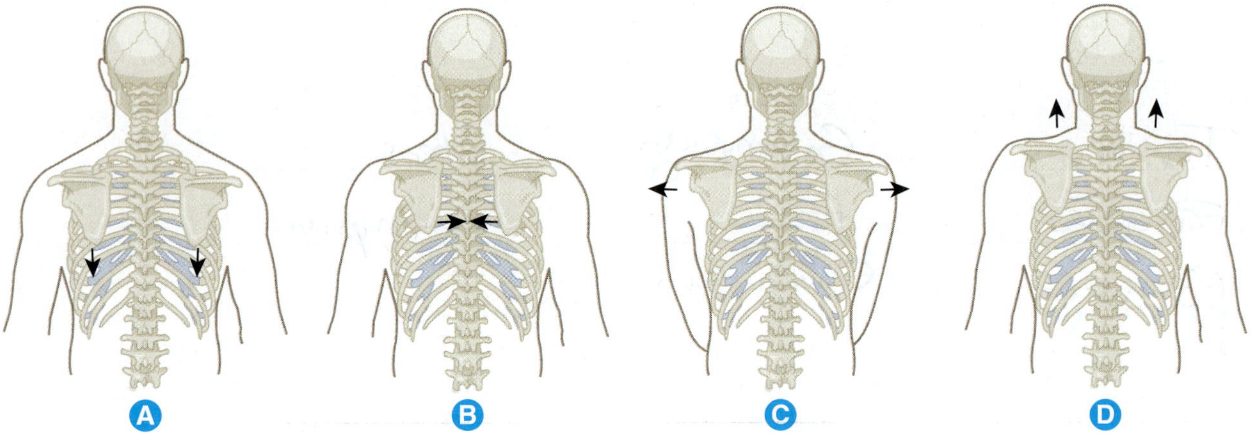

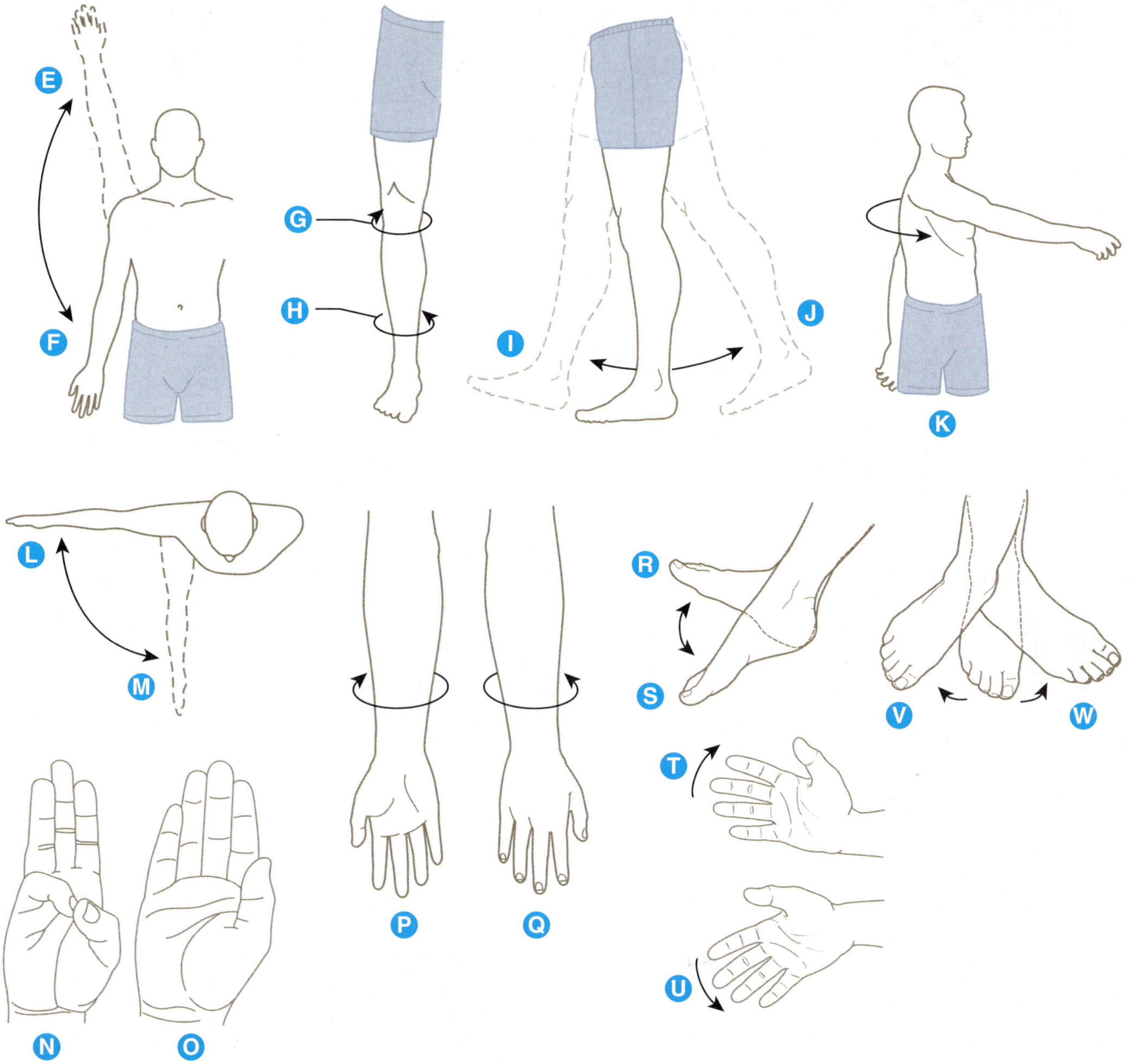

A. ____________________ B. ____________________ C. ____________________

D. ____________________ E. ____________________ F. ____________________

G. ____________________ H. ____________________ I. ____________________

J. ____________________ K. ____________________ L. ____________________

M. ____________________ N. ____________________ O. ____________________

P. ____________________ Q. ____________________ R. ____________________

S. ____________________ T. ____________________ U. ____________________

V. ____________________ W. ____________________

Workbook 4

Name ______________________ Section ________ Date ______________

The Nervous System

Using the word banks in each item, record the anatomical term in the blank that corresponds with what is being pointed at in the image(s) of each item.

1. Basic anatomy of a typical neuron.

axon   terminal end fibers   dendrites   soma

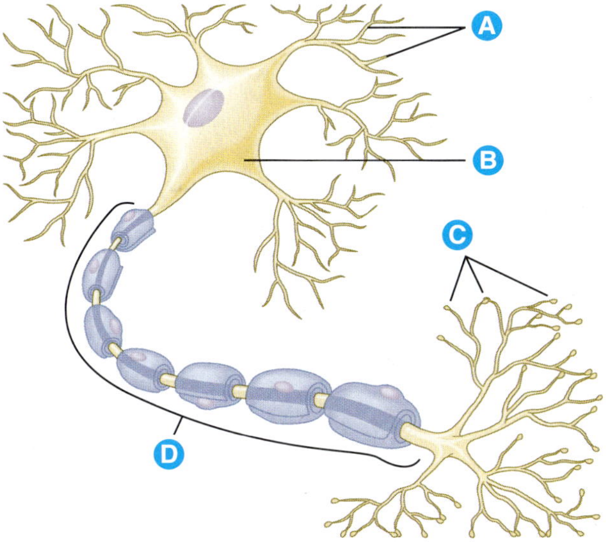

A. ______________________

B. ______________________

C. ______________________

D. ______________________

2. Anatomy of a synapse.

neurotransmitters   synaptic knob   vesicles   synaptic cleft

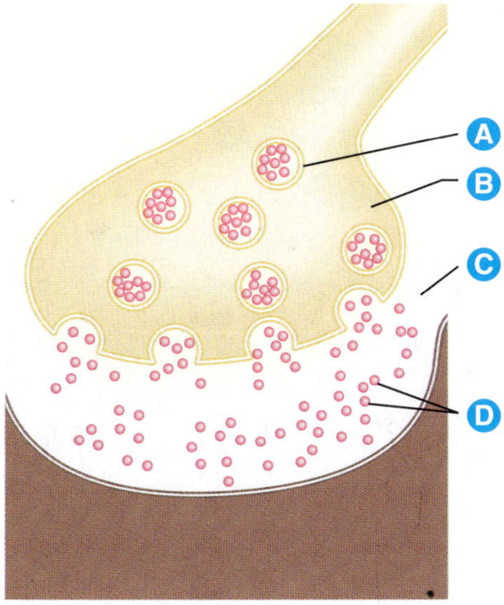

A. ______________________

B. ______________________

C. ______________________

D. ______________________

3. Main parts of the brain.

brainstem cerebellum cerebrum

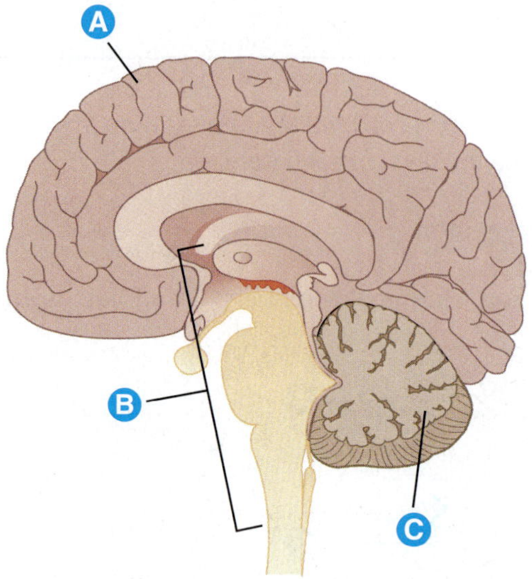

A. ____________________ B. ____________________

C. ____________________

4. Main parts of the brain.

parietal lobe temporal lobe occipital lobe frontal lobe

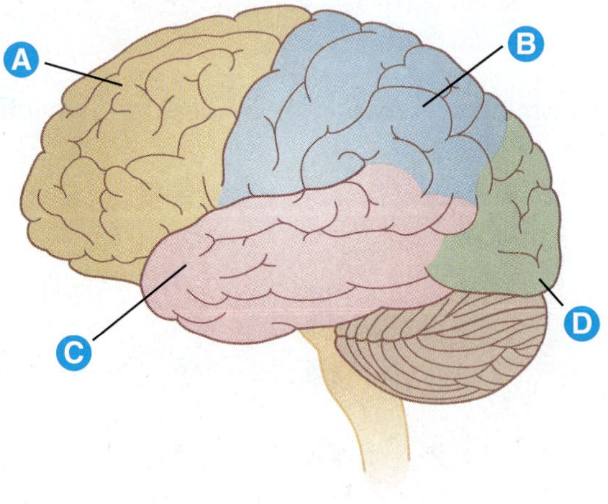

A. ____________________ B. ____________________

C. ____________________ D. ____________________

5. Main parts of the brainstem.

pons thalamus medulla oblongata midbrain

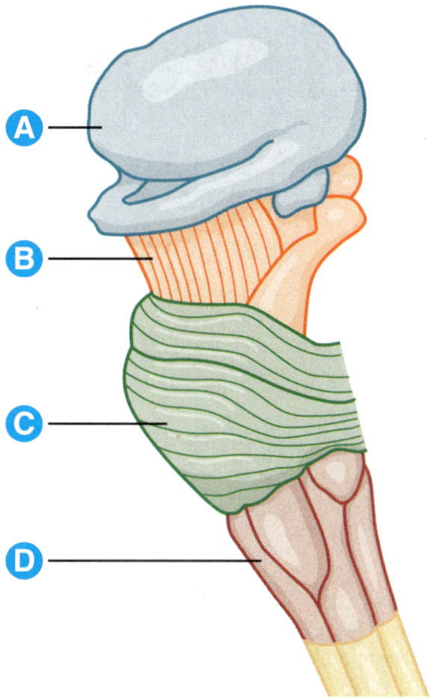

A. ____________________ B. ____________________

C. ____________________ D. ____________________

6. The connection between the spinal cord and nerves.

dorsal root spinal cord anterior ramus ventral root posterior ramus

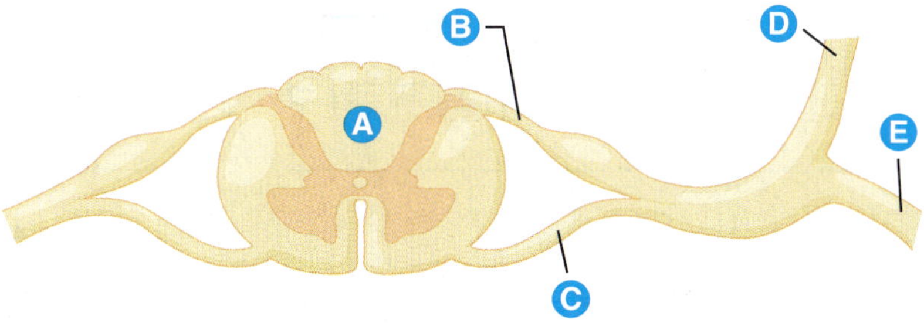

A. ____________________ B. ____________________

C. ____________________ D. ____________________

E. ____________________

Name ______________________________ Section ________ Date ____________

The Bones of the Axial Skeleton

Using the word banks in each item, record the anatomical term in the blank that corresponds with what is being pointed at in the image(s) of each item.

1. The bones of the skull.

maxilla temporal zygomatic frontal mandible
occipital sphenoid parietal nasal

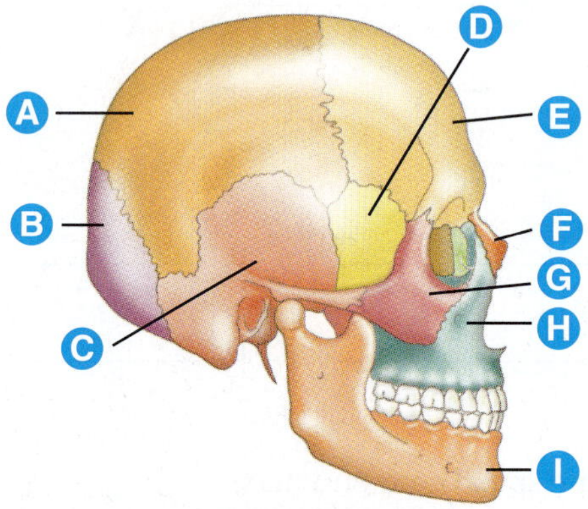

Lateral view

A. ____________ B. ____________ C. ____________

D. ____________ E. ____________ F. ____________

G. ____________ H. ____________ I. ____________

2. Anatomical features of the occipital bone.

external occipital protuberance   condyles   inferior nuchal line
superior nuchal line   foramen magnum

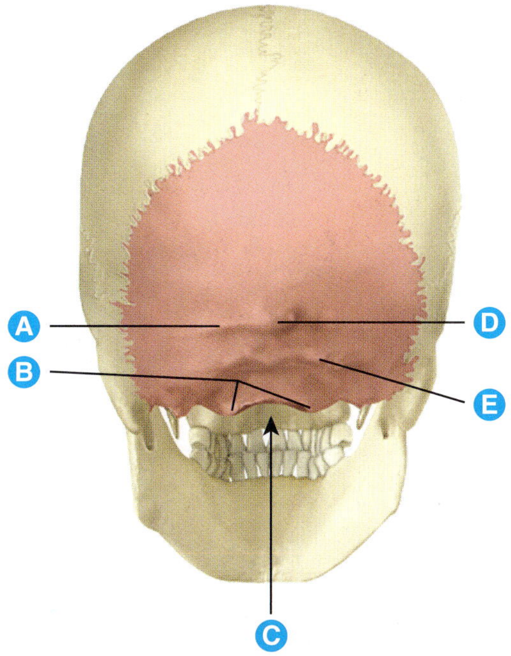

A. ____________________   B. ____________________

C. ____________________   D. ____________________

E. ____________________

3. Bones of the vertebral column.

sacrum   cervical vertebrae   lumbar vertebrae   coccyx   thoracic vertebrae

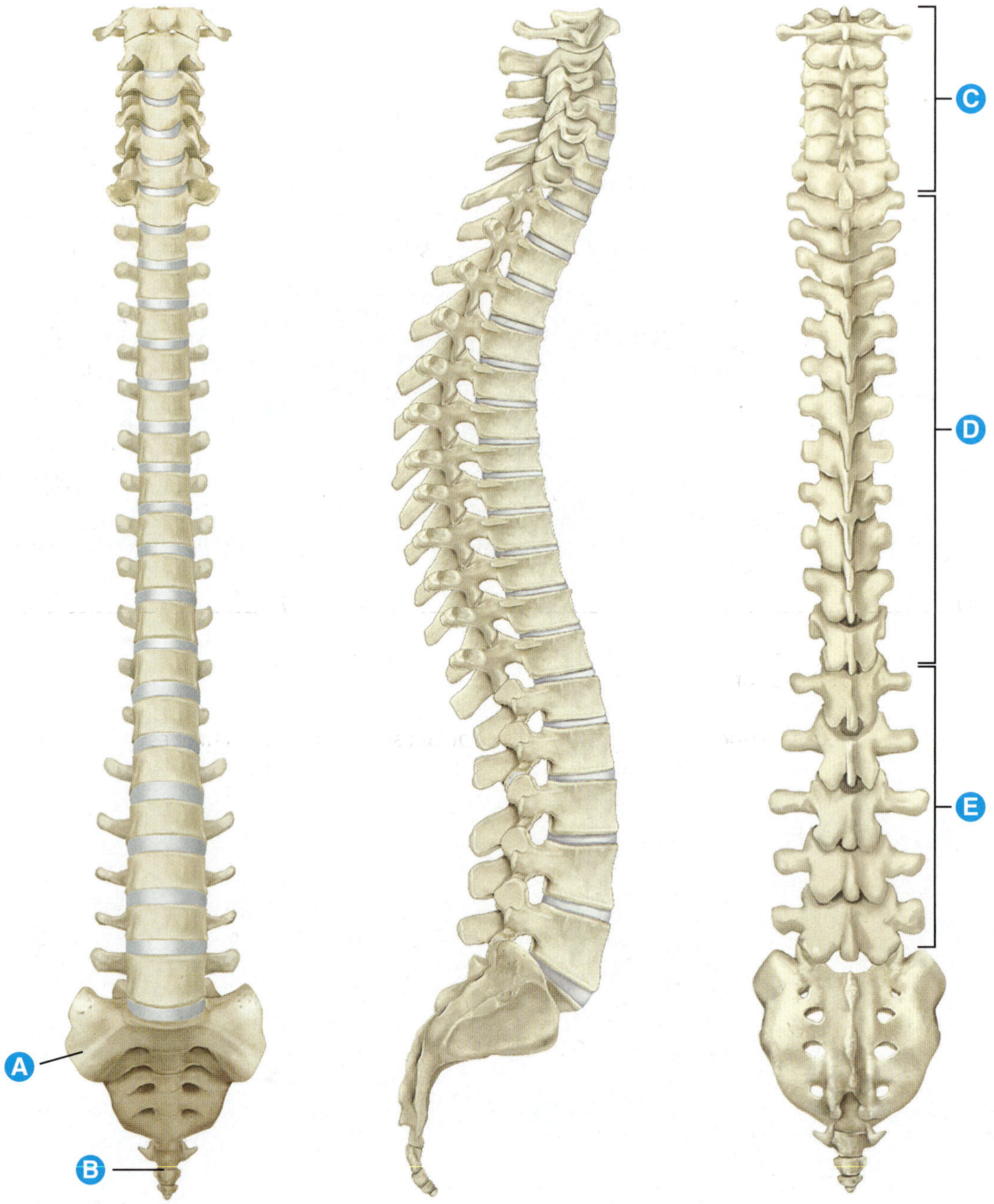

A. ______________________ B. ______________________

C. ______________________ D. ______________________

E. ______________________

4. Anatomical features of a vertebra.

body transverse process superior articular facet superior articular process spinous process vertebral arch

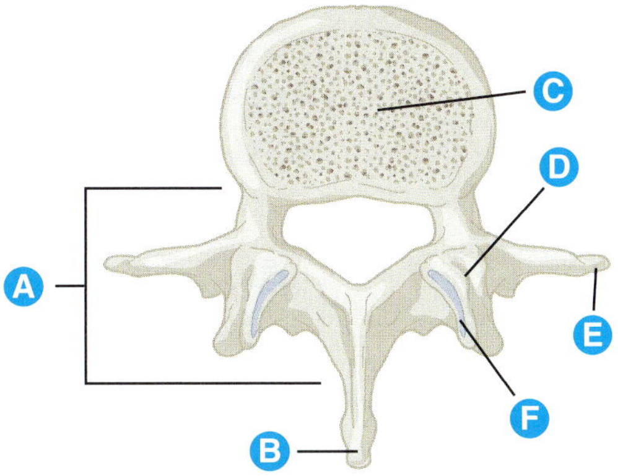

Superior View

A. ____________ B. ____________ C. ____________

D. ____________ E. ____________ F. ____________

5. Anatomical features of the sacrum.

base superior articular process sacral canal median sacral crest apex sacral foramen lateral sacral crest superior articular facet

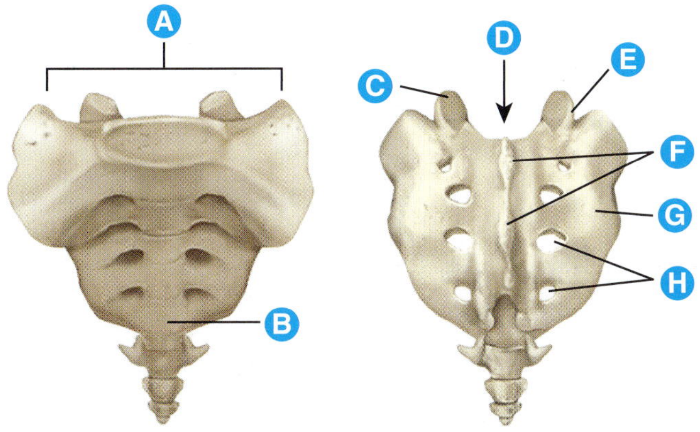

Anterior View **Posterior View**

A. ____________ B. ____________ C. ____________

D. ____________ E. ____________ F. ____________

G. ____________ H. ____________

6. The thoracic cage.

manubrium true rib clavicle false floating rib
false rib sternal body xiphoid process

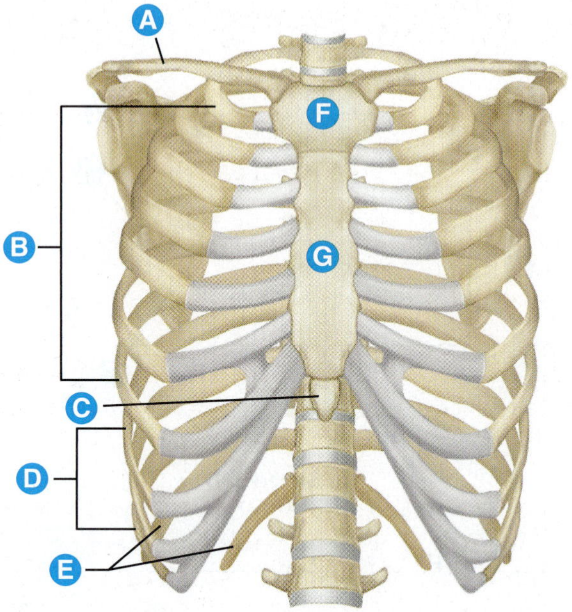

A. ____________________ B. ____________________ C. ____________________

D. ____________________ E. ____________________ F. ____________________

G. ____________________

Workbook 6

Name ______________________ Section ________ Date ______________

The Bones of the Upper Extremities

Using the word banks in each item, record the anatomical term in the blank that corresponds with what is being pointed at in the image(s) of each item.

1. Anatomical features of the scapula.

scapular spine	infraspinous fossa	glenoid cavity	medial border
inferior angle	subscapular fossa	superior border	acromial process
lateral border	coracoid process	supraspinous fossa	superior angle

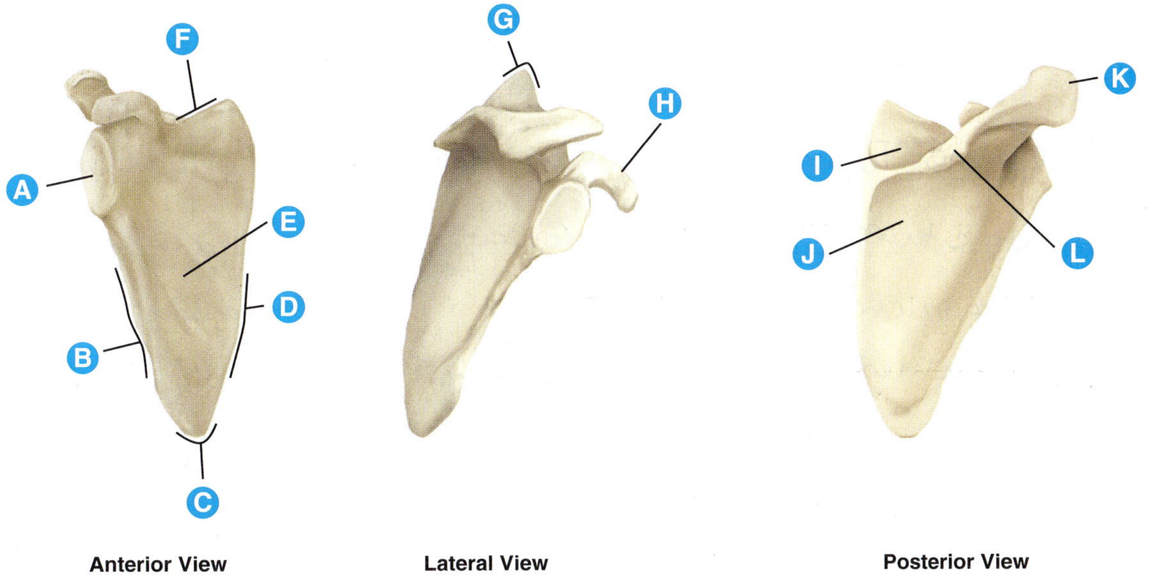

A. Glenoid Cavity B. Lateral Border C. Inferior Angle

D. Medial Border E. Subscapular fossa F. Superior Border

G. Superior Angle H. Coracoid Process I. Supraspinous fossa

J. Infraspinous fossa K. Acromial process L. Scapular Spine

2. Anatomical features of the humerus.

capitulum coronoid fossa olecranon fossa deltoid tuberosity greater tubercle
head lesser tubercle trochlea intertubercular groove radial fossa epicondyle

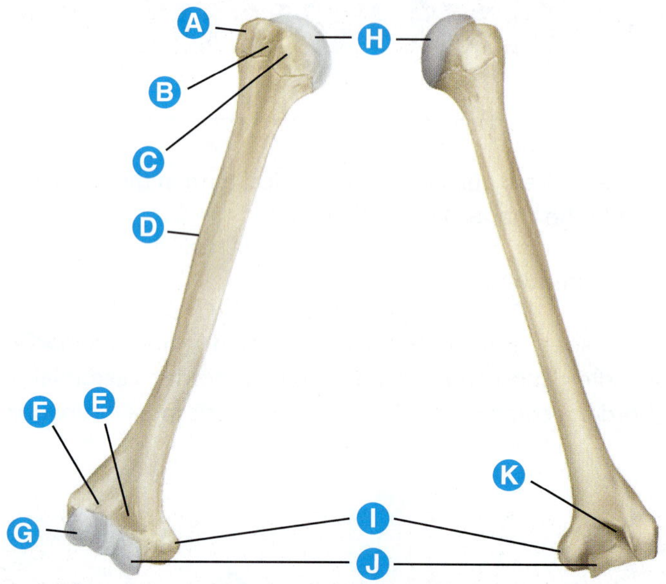

Pg 49

A. Greater Tubercle B. Intertuberculear Groove C. Lesser Tuberculer

D. Deltoid tuberosity E. Coronod fossa F. Epicondyle

G. Capitulum H. Head I. Epicondyle

J. Trochlea K. Olecranon fossa

3. Anatomical features of the ulna.

medial styloid process coronoid process olecranon process trochlear notch radial notch

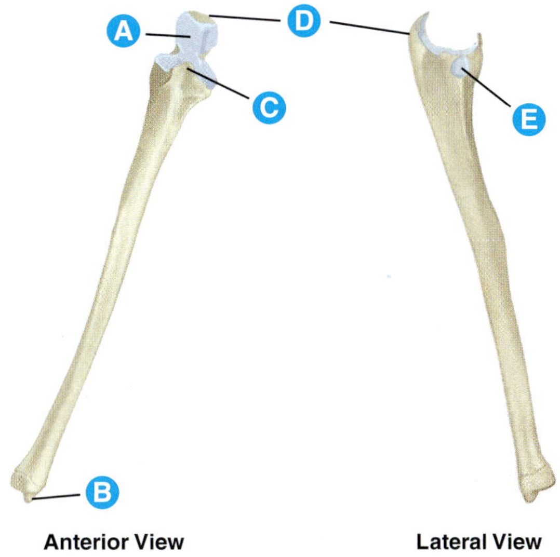

A. Trochlear Notch B. Medial Styloid Process C. Coronoid process

D. Olecranon Process E. Radial Notch

4. Anatomical features of the radius.

neck lateral styloid process radial tuberosity head

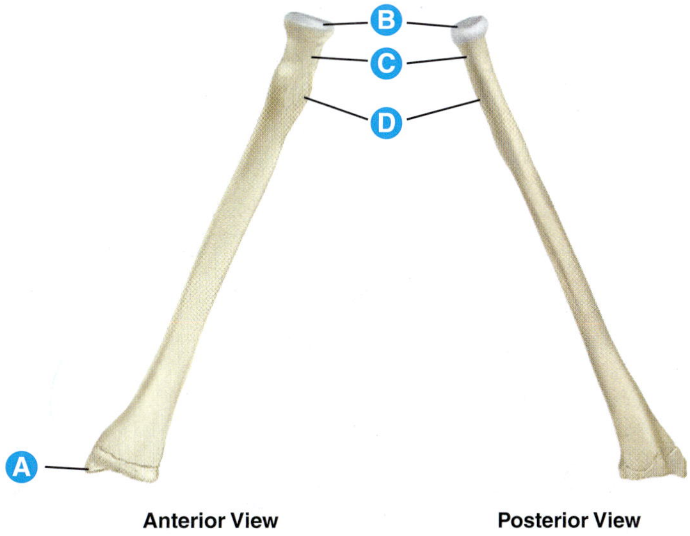

A. Lateral Styloid Process B. Head

C. Neck D. Radial Tuberosity

5. Bones of the hand.

carpals finger I finger II finger III finger IV finger V
proximal phalange intermediate phalange distal phalange metacarpal

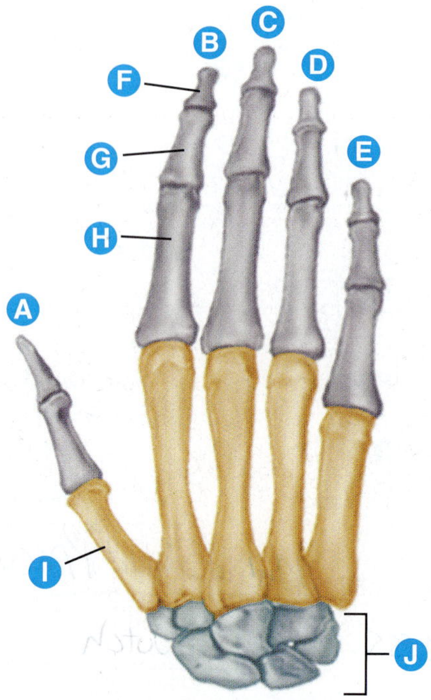

Posterior View

A. Finger I
B. Finger II
C. Finger III
D. Finger IV
E. Finger V
F. Distal Phalange
G. Intermediate Phalange
H. Proximal Phalange
I. ~~Proximal Phalange~~ MetaCarpal
J. Carpal

Name ______________________ Section ________ Date ______________

The Bones of the Lower Extremities

Using the word banks in each item, record the anatomical term in the blank that corresponds with what is being pointed at in the image(s) of each item.

1. Anatomical features of the coxal bone.

pubic ramus   ilium   greater sciatic notch   anterior inferior iliac spine
iliac spine   pubis   posterior superior iliac spine   ischial ramus
ischial spine   superior ramus of pubis   ischial tuberosity   lesser sciatic notch
inferior ramus of pubis   anterior superior iliac spine   pubic body   ischial body
iliac fossa   posterior inferior iliac spine   ischium   ischial spine

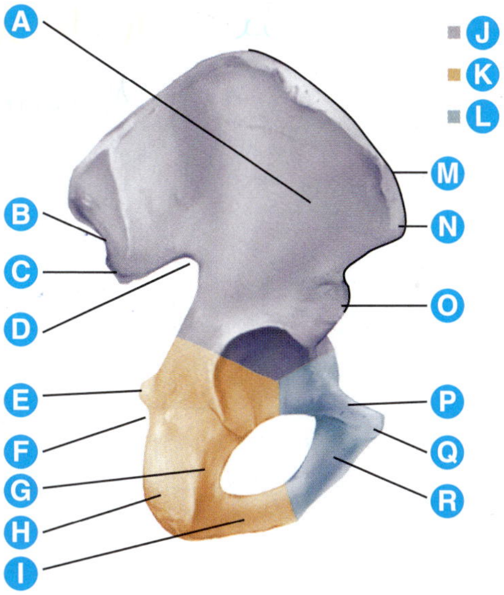

Lateral View

A. ______________________

B. ______________________

C. ______________________

D. ______________________

E. ______________________

F. ______________________ G. ______________________ H. ______________________

I. ______________________ J. ______________________ K. ______________________

L. ______________________ M. ______________________ N. ______________________

O. ______________________ P. ______________________ Q. ______________________

R. ______________________

2. Anatomical features of the femur.

lateral condyle pectineal line linea aspera medial epicondyle
adductor tubercle head intertrochanteric crest intertrochanteric line
lesser trochanter intercondylar fossa fovea capitis neck
greater trochanter gluteal tuberosity medial condyle lateral epicondyle

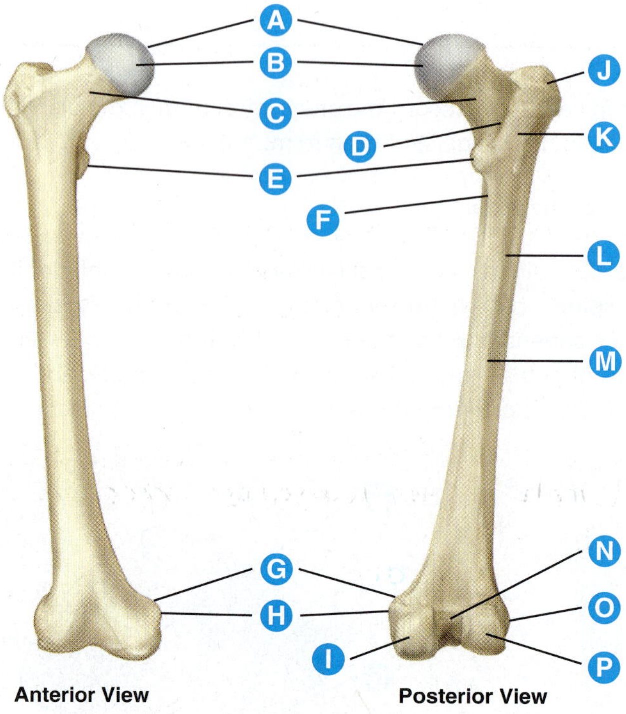

A. ____________ B. ____________ C. ____________

D. ____________ E. ____________ F. ____________

G. ____________ H. ____________ I. ____________

J. ____________ K. ____________ L. ____________

M. ____________ N. ____________ O. ____________

P. ____________

3. Anatomical features of the patella.

apex   articular facets   base

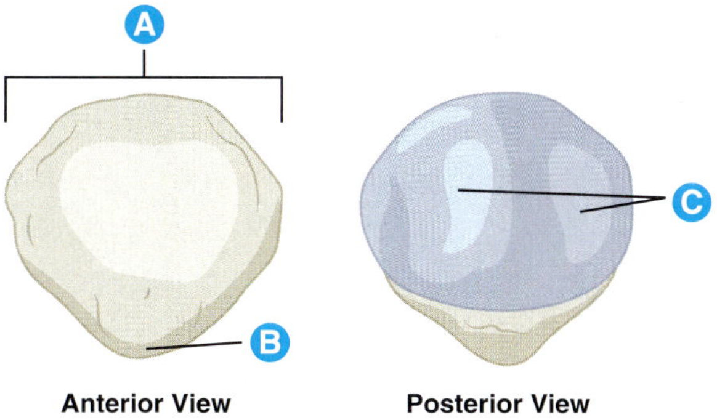

A. ____________________   B. ____________________   C. ____________________

4. Anatomical features of the tibia.

tibial tuberosity   intercondylar eminence   anterior crest   medial condyle
medial malleolus   lateral condyle   tibial plateau

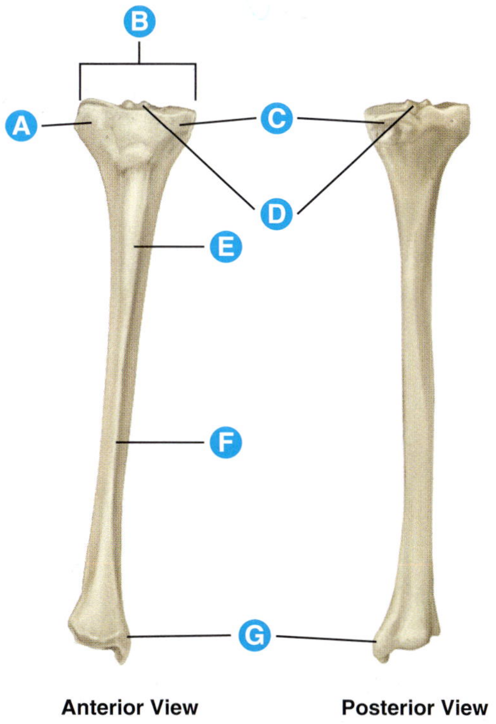

A. ____________________   B. ____________________   C. ____________________

D. ____________________   E. ____________________   F. ____________________

G. ____________________

5. Anatomical features of the fibula.

lateral malleolus   head

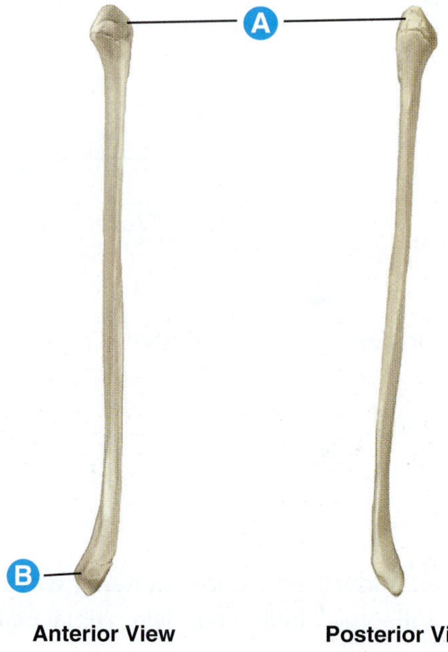

A. ____________________   B. ____________________

6. Bones of the foot.

calcaneus toe I toe II medial cuneiform toe III cuboid phalanges
talus intermediate phalange lateral cuneiform toe IV toe V tarsals
middle cuneiform distal phalange metatarsals navicular proximal phalange

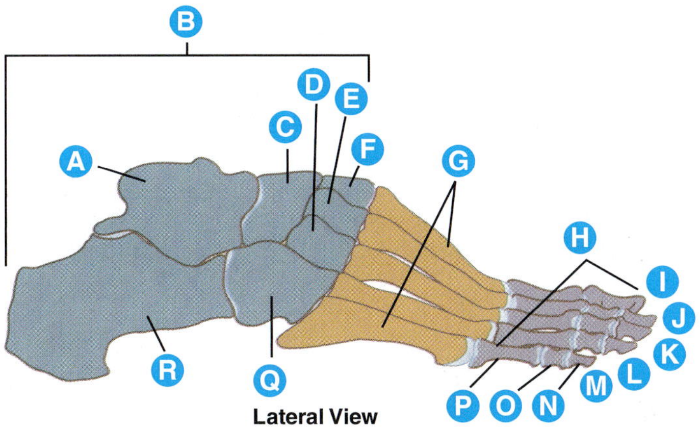

Lateral View

A. ____________ B. ____________ C. ____________

D. ____________ E. ____________ F. ____________

G. ____________ H. ____________ I. ____________

J. ____________ K. ____________ L. ____________

M. ____________ N. ____________ O. ____________

P. ____________ Q. ____________ R. ____________

Name ______________________ Section ________ Date ____________

The Foot

Using the word banks in each item, record the anatomical term in the blank that corresponds with what is being pointed at in the image(s) of each item.

1. Joints of the foot.

tarsometatarsal   metatarsophalangeal   interphalangeal
proximal interphalangeal   distal interphalangeal

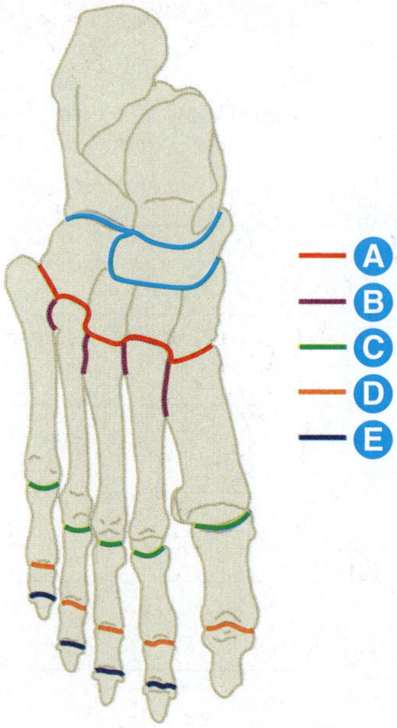

A. ____________________   B. ____________________   C. ____________________

D. ____________________   E. ____________________

2. Motions of the toes.

extension   flexion   abduction   adduction

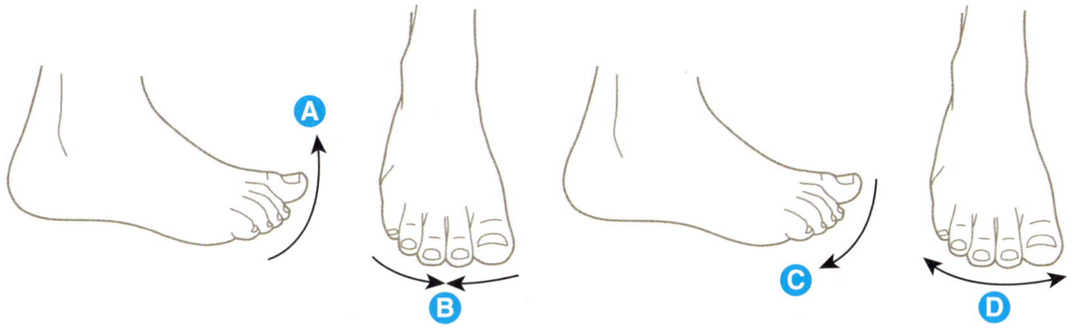

A. ____________________ B. ____________________

C. ____________________ D. ____________________

3. Muscles that flex the toes.

flexor hallucis brevis   flexor hallucis longus   flexor digitorum brevis
flexor digitorum longus   flexor digiti minimi   quadratus plantae   lumbricals

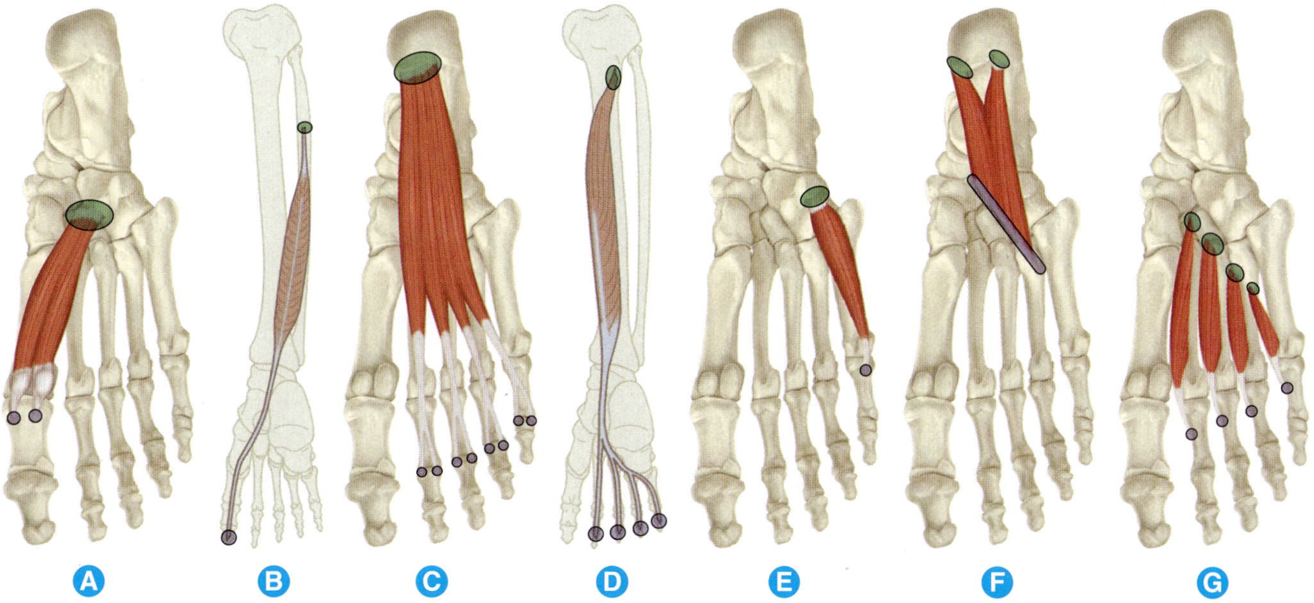

A. ____________________ B. ____________________ C. ____________________

D. ____________________ E. ____________________ F. ____________________

G. ____________________

4. Muscles that extend the toes.

extensor hallucis brevis | extensor hallucis longus
extensor digitorum brevis | extensor digitorum longus

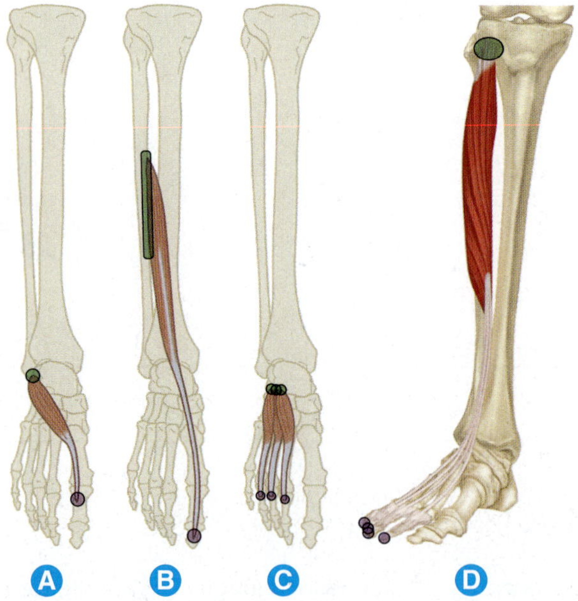

A. ______________________ B. ______________________

C. ______________________ D. ______________________

5. Muscles that abduct and adduct the toes.

abductor hallucis dorsal interossei abductor digiti minimi
adductor hallucis plantar interossei

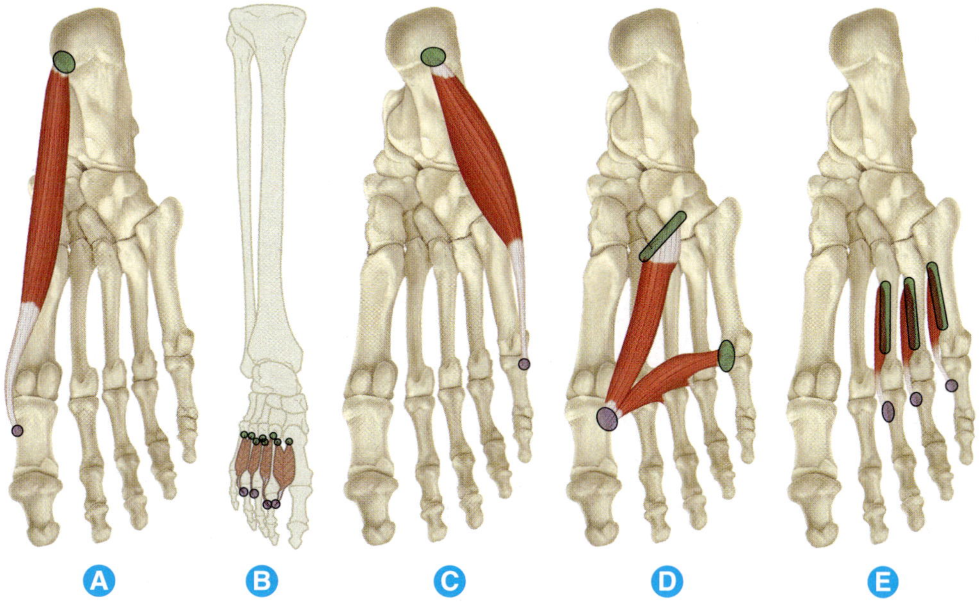

A. ____________________ B. ____________________ C. ____________________

D. ____________________ E. ____________________

Name ______________________ Section ________ Date ______________

The Ankle

Using the word banks in each item, record the anatomical term in the blank that corresponds with what is being pointed at in the image(s) of each item.

1. Joints of the ankle.

distal tibiofibular   talocrural   subtalar

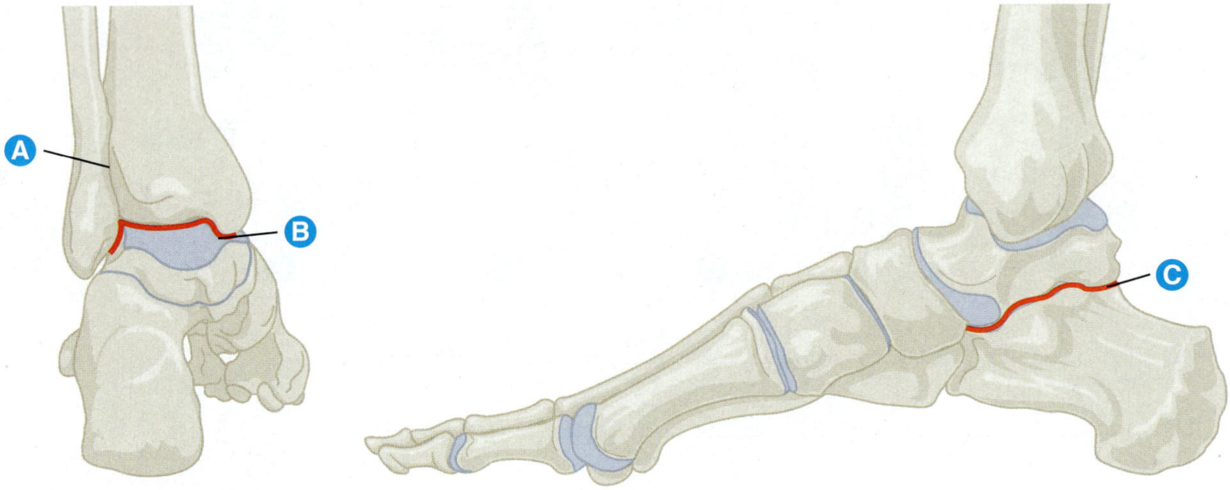

A. ______________ B. ______________ C. ______________

2. Motions of the ankle.

plantar flexion dorsiflexion inversion eversion

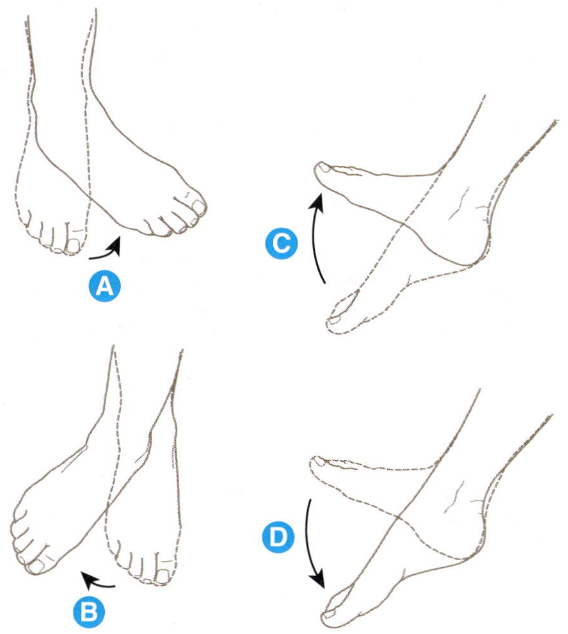

A. ____________________ B. ____________________

C. ____________________ D. ____________________

3. Muscles of the ankle.

tibialis anterior extensor hallucis longus extensor digitorum longus
fibularis brevis fibularis longus fibularis tertius gastrocnemius soleus
plantaris posterior tibialis extensor digitorum longus extensor hallucis longus

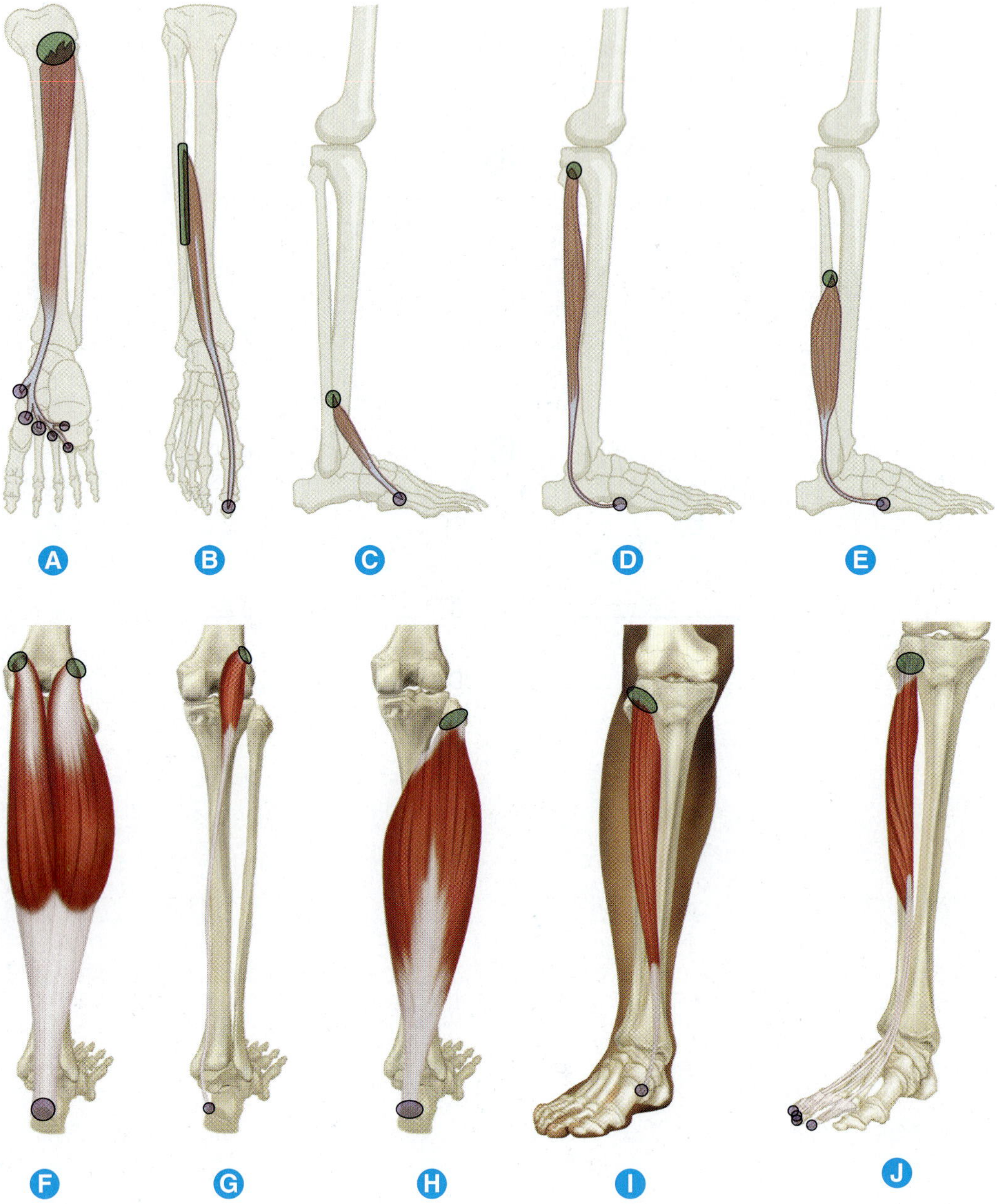

A. ______________________ B. ______________________ C. ______________________

D. ______________________ E. ______________________ F. ______________________

G. ______________________ H. ______________________ I. ______________________

J. ______________________

Workbook 10

Name ______________________ Section ________ Date ______________

The Knee

Using the word banks in each item, record the anatomical term in the blank that corresponds with what is being pointed at in the image(s) of each item.

1. Joints of the knee.

tibiofemoral joint   patellofemoral joint   tibiofibular joint

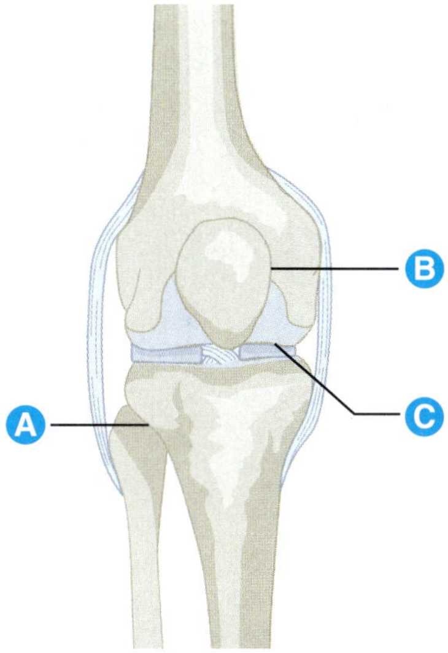

Right Knee

A. ______________ B. ______________ C. ______________

2. Motions of the knee.

flexion extension

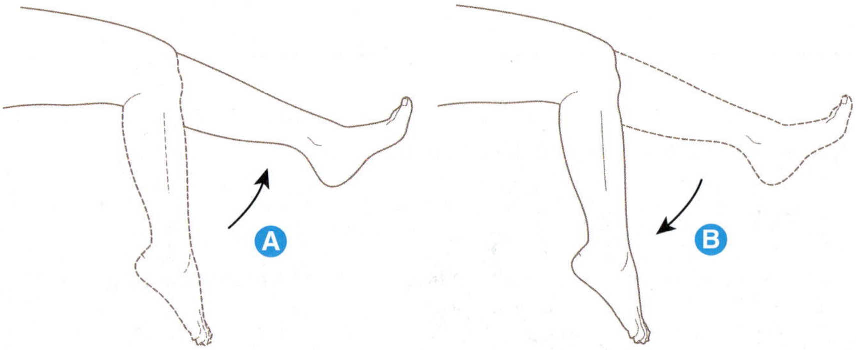

A. ______________________ B. ______________________

3. Muscles of the knee.

vastus intermedius biceps femoris rectus femoris popliteus
semimembranosus vastus medialis semitendinosus vastus lateralis

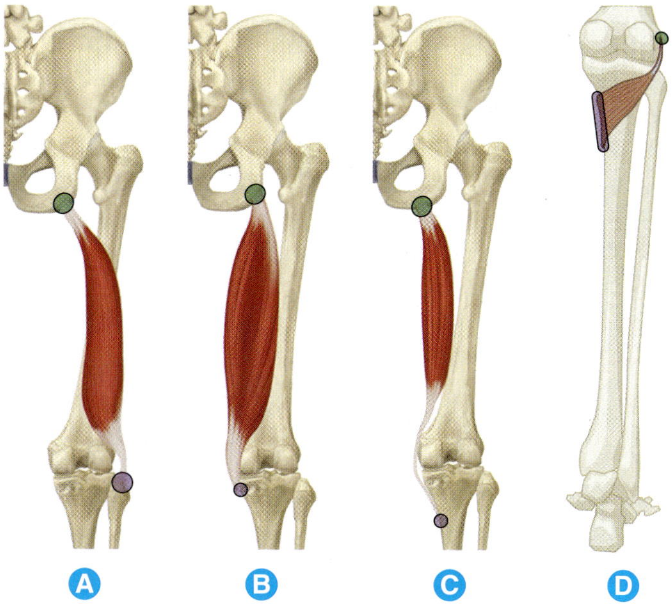

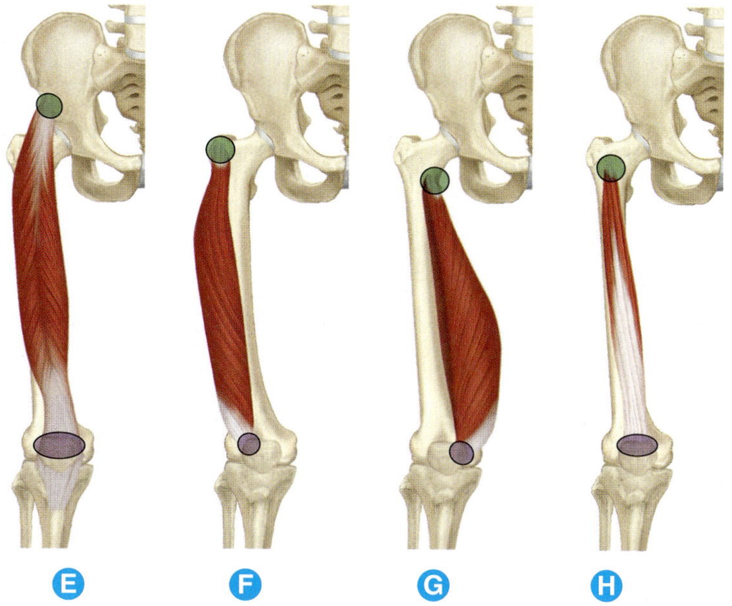

A. ____________ B. ____________ C. ____________

D. ____________ E. ____________ F. ____________

G. ____________ H. ____________

Workbook 11

Name ______________________ Section ________ Date ____________

The Hip

Using the word banks in each item, record the anatomical term in the blank that corresponds with what is being pointed at in the image(s) of each item.

1. Motions of the hip.

flexion extension abduction adduction lateral rotation medial rotation

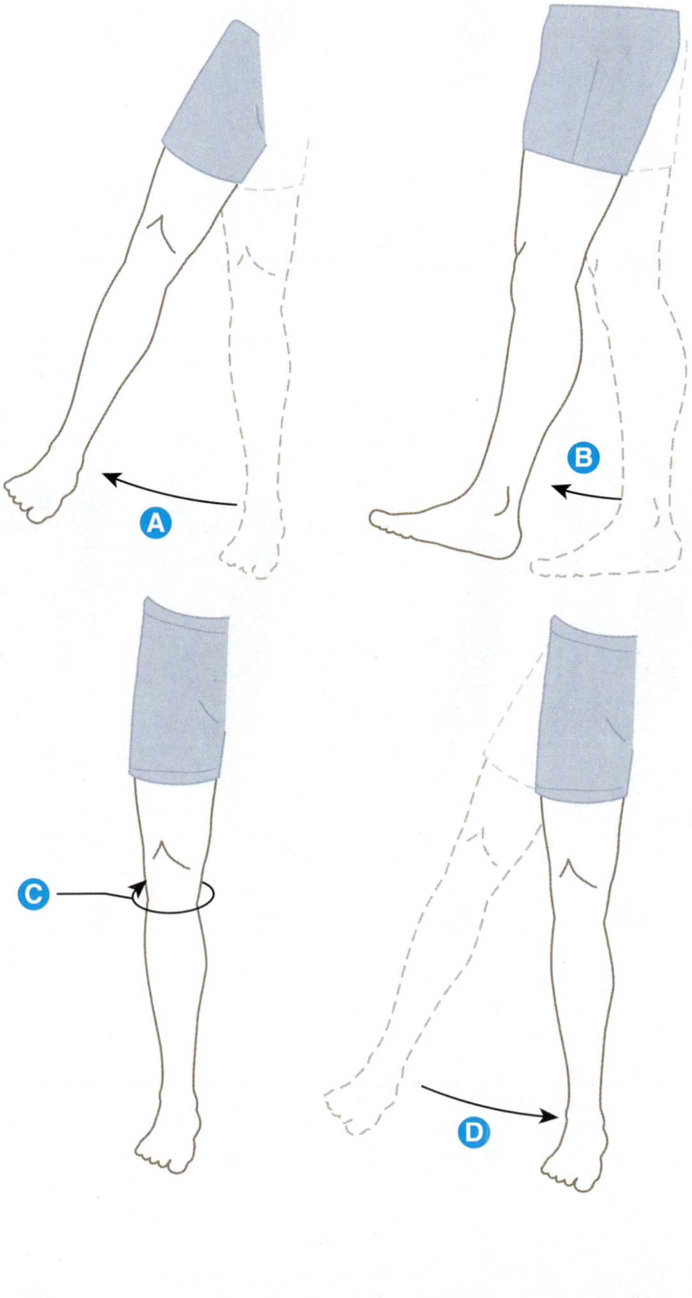

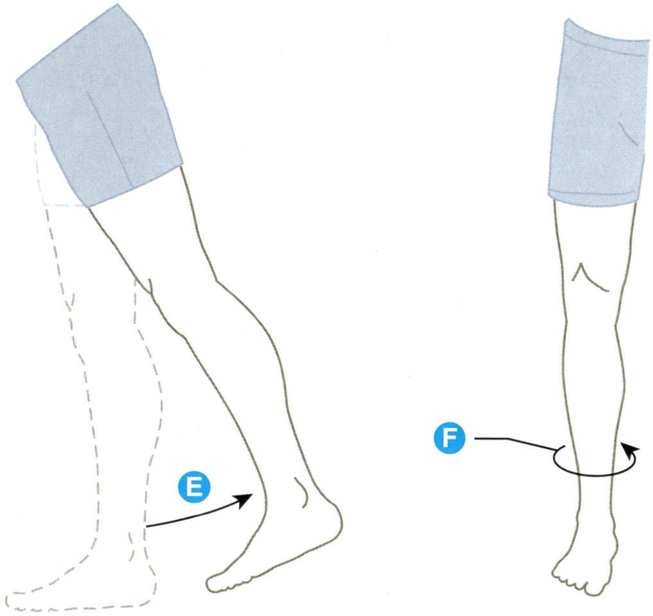

A. ____________________ B. ____________________

C. ____________________ D. ____________________

E. ____________________ F. ____________________

2. Four muscles of the hip.

sartorius psoas major iliacus tensor fasciae latae

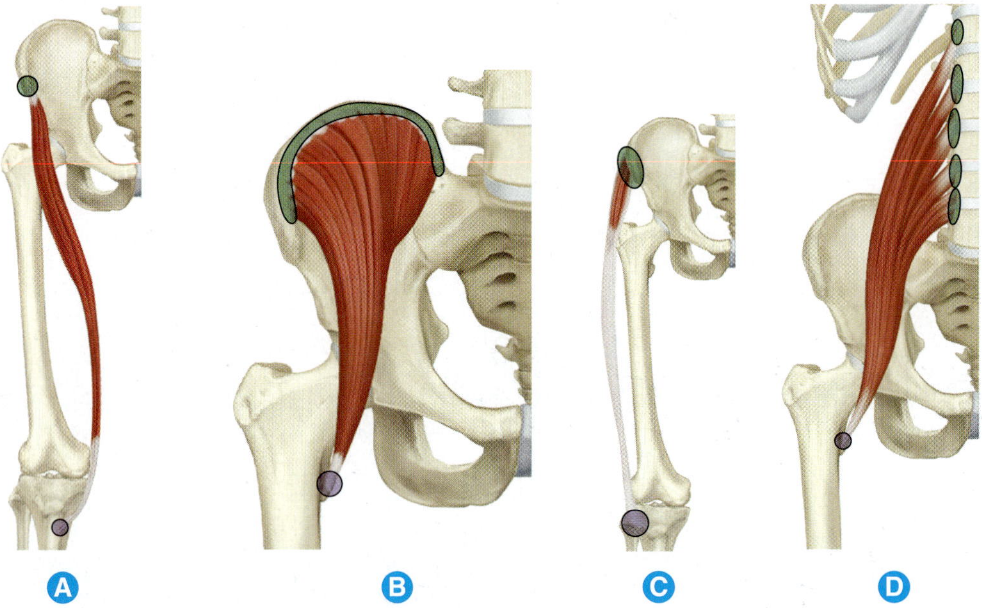

A. ____________________ B. ____________________

C. ____________________ D. ____________________

3. Four more muscles of the hip.

adductor brevis   gluteus minimus   gracilis   gluteus medius

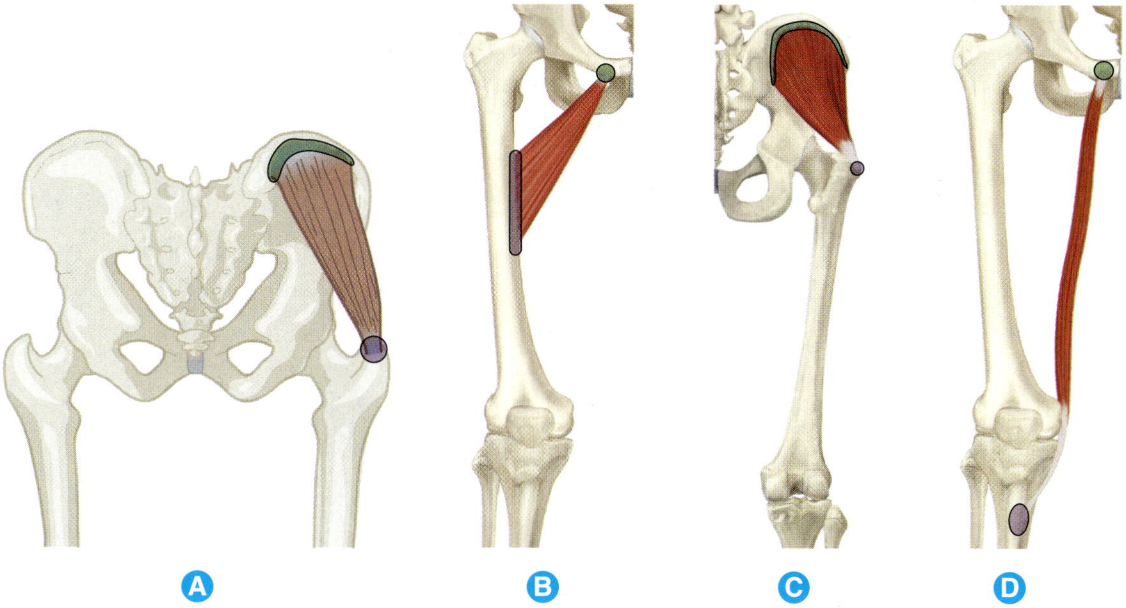

A. ______________________ B. ______________________

C. ______________________ D. ______________________

4. Five more muscles of the hip.

rectus femoris   semitendinosus   semimembranosus   biceps femoris   adductor longus

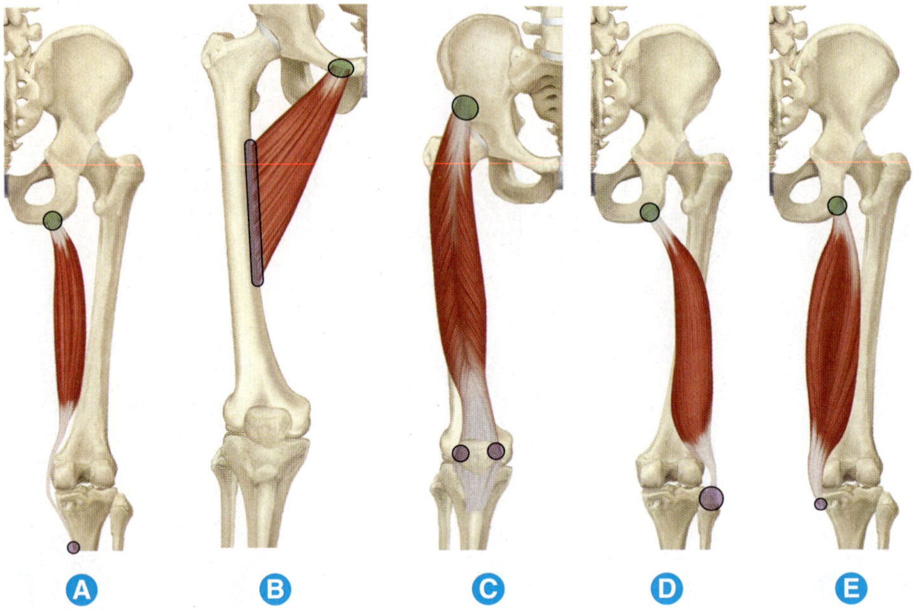

A. ____________________   B. ____________________   C. ____________________

D. ____________________   E. ____________________

5. Five more muscles of the hip.

piriformis   pectineus   obturator externus   quadratus femoris   obturator internus

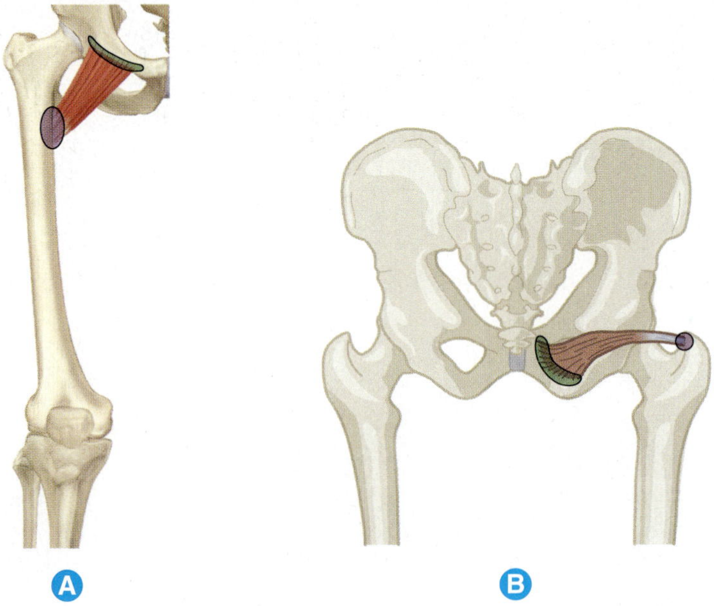

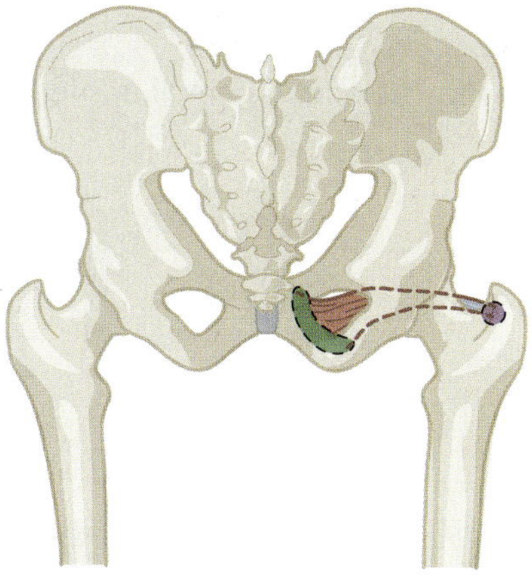

C

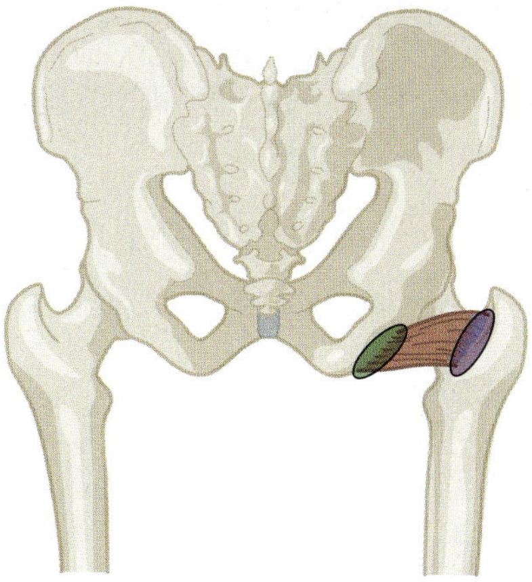

D

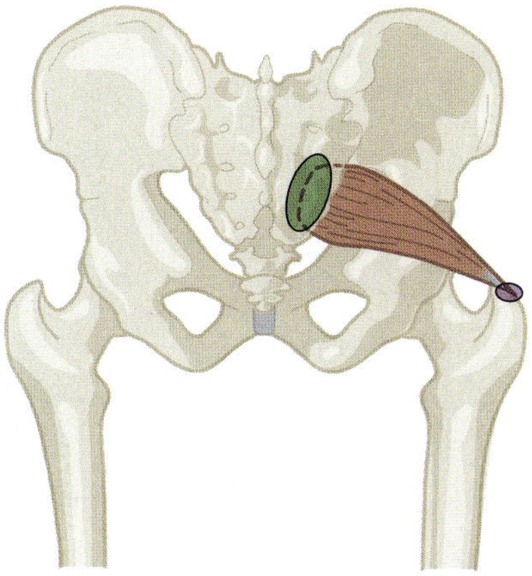

E

A. ______________________ B. ______________________ C. ______________________

D. ______________________ E. ______________________

6. The last four muscles of the hip.

gluteus maximus   adductor magnus   gemellus superior   gemellus inferior

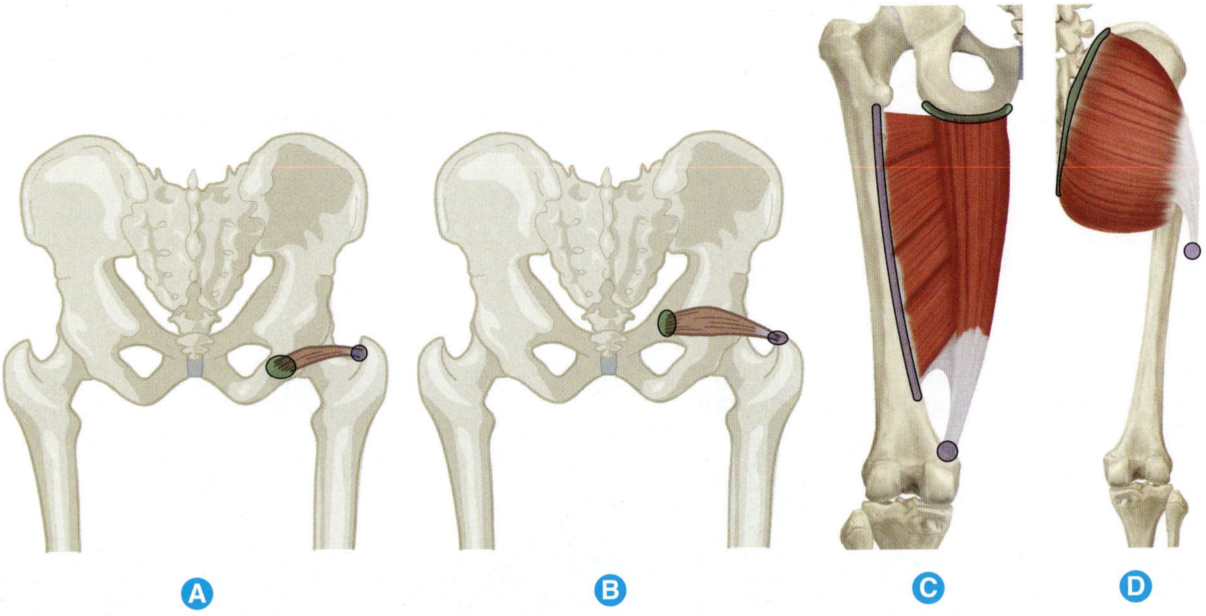

A. ____________________ B. ____________________

C. ____________________ D. ____________________

Workbook 12

Name ____________________ Section ________ Date ____________

The Trunk

Using the word banks in each item, record the anatomical term in the blank that corresponds with what is being pointed at in the image(s) of each item.

1. Joints of the trunk.

intervertebral lumbosacral

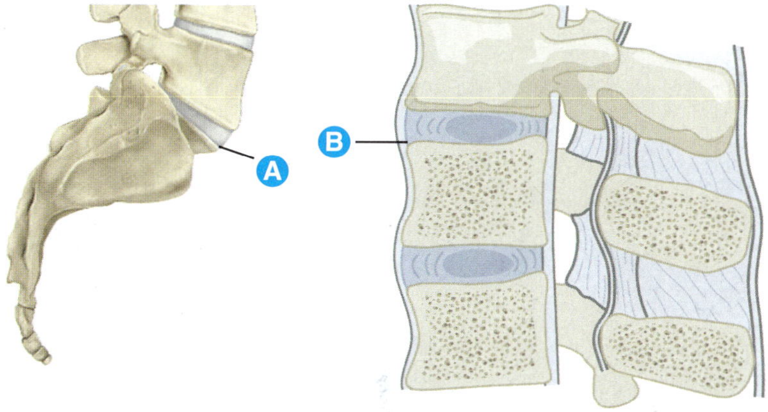

A. ____________________

B. ____________________

2. Motions of the trunk.

flexion extension rotation lateral flexion

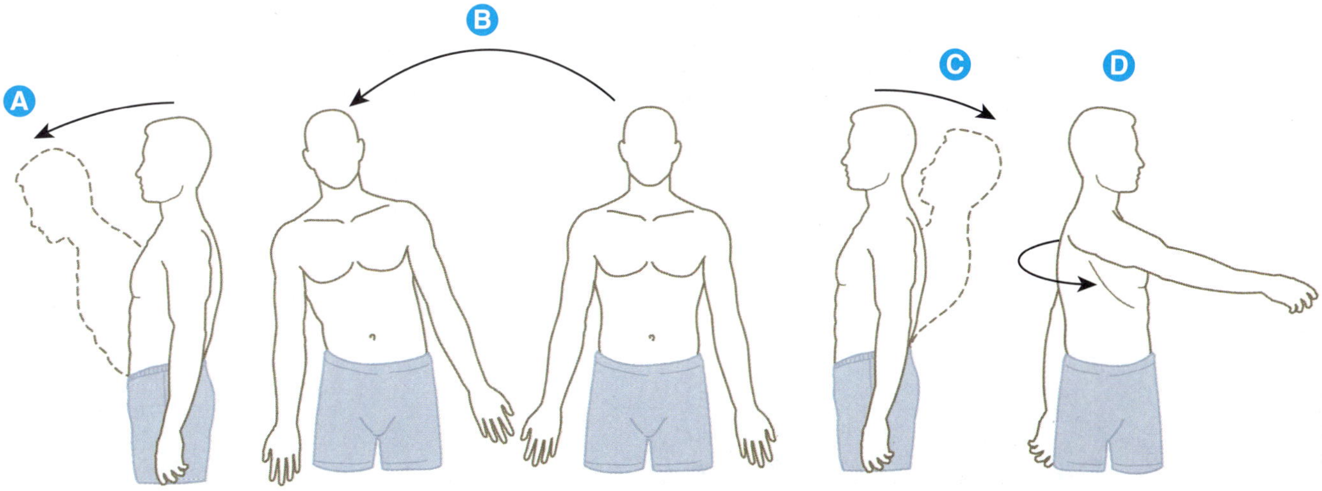

A. ____________________ B. ____________________

C. ____________________ D. ____________________

3. Muscles of the trunk.

spinalis longissimus iliocostalis rectus abdominis
external abdominal oblique internal abdominal oblique quadratus lumborum

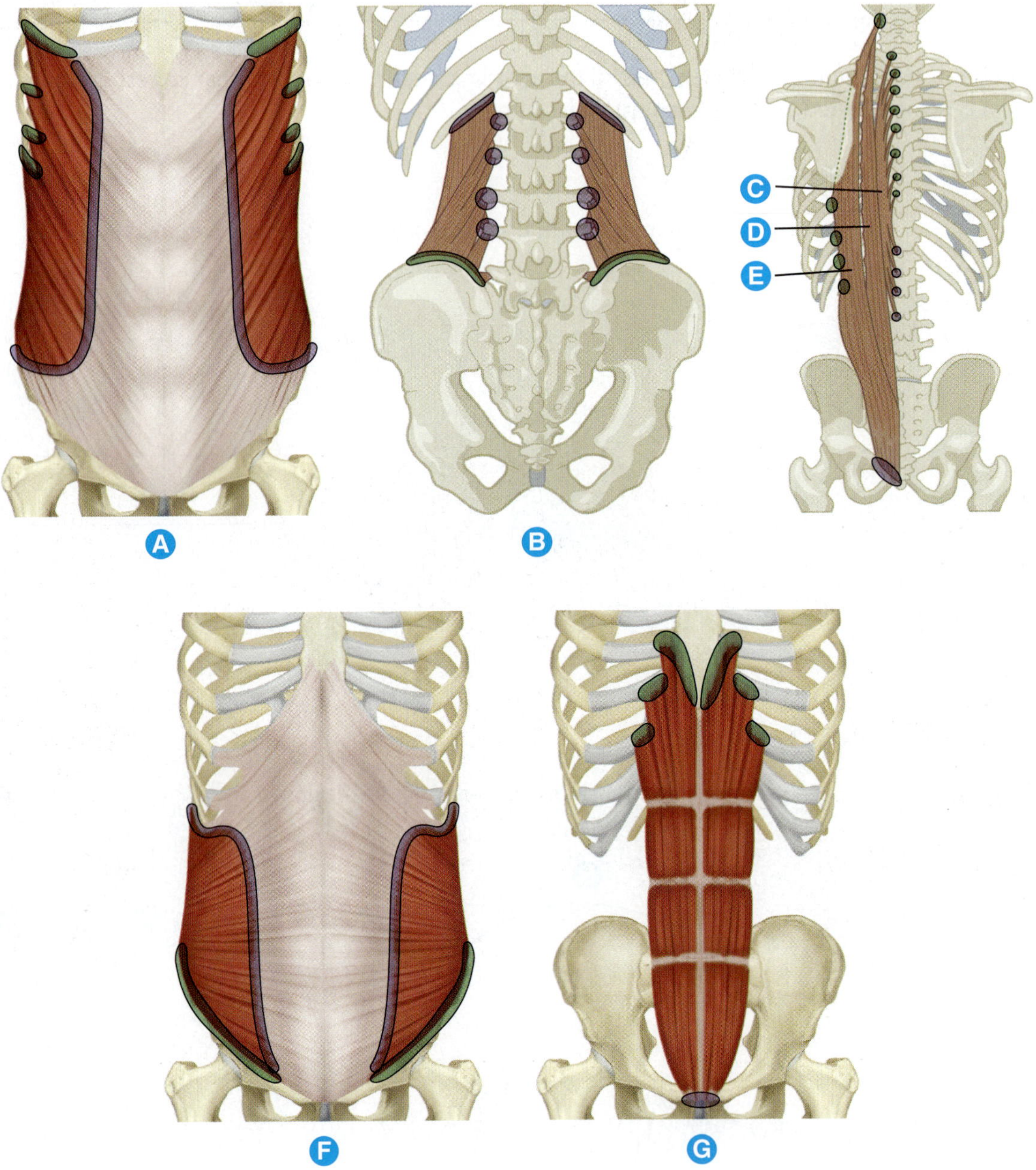

A. ______________ B. ______________ C. ______________

D. ______________ E. ______________ F. ______________

G. ______________

Workbook 13

Name ______________________ Section ________ Date ____________

The Neck

Using the word banks in each item, record the anatomical term in the blank that corresponds with what is being pointed at in the image(s) of each item.

1. Joints of the neck.

atlantoaxial joint intervertebral joints atlanto-occipital joint

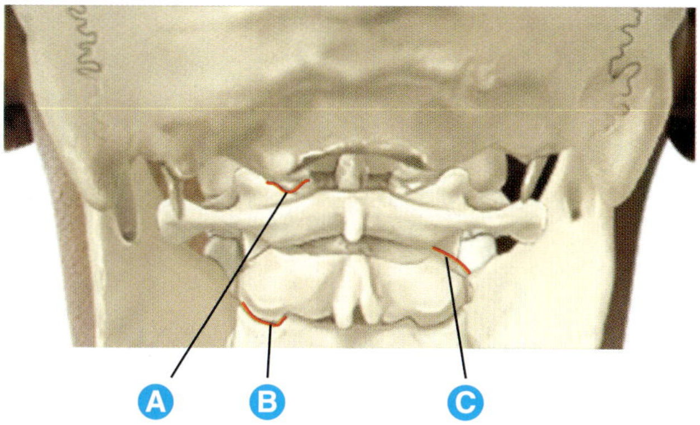

A. ______________ B. ______________ C. ______________

2. Motions of the neck.

rotation extension lateral flexion flexion

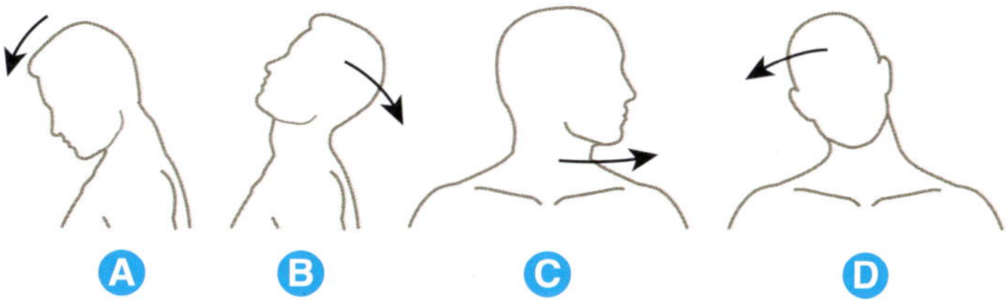

A. ______________ B. ______________

C. ______________ D. ______________

3. Muscles of the neck.

sternocleidomastoid scalenes trapezius splenius capitis splenis cervicis
semispinalis capitis semispinalis cervicis

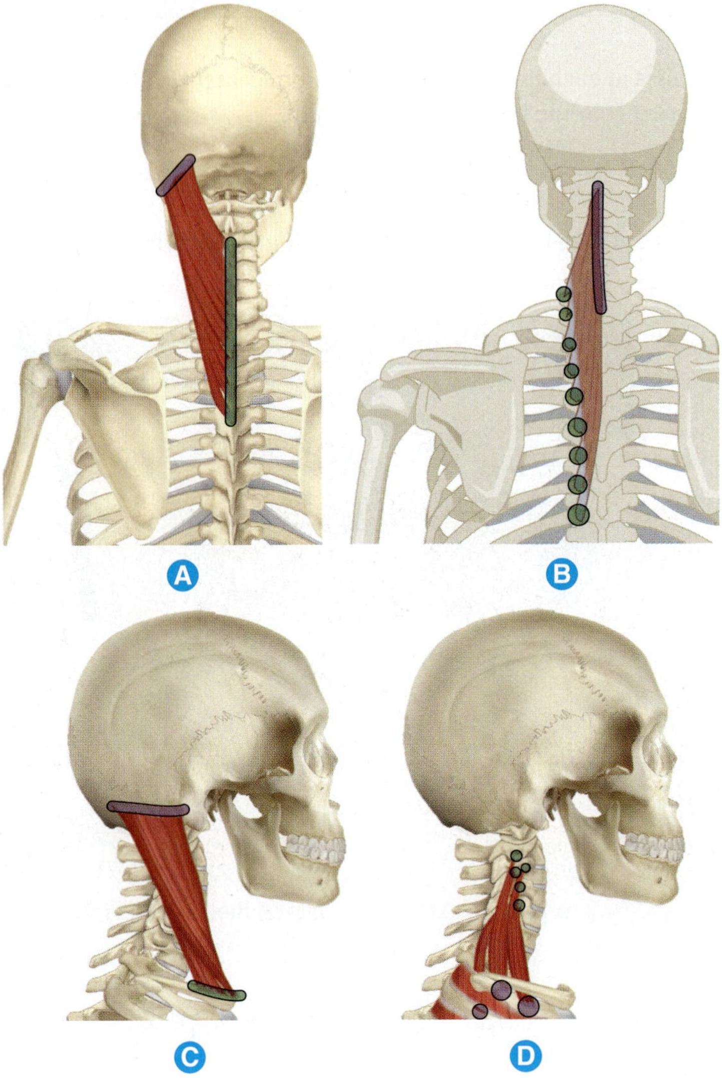

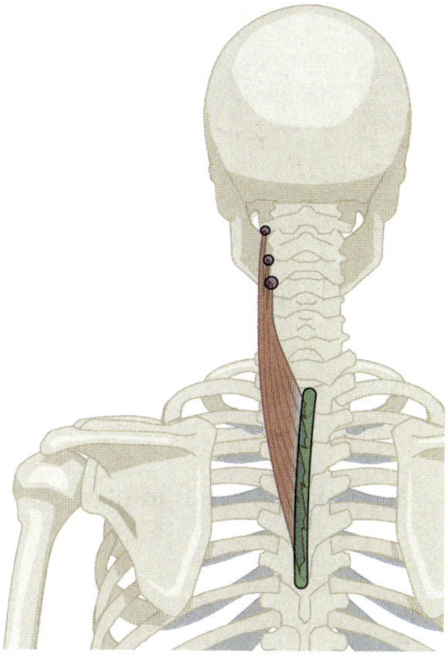

E

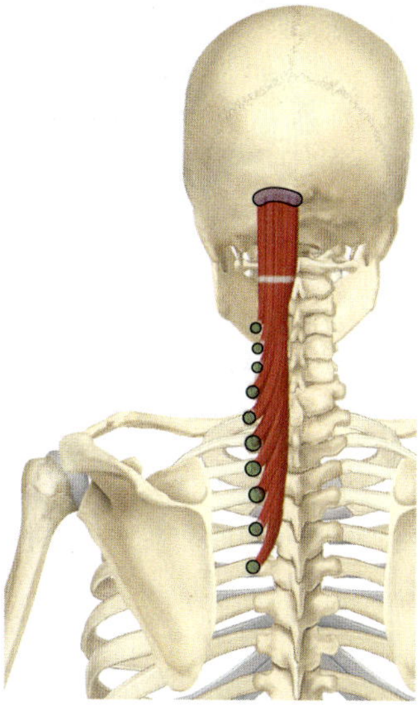

F

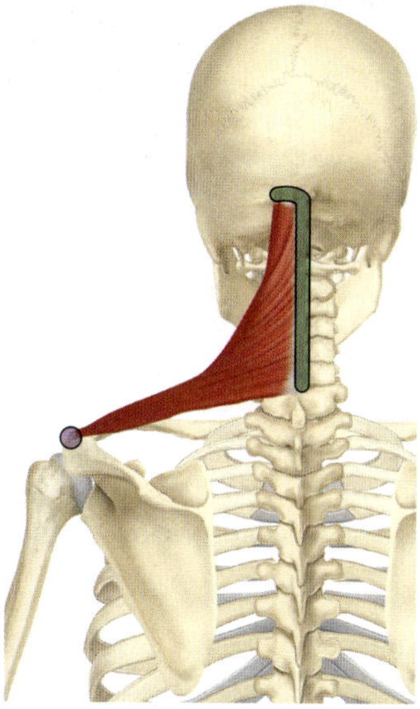

G

A. ____________________

B. ____________________

C. ____________________

D. ____________________

E. ____________________

F. ____________________

G. ____________________

Name ______________________ Section ________ Date ______________

The Shoulder Girdle

Using the word banks in each item, record the anatomical term in the blank that corresponds with what is being pointed at in the image(s) of each item.

1. Joints of the shoulder girdle.

sternoclavicular joint   scapulothoracic articulation   acromioclavicular joint

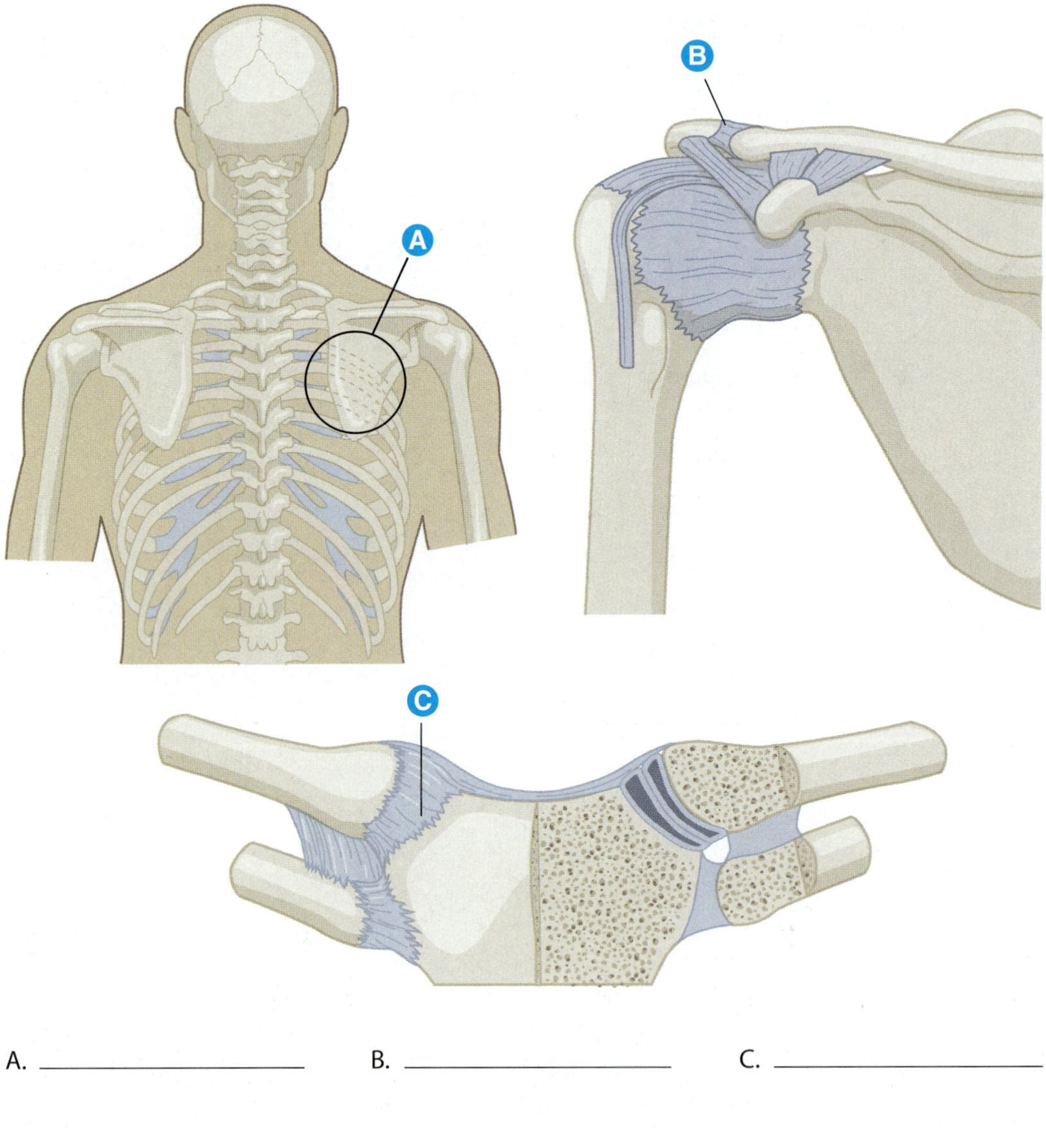

A. ______________ B. ______________ C. ______________

2. Motions of the shoulder girdle.

elevation depression retraction protraction

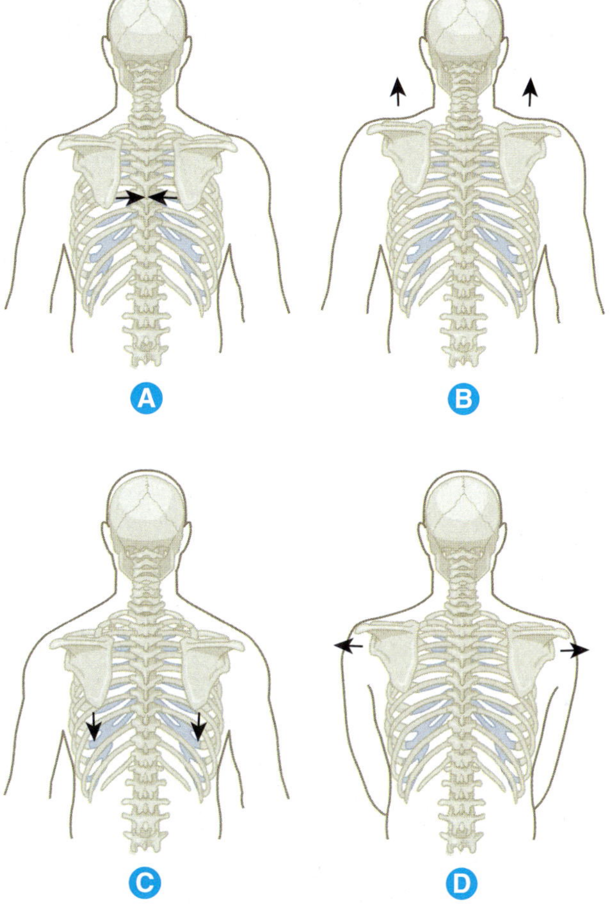

A. ____________________ B. ____________________

C. ____________________ D. ____________________

3. Muscles of the shoulder girdle.

trapezius levator scapula rhomboideus major
rhomboideus minor serratus anterior pectoralis minor

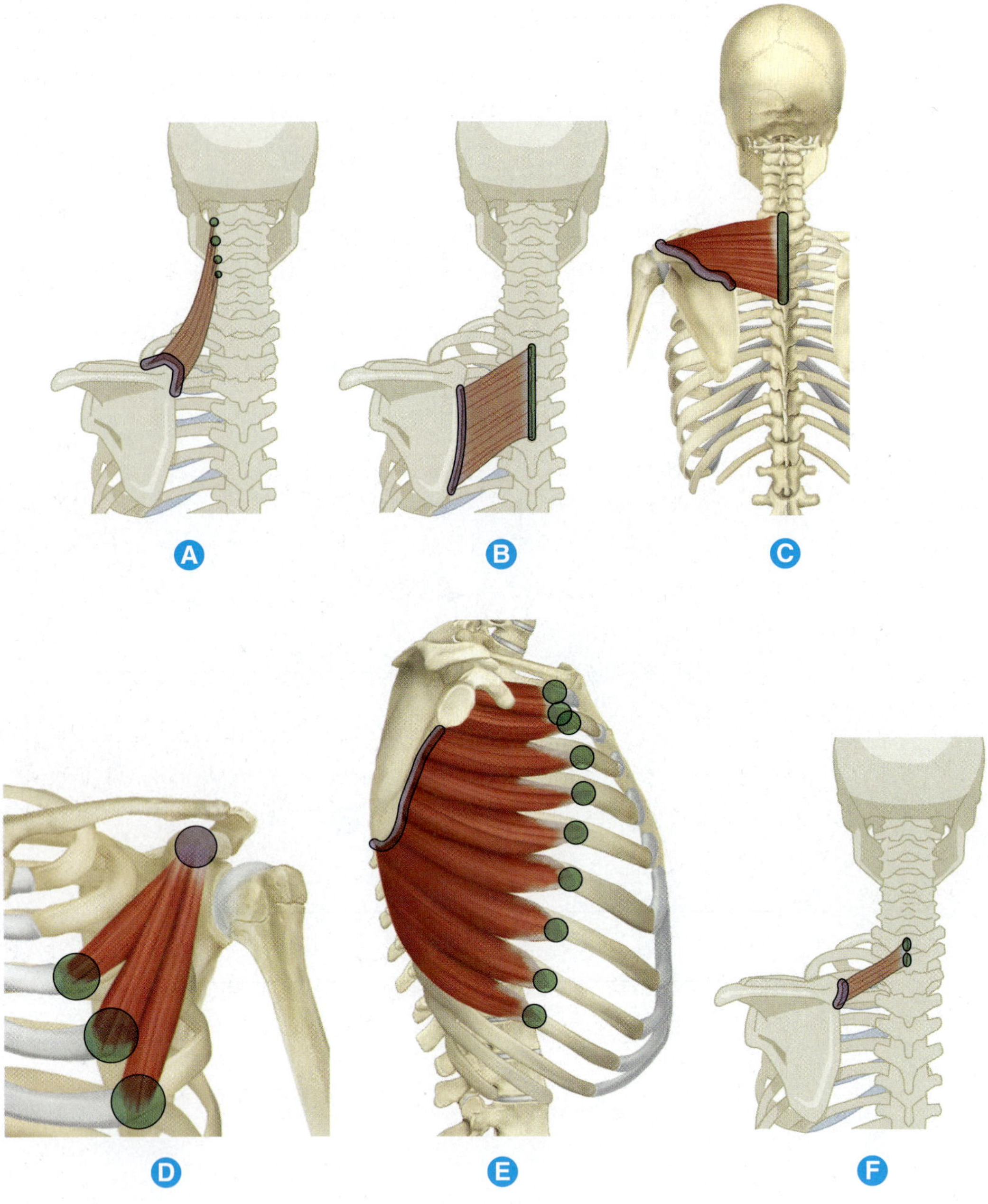

A. ______________________ B. ______________________ C. ______________________

D. ______________________ E. ______________________ F. ______________________

Workbook 15

Name ______________________________ Section ________ Date ______________

The Shoulder

Using the word banks in each item, record the anatomical term in the blank that corresponds with what is being pointed at in the image(s) of each item.

1. Motions of the glenohumeral joint.

extension flexion abduction adduction
lateral rotation medial rotation horizontal abduction horizontal adduction

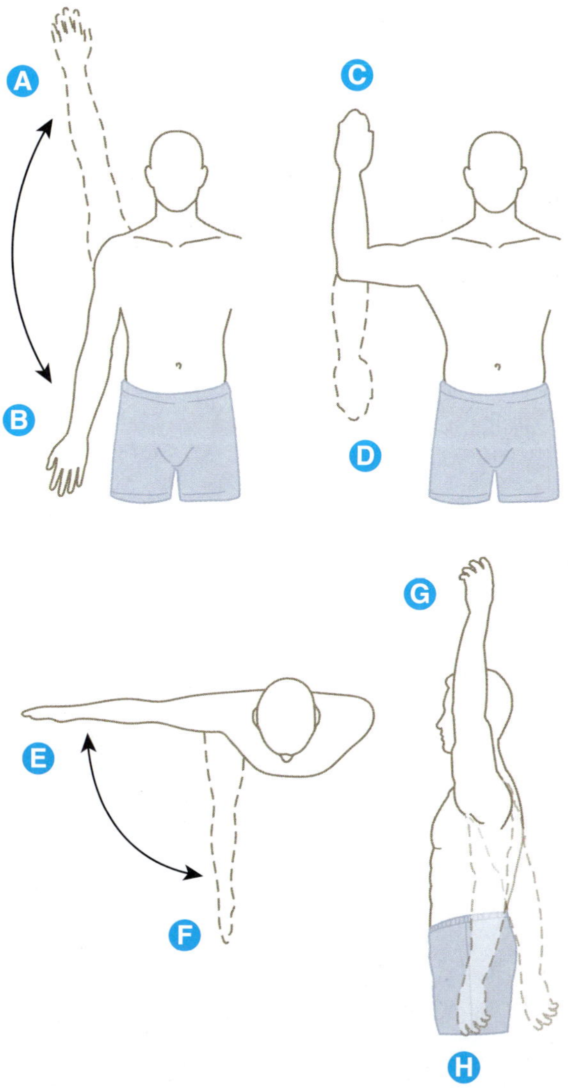

A. ______________________

B. ______________________

C. ______________________

D. ______________________

E. ______________________

F. ______________________

G. ______________________

H. ______________________

2. Muscles of the shoulder.

deltoid pectoralis major coracobrachialis biceps brachii triceps brachii
supraspinatus latissimus dorsi teres major infraspinatus teres minor subscapularis

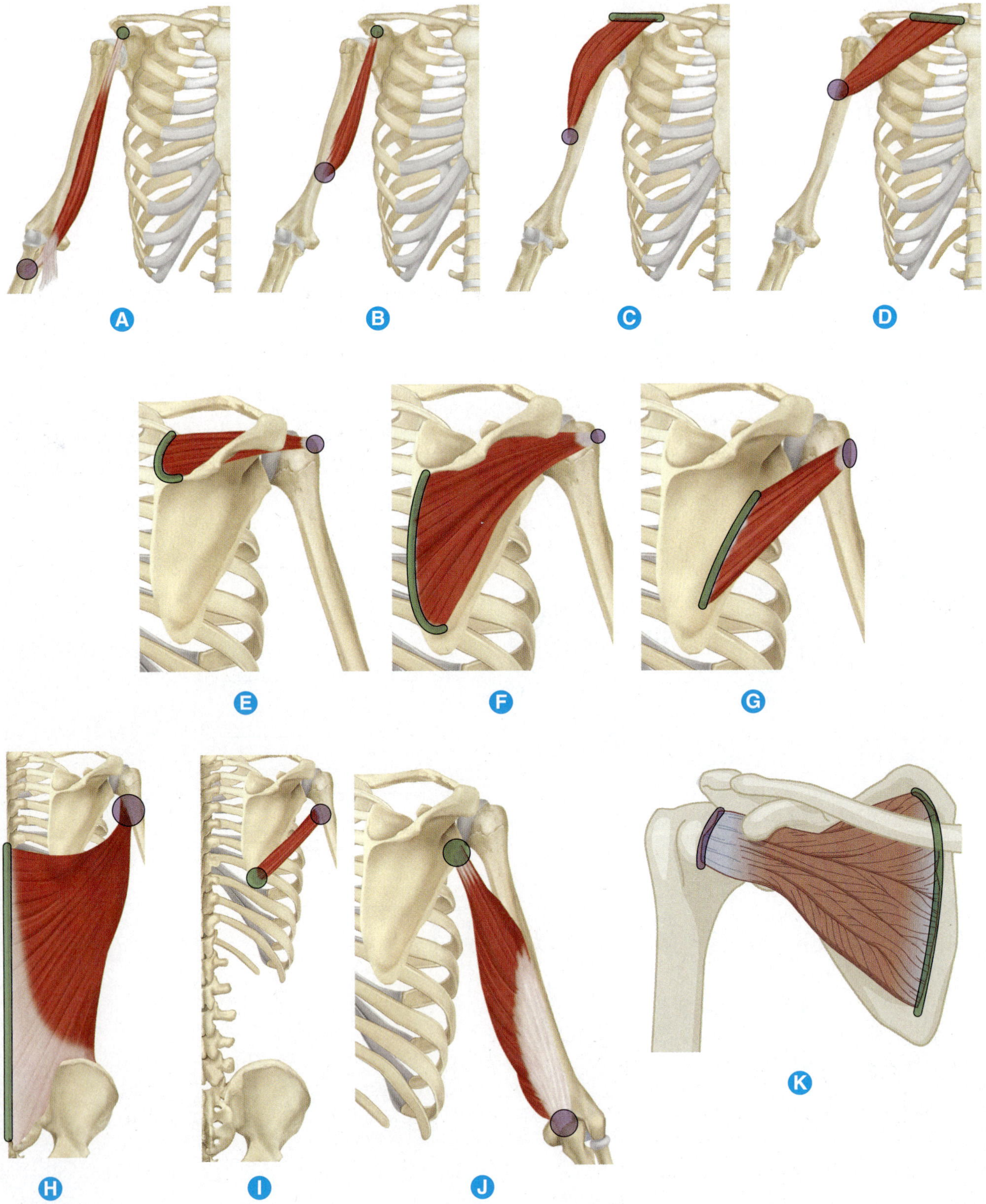

A. ______________________ B. ______________________ C. ______________________

D. ______________________ E. ______________________ F. ______________________

G. ______________________ H. ______________________ I. ______________________

J. ______________________ K. ______________________

Workbook 16

Name ______________________ Section ________ Date ____________

The Elbow

Using the word banks in each item, record the anatomical term in the blank that corresponds with what is being pointed at in the image(s) of each item.

1. Joints of the elbow.

humeroradial joint humeroulnar joint radioulnar joint

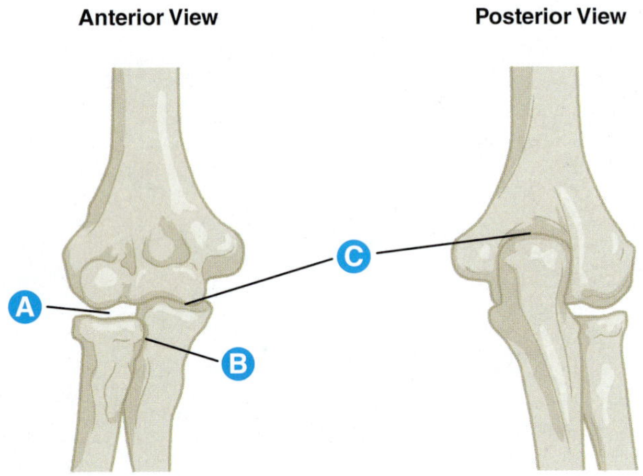

A. ____________________

B. ____________________

C. ____________________

2. Motions of the elbow.

flexion extension pronation supination

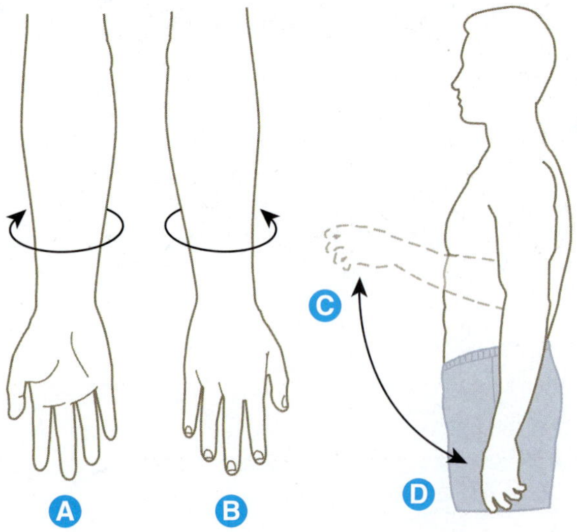

A. ____________________

B. ____________________

C. ____________________

D. ____________________

3. Muscles of the elbow.

biceps brachii   brachialis   brachioradialis   triceps brachii
pronator teres   pronator quadratus   supinator

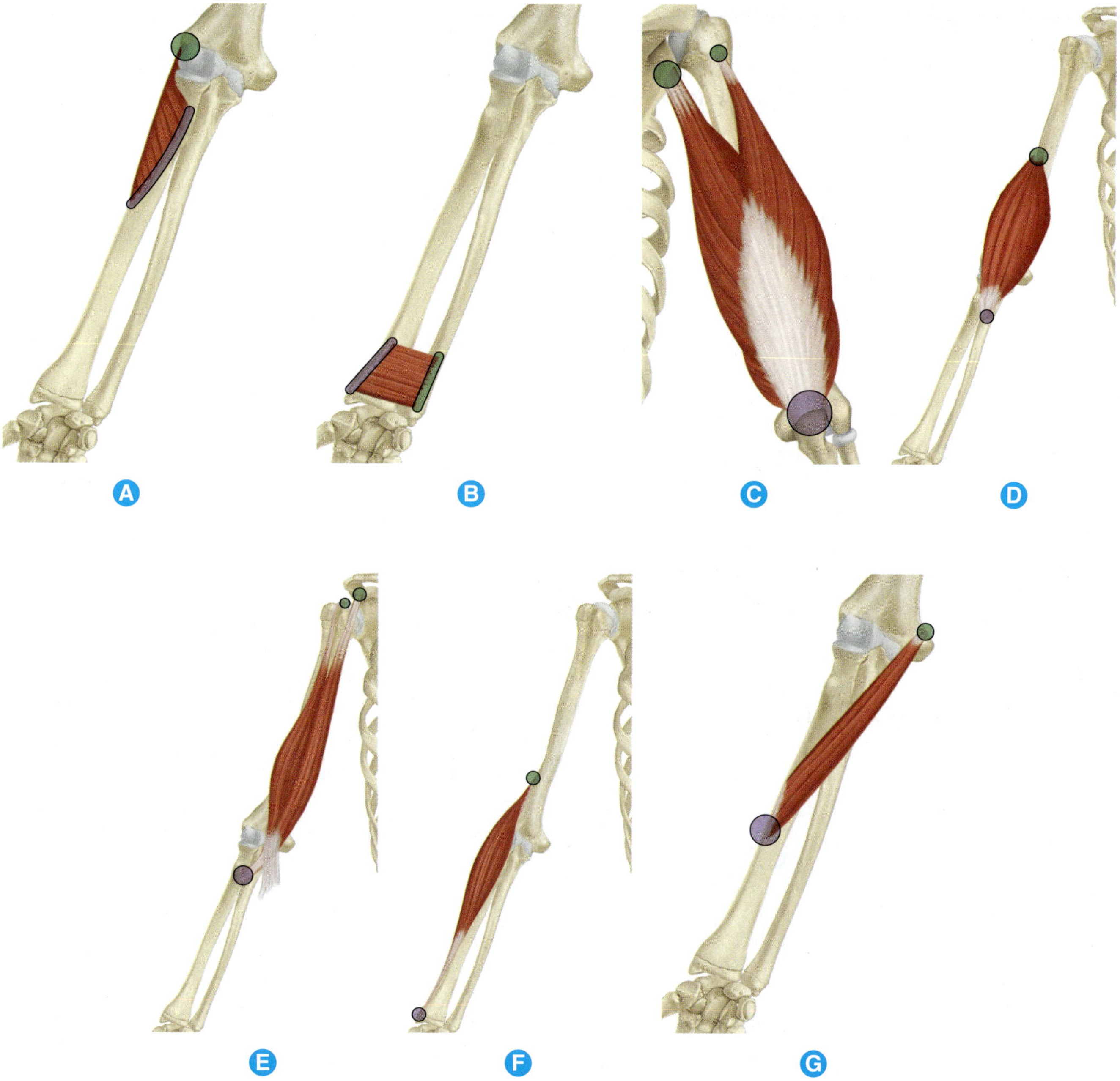

A. ____________________   B. ____________________   C. ____________________

D. ____________________   E. ____________________   F. ____________________

G. ____________________

Name ______________________ Section ________ Date ____________

The Wrist

Using the word banks in each item, record the anatomical term in the blank that corresponds with what is being pointed at in the image(s) of each item.

1. Joints of the wrist.

distal radioulnar joint   radiocarpal joint (radioulnar-carpal joint)

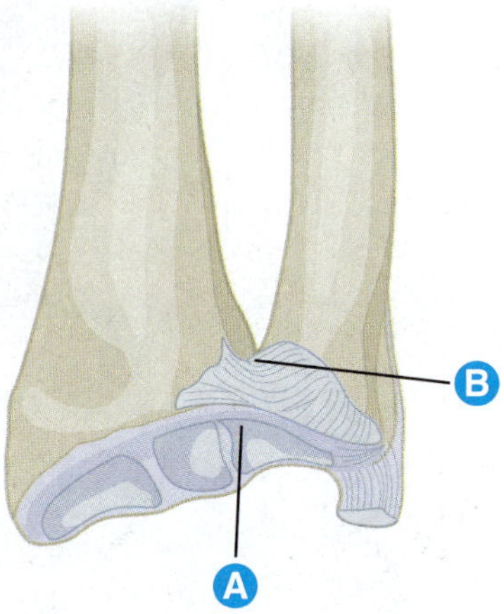

A. ____________________

B. ____________________

2. Motions of the wrist.

flexion   extension   radial deviation   ulnar deviation

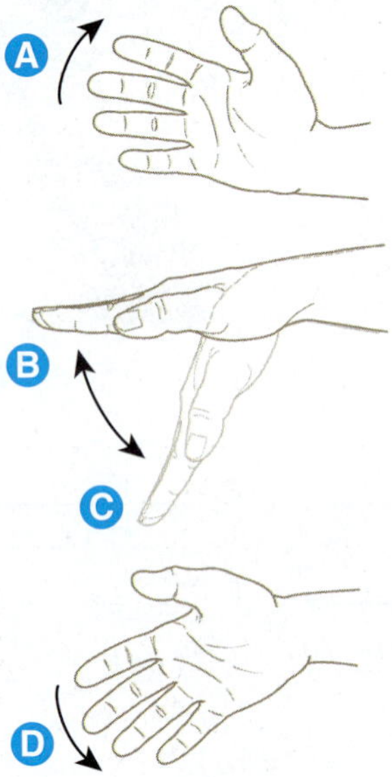

A. ____________________

B. ____________________

C. ____________________

D. ____________________

3. Muscles of the wrist.

flexor carpi radialis   palmaris longus   flexor carpi ulnaris
extensor carpi radialis brevis   extensor carpi radialis longus   extensor carpi ulnaris

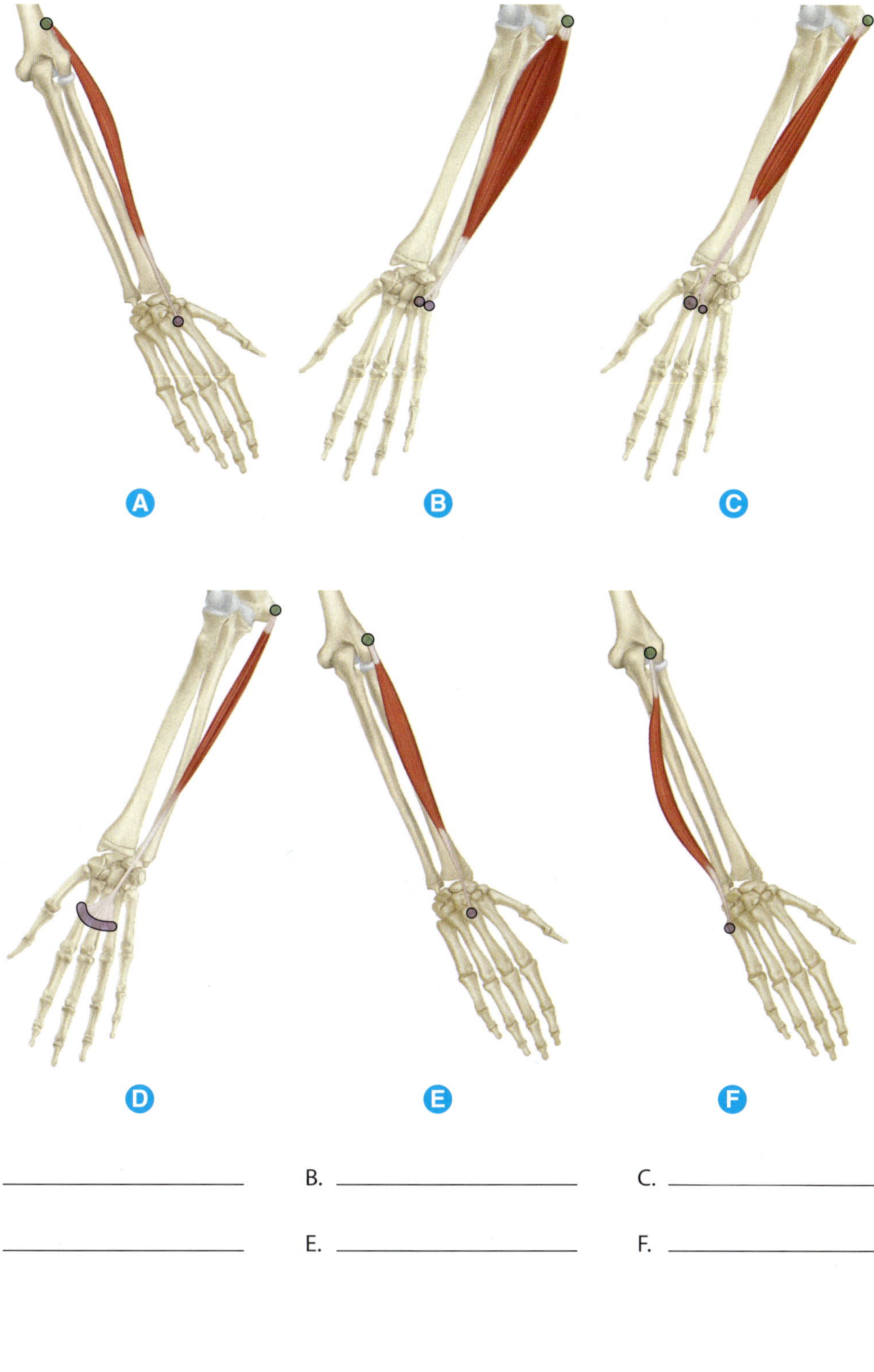

A. ____________________ B. ____________________ C. ____________________

D. ____________________ E. ____________________ F. ____________________

Name ______________________ Section __________ Date ______________

The Hand

Using the word banks in each item, record the anatomical term in the blank that corresponds with what is being pointed at in the image(s) of each item.

1. Joints of the hand.

carpometacarpal joint   metatarsophalangeal joint
interphalangeal joint   proximal interphalangeal joint   distal interphalangeal joint

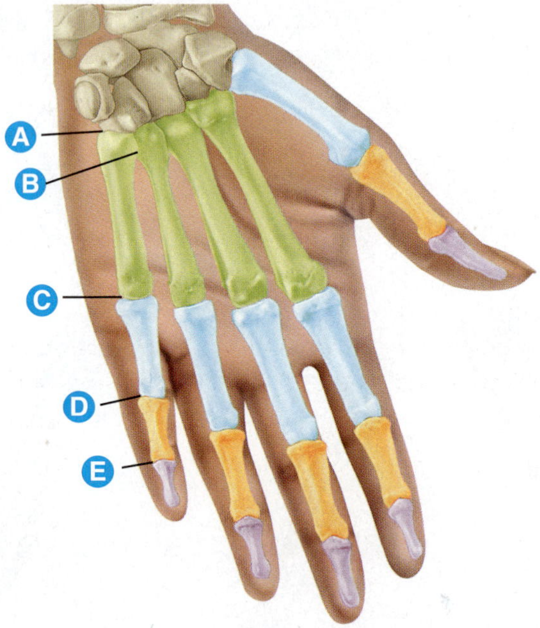

A. ______________________ B. ______________________ C. ______________________

D. ______________________ E. ______________________

2. Motions of the fingers.

flexion extension abduction adduction opposition reposition

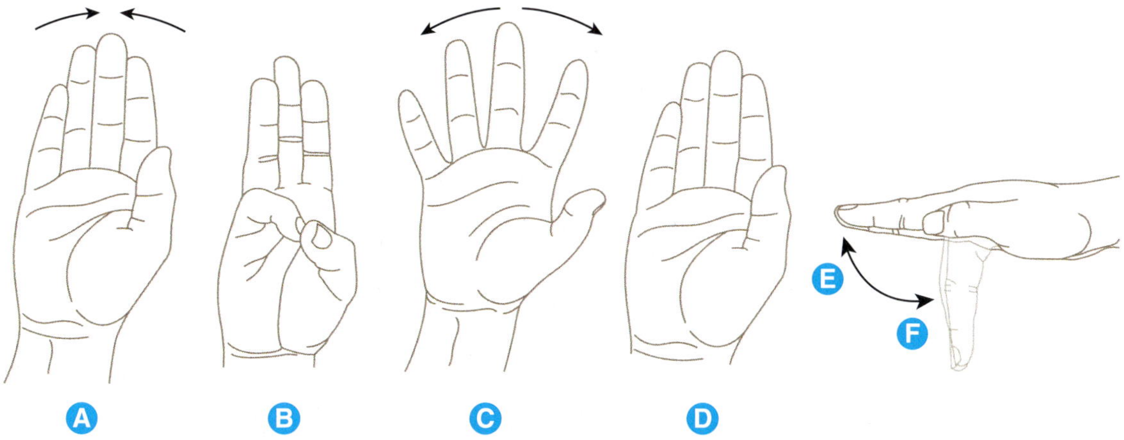

A. ____________________ B. ____________________ C. ____________________

D. ____________________ E. ____________________ F. ____________________

3. Muscles that flex fingers.

flexor pollicis brevis flexor pollicis longus flexor digitorum superficialis
flexor digitorum profundus flexor digiti minimi lumbricals

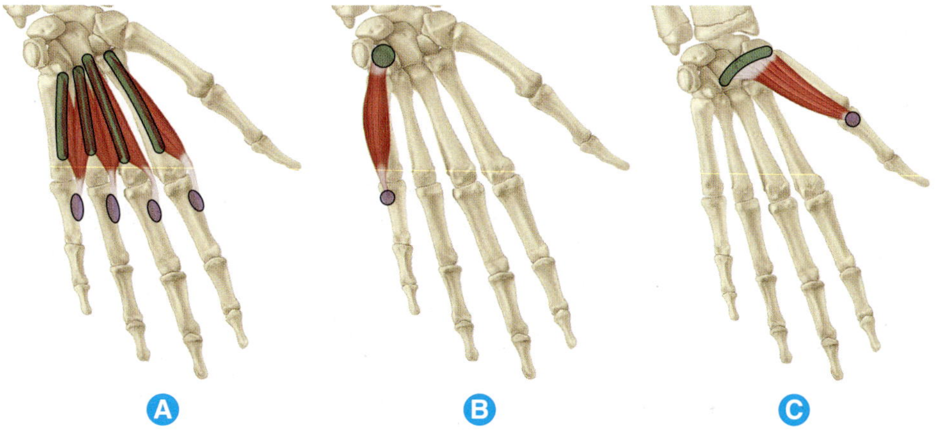

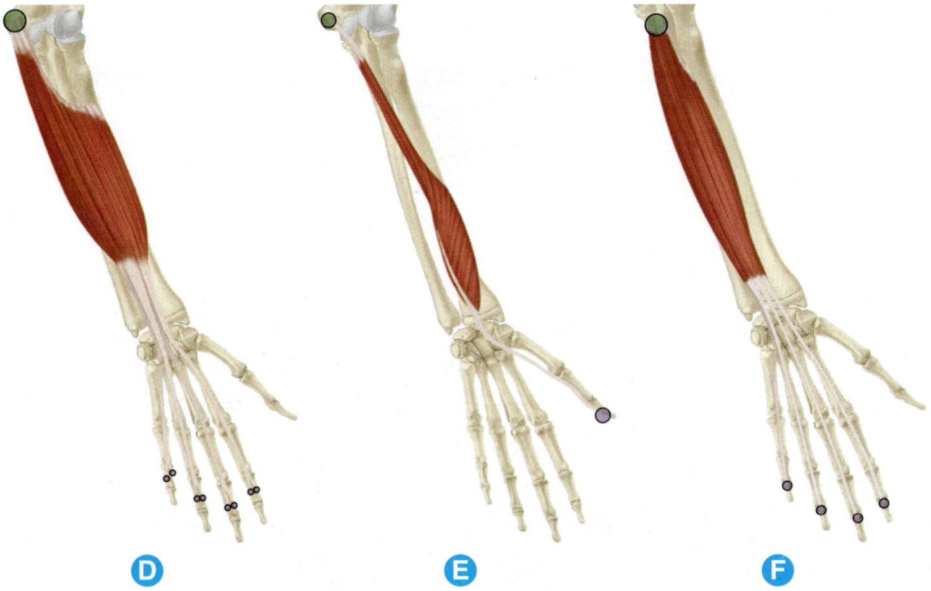

A. ______________________ B. ______________________ C. ______________________

D. ______________________ E. ______________________ F. ______________________

4. Muscles that extend fingers.

extensor pollicis brevis extensor pollicis longus extensor indicis
extensor digitorum extensor digiti minimi

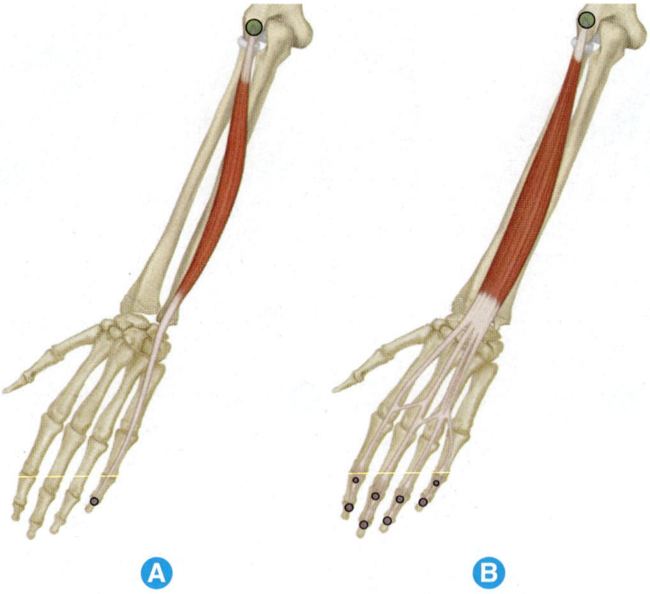

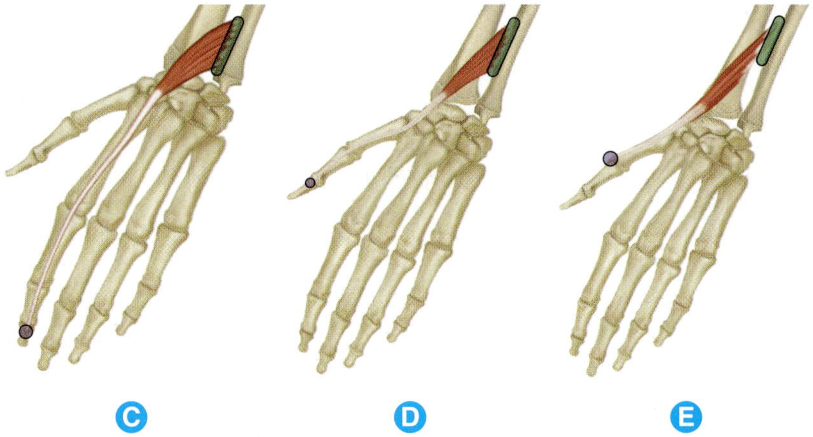

A. ______________________ B. ______________________ C. ______________________

D. ______________________ E. ______________________

5. Muscles that abduct fingers.

abductor pollicis brevis   abductor pollicis longus   abductor digiti minimi   dorsal interossei

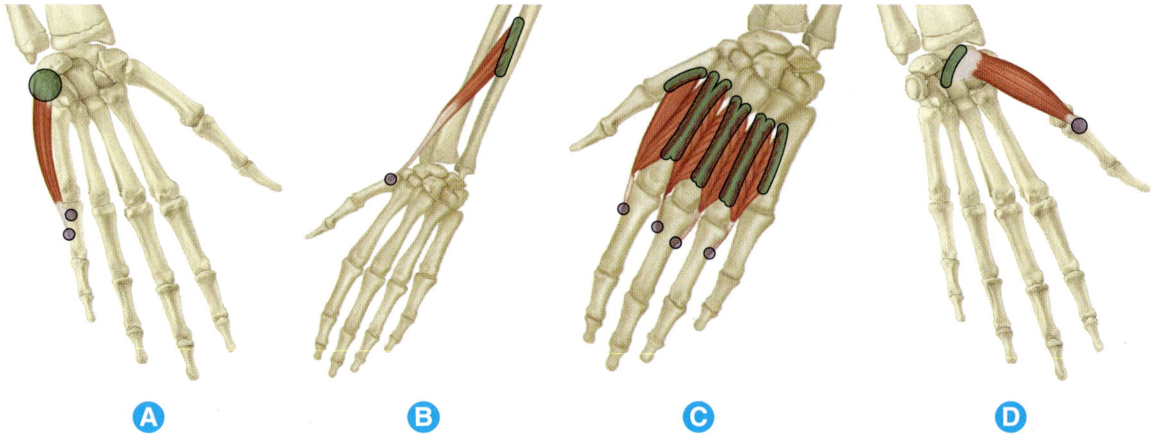

A. ______________________ B. ______________________

C. ______________________ D. ______________________

6. Muscles that adduct and oppose the fingers.

adductor pollicis palmar interossei opponens pollicis opponens digiti minimi

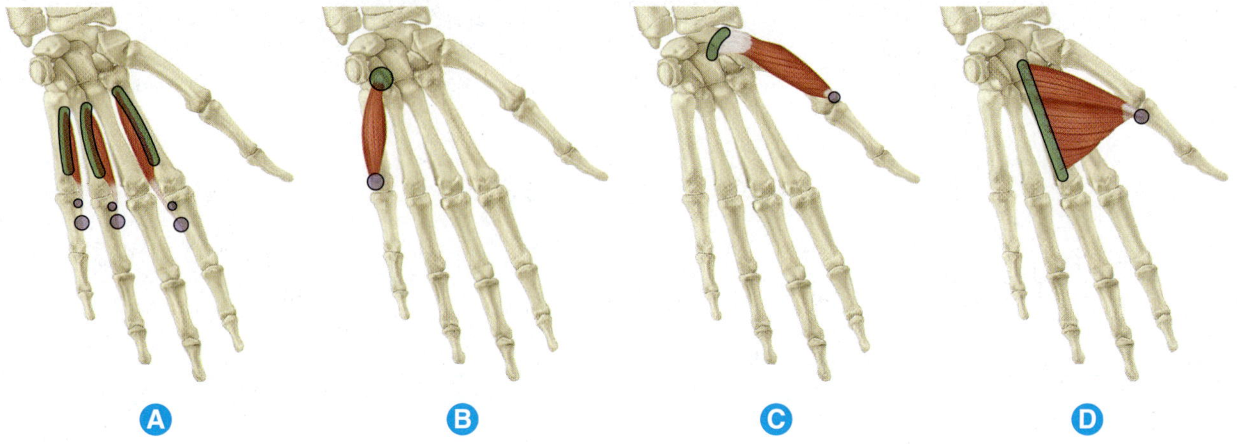

A. ______________ B. ______________

C. ______________ D. ______________

© BraunS/E+/Getty Images

APPENDIX

Bibliography

1. Agur, A. M. R. (1991). *Grant's atlas of anatomy* (9th ed.). Baltimore, MD: Williams, & Wilkins.
2. American Academy of Orthopedic Surgeons. (1991). *Athletic training and sports medicine* (2nd ed.). Rosemont, IL: AAOS.
3. Anderson, M. K., Hall, S. J., & Martin, M. (2000). *Sports injury management* (2nd ed.). Philadelphia, PA: Lippincott, Williams, & Wilkins.
4. Clarkson, H. M. (2000). *Musculoskeletal assessment: Joint range of motion and manual muscle strength* (2nd ed.). Philadelphia, PA: Lippincott, Williams, & Wilkins.
5. Ehrlich, A. & Schroeder, C. L. (2009). *Introduction to medical terminology* (2nd ed.). Clifton Park, NY: Delmar Learning.
6. Floyd, R. T. (2012). *Manual of structural kinesiology* (18th ed.). New York, NY: McGraw-Hill.
7. Hall, S. (2012). *Basic biomechanics* (6th ed.). New York, NY: McGraw-Hill.
8. Hamill, J. & Knutzen, K. M. (2009). *Biomechanical basis of human movement* (3rd ed.). Philadelphia, PA: Lippincott, Williams, & Wilkins.
9. Hamilton, N., Weimar, W., & Luttgens, K. (2012). *Kinesiology: Scientific basis for human movement* (12th ed.). New York, NY: McGraw-Hill.
10. Houglum, P. A. & Bertoti, D. B. (2012). *Brunnstrom's clinical kinesiology* (6th ed.). Philadelphia, PA: FA Davis.
11. Hole, J. W. (1990). *Human anatomy and physiology* (5th ed.). Dubuque, IA: Wm. C. Brown Publishers.
12. Hoppenfeld, S. (1976). *Physical examination of the spine and extremities*. Norwalk, CT: Appleton & Lange.
13. Kenney, W. L., Wilmore, J. H., & Costill, D. L. (2015). *Physiology of sport and exercise*. Champaign, IL: Human Kinetics.
14. Lippert, L. S. (2011). *Clinical kinesiology and anatomy* (5th ed.). Philadelphia, PA: F.A. Davis.
15. Loudon, J. K., Manske, R. C., & Reiman, M. P. (2013). *Clinical mechanics and kinesiology*. Champaign, IL: Human Kinetics.
16. Magee, D. J. (1992). *Orthopedic physical assessment*. Philadelphia, PA: W. B. Saunders Company.
17. Marieb, E. N. (2009). *Anatomy and physiology coloring book* (9th ed.). San Francisco, CA: Pearson Benjamin Cummings.
18. Martini, F. H. & Nath, J. L. (2009). *Fundamentals of anatomy and physiology* (8th ed.). San Francisco, CA: Pearson Benjamin Cummings.
19. Muscolino, J. E. (2017). *The muscular system manual* (4th ed.). St. Louis, MO: Elsevier.
20. Norkin, C. C. & White, D. J. (2016). *Measurement of joint motion: A guide to goniometry* (5th ed.). Philadelphia, PA: FA Davis.
21. Powers, S. K. & Howley, E. T. (2012). *Exercise physiology: Theory and application to fitness and performance* (8th ed.). Boston, MA: McGraw-Hill.
22. Prentice, W. E. (2009). *Arnheim's principles of athletic training: A competency based approach* (13th ed.). New York, NY: McGraw-Hill.
23. Saladin, K. S. (2012). *Anatomy & physiology: The unity of form and function* (6th ed.). New York, NY: McGraw-Hill.
24. Starkey, C. & Ryan, J. (2002). *Evaluation of orthopedic and athletic injuries* (2nd ed.). Philadelphia, PA: FA Davis.
25. Tortora, G. J. & Derrickson, B. (2012). *Principles of anatomy & physiology* (13th ed.). Hoboken, NJ: John Wiley & Sons.
26. Venes, D. (Ed.) (2017). *Taber's cyclopedic medical dictionary* (23rd ed.). Philadelphia, PA: FA Davis.

Glossary

A

Abductor digiti minimi Muscle which lies along the lateral border of the foot or hand.

Abductor pollicis brevis The largest muscle of the thenar eminence muscle group.

Abductor pollicis longus An extrinsic forearm muscle that acts on the thumb (digit I).

Acromioclavicular (AC) joint A joint at the top of the shoulder. It is the junction between the acromion (part of the scapula that forms the highest point of the shoulder) and the clavicle.

Acromioclavicular ligament This ligament is a quadrilateral band, covering the superior part of the articulation, and extending between the upper part of the lateral end of the clavicle and the adjoining part of the upper surface of the acromion.

Adductor brevis Muscle that lies inferior to the pectineus in the groin. It originates on the body and inferior ramus of the pubis and inserts on the pectineal line and the proximal end of the linea aspera.

Adductor longus Muscle that lies just inferior to the adductor brevis muscle. It originates on the body and inferior ramus of the pubis just inferior to the adductor brevis origin. It then passes through the groin and inserts on the middle one-third of the linea aspera.

Adductor magnus A large triangular muscle situated on the thigh.

Adductor pollicis Muscle that acts on the thumb and whose fibers pass transversely across the palmar hand from the pollicis muscles; it acts on the middle of the hand to the thumb.

Afferent A term that means, literally, "carry towards."

Agonist Sometimes called primary mover, this describes the muscle or group of muscles primarily responsible for a given motion.

Amphiarthroses Joints which are slightly movable.

Anatomical kinesiology The study of human anatomy and body movement in relation to health and wellness.

Anatomical position The description of any part of the body in a specific stance.

Annular ligament A strong band of fibers that encircles the head of the radius.

Antagonists Muscles that regulate motion from the opposite side of the agonists and serve as the brake when it is time to slow or stop the motion.

Anterior The area toward the head of the body.

Anterior and middle scalene Two muscles that lie deep in the posterolateral neck and insert on the first rib.

Anterior atlanto-occipital membrane Located in the front of the spine, a membranous ligament that stabilizes the atlanto-occipital joint.

Anterior compartment Contains the muscles that dorsiflex the ankle.

Anterior cruciate ligament One of four key ligaments that help stabilize knee joint, it extends from the inferior posterior femur to the superior anterior intercondylar eminence of the tibia. Its purpose is to restrict anterior translation of the tibia from the femur.

Anterior deltoid A large triangular muscle occupying the upper arm.

Anterior longitudinal ligament A ligament that runs down the anterior surface of the spine.

Anterior ramus One of the primary branches of a spinal nerve.

Anterior talofibular ligament A ligament in the ankle that connects the fibula and talus on the anterolateral side.

Apex The narrower distal end of the patella.

Appendicular region The four extremities of the body, which attach to the axial region.

Articular cartilage The smooth white tissue that allows the bones to glide over each other with very little friction.

Articulations Features of bones related to the structure of a joint.

Ascending neurons Also called afferent neurons, these conduct messages toward the CNS.

Ascending tracts Bundles of sensory nerve fibers that transmit information from PNS sensory neurons to the brain.

Association tracts Tracts that connect different areas within each hemisphere of the brain.

Associative or connecting neurons Also called interneurons, they connect the sensory neurons and the motor neurons and are only found in the CNS.

Atlanto-occipital joint Articulation between the atlas and the occipital bone, is a synovial joint.

Atlanto-occipital joint capsule Joint which is the articulation between the occipital bone and the first cervical vertebra which is also named atlas.

Atlantoaxial joint A joint in the upper part of the neck between the first and second cervical vertebrae; the atlas and axis.

Atlas The first vertebra in the neck.

Axial region The main axis of the human body, includes head, neck, chest, and trunk.

Axon The long thread-like part of a nerve cell.

B

Ball and socket joints A joint in which the round end of a bone fits into the cavity of another bone.

Base The lowest or proximal end of something; the wider proximal end of the patella.

Biceps brachii Muscle of the upper arm that is divided into three sections called the long head, the medial head, and the lateral head.

Biceps femoris A muscle that resides on the lateral side of the posterior thigh; it has two heads and therefore two origins.

Bicipital ligament A ligament that holds the long head of the biceps tendon within the bicipital groove.

Bilateral Involving both sides of the body or brain.

Bipennate The fibers arranged obliquely and inserting on both sides into a central tendon.

Body motion A simultaneous displacement of more than one body segment.

Bone cells Cells that are responsible for the growing, shaping, and maintenance of the skeleton.

Bone marrow A soft tissue that occupies the spaces of spongy osseous tissue and the passageways of compact osseous tissue.

Brachial plexus A network of nerves. It is formed by C5–T1 nerve roots. Thirteen (13) pairs of nerves emerge from these nerve roots and/or this plexus.

Brachialis A muscle in the upper arm that flexes the elbow joint.

Brainstem The posterior part of the brain.

C

Calcaneofibular ligament A narrow rounded cord running from the fibula downward.

Calcaneus Also called the heel bone, is a large bone that forms the foundation of the rear part of the foot. It connects with the talus and cuboid bones.

Cardiac muscles Striped muscles in the walls of the heart.

Carpal bones The eight small bones that make up the wrist.

Carpometacarpal joints Five joints in the wrist.

Cartilaginous joints Joints joined by cartilage, which allows more movement between the bones than a fibrous joint.

Caudal The posterior part of the body.

Cell membrane A biological membrane that protects the cell.

Central nervous system Formed by the brain and spinal cord, this controls most functions of the body and mind.

Cephalic The head or the head end of the body.

Cerebellum The part of the brain that coordinates voluntary movements.

Cerebrum The uppermost region of the central nervous system.

Cervical Pertaining to the neck.

Cervical plexus A network of nerve fibers located in the posterior triangle of the neck.

Cervical vertebrae The seven small vertebrae that form the neck.

Clavicle A long bone that serves as a strut between the shoulder blade and the breastbone.

Clavicular section Portion of the pectoralis major that runs obliquely originating from the proximal one-third of the clavicle and inserts on the lateral lip of the intertubercular (bicipital) groove.

Coccygeal Describes the vertebrae that form the small triangular bone forming the lower extremity of the spinal column in humans.

Coccygeal plexus A plexus of nerves near the coccyx bone.

Commissural tracts Tracts that connect each cerebral hemisphere (side of the brain) to the other.

Compact osseous tissue Tissue that is arranged in cylindrical concentric lamellae and is present in every bone.

Complex body motion The simultaneous displacement of more than one body segment in which the entire body is displaced.

Complex joint motion The displacement of a single body segment in one direction; however, it is moved through more than one body plane and therefore more than one axis of rotation.

Concentric muscular action A motion that causes the length of a muscle to decrease; used to lift objects.

Conductivity The ability to transmit (carry) electrical energy.

Condyloid joints One in which an ovoid head of one bone moves in an elliptical cavity of another, permitting all movements except axial rotation.

Connective tissue Type of body tissue that performs varied functions; it supports and connects internal organs, forms bones and the walls of blood vessels, attaches muscles to bones, and replaces tissues of other types following injury.

Contralateral Pertaining to the opposite side.

Convergent A type of muscle in which the fascicles merge at the end of a tendon. They are broad at one end and narrow at the other creating a triangular shape.

Coracoacromial ligament A strong triangular band, extending between the coracoid process and the acromion.

Coracobrachialis A long, slender muscle of the shoulder joint.

Coracoclavicular ligament Ligament that connects the clavicle with the coracoid process of the scapula.

Coracohumeral ligament A broad ligament which strengthens the upper part of the capsule of the shoulder joint.

Coronal plane An imaginary plane dividing the body into dorsal and ventral parts.

Coxal A large irregular bone, constricted in the center and expanded above and below that forms the hip.

Coxal joint The joint between the femur and acetabulum of the pelvis.

Cranial bones The eight bones of the skull: one frontal, two parietal, one occipital, two temporal, one sphenoid, and one ethmoid.

Cranial nerves Nerves that bring information from the sense organs to the brain.

Cuboid A wedge-shaped foot bone, widest at its medial edge and narrow at its lateral edge.

Cytoplasm A thick solution that fills each cell and is enclosed by the cell membrane.

Cytoskeleton A structure that helps cells maintain their shape and internal organization.

Cytosol The aqueous component of the cytoplasm of a cell.

D

Deep Farther from the body's surface.

Deep posterior compartment Contains the muscles that invert the ankle.

Deltoid Triangular muscle located on the uppermost part of the arm and the top of the shoulder.

Deltoid ligament A triangular-shaped ligament that attaches from the medical malleolus to various location on the talus and calcaneus.

Dendrites Short arm-like protuberances from a nerve cell (a neuron).

Depressions Concavities that dip inwardly from the main surface of a bone.

Descending neurons Motor neurons conduct messages from the CNS to motors (muscles or glands) to cause an effect. They are also sometimes called efferent neurons.

Descending tracts Bundles of nerve fibers that carry nerve impulses in a direction away from the central nervous system.

Diarthroses Joints that are freely movable.

DIP joints Joints between the second (intermediate) and third (distal) phalanges.

Distal Situated away from the point of origin or attachment, as of a limb or bone.

Distal anterior tibiofibular ligament Anterior ligament that stabilizes the tibia and fibula at the ankle.

Distal interphalangeal joints Joints between the second (intermediate) and third (distal) phalanges.

Distal phalange The bone at the tip of the fingers or toes.

Distal posterior tibiofibular ligament Posterior ligament that stabilizes the tibia and fibula at the ankle.

Distal radial collateral ligament Ligament that stabilizes the lateral wrist connecting the radius to the scaphoid bone.

Distal radioulnar joint The articulation between the distal ends of the radius and ulna.

Distal tibiofibular joint Joint formed by the rough, convex surface of the medial side of the distal end of the fibula, and a rough concave surface on the lateral side of the tibia.

Distal ulnar collateral ligament Ligament that stabilizes the medial wrist attaching at the ulnar styloid process to some carpal bones.

Dorsal Relating to the back or posterior of a structure; the opposite of the ventral, or front, of the structure.

Dorsal interossei Four muscles in the back of the hand that act to abduct (spread) the index, middle, and ring fingers away from hand's midline. In the foot, the four muscles lying between the metatarsals.

Dorsal radiocarpal ligament The ligament that extends from the distal end of the radius posteriorly to the proximal row of carpal bones.

Dorsal root The root that extends from the posterior spinal cord and contains sensory neurons bound for an ascending tract.

Dorsiflexion Ankle movement which straightens and bends the ankle in the sagittal plane rotating about the frontal axis.

Downward scapular rotation Motion of scapula in clockwise direction.

E

Eccentric muscular action A motion that causes the length of a muscle to increase; used to lower objects.

Efferent neurons Motor neurons conduct messages from the CNS to motors (muscles or glands) to cause an effect. They are also sometimes called descending neurons.

Elbow extension and flexion Straightening motion of elbow.

Elbow pronation Turning of forearm toward the midline.

Elevation and depression The motion of lifting (shoulder shrug) and lowering, respectively, of the shoulder girdle in the frontal plane rotating about the sagittal axis.

Endomysium A connective tissue that wraps around individual muscle fibers (cells).

Endosteum A thin vascular membrane of connective tissue that lines the inner surface of the bony tissue that forms the medullary cavity of long bones.

Epimysium A sheath of fibrous elastic tissue surrounding a muscle.

Epiphyses The rounded end of a long bone.

Epithelial tissue The outer surfaces of organs and blood vessels throughout the body.

Erector spinae A group of muscles and tendons that run the length of the spine and hips to the base of the skull.

Ethmoid A square bone at the root of the nose.

Eversion Rotation of the ankle in the frontal plane around the sagittal axis away from the midline.

Excitability The ability to respond to stimuli.

Extensor carpi radialis brevis A muscle in the forearm that acts to extend and abduct the wrist.

Extensor carpi radialis longus A wrist extensor muscle originates from the distal end of the lateral supracondylar ridge of the humerus

Extensor carpi ulnaris A skeletal muscle located on the ulnar side of the forearm.

Extensor digiti minimi A muscle located in the forearm of the human body.

Extensor digitorum A muscle of the posterior forearm present in humans and animals.

Extensor digitorum longus A pennate muscle situated at the lateral part for extension of toes and dorsiflexion of ankle.

Extensor hallucis brevis A thin muscle situated between the tibialis anterior and the extensor digitorum longus.

Extensor hallucis longus A thin muscle situated between the tibialis anterior and the extensor digitorum longus.

Extensor indicis An extensor muscle that acts upon the second digit (index finger).

Extensor pollicis brevis A skeletal muscle on the dorsal side of the forearm.

Extensor pollicis longus A skeletal muscle located dorsally on the forearm, much larger than the extensor pollicis brevis.

External abdominal oblique A pair of broad, thin, superficial muscles that lie on the lateral sides of the abdominal region of the body.

External rotation Rotation away from the center of the body.

Extrinsic Operating or coming from without; describes muscles whose origin and belly are outside the region of the joint it acts upon.

Extrinsic muscles A group of muscles lying superficially on a structure.

F

Facial bones A set of bones that make up the face.

False floating ribs The last two false ribs, having no ventral attachment.

False ribs Attach through coastal cartilage.

Fascia A sheet of connective tissue.

Fascicles Bundles of skeletal muscle fibers surrounded by perimysium (a type of connective tissue).

Femur The longest bone in the human body.

Fibrocartilage Very strong tissue found predominantly in the intervertebral disks and at the insertions of ligaments and tendons.

Fibrous joint Connected by dense fibrous connective tissue, have no joints.

Fibula The long, thin and lateral bone of the lower leg.

Fibularis brevis A muscle of the lower leg, and aids in plantarflexion and eversion of the foot.

Fibularis longus A muscle inside the outer area of the human leg, which everts and flexes the ankle.

Fibularis tertius A muscle that originates from the medial side of the distal one-third of the fibula and inserts on the base of the fifth metatarsal superior to the fibularis brevis attachment.

Fifth metacarpal Bone of the little finger.

Fifth metatarsal The long bone on the outside of the foot that connects to the little toe.

First distal phalange The bone at the tip of the first finger.

First IP joint The first joint of the finger and is located between the first two bones of the finger.

First metacarpal The bone proximal to the thumb.

First metatarsal The thickest, strongest, and the shortest of the metatarsal bones.

First TM joint (TM joint I) The joint between the first metatarsal (metatarsal I) and the medial cuneiform.

Flat bones Bones that are expanded into broad, flat plates, as in the skull, the pelvis, and the rib cage.

Flexor carpi radialis A muscle of the forearm that acts to flex and abduct the hand.

Flexor carpi ulnaris A superficial flexor muscle of the forearm that flexes adducts the hand.

Flexor digiti minimi A muscle that originates from the plantar side of the base of the fifth metatarsal and inserts on the plantar side of the base of the fifth proximal phalange.

Flexor digitorum brevis A muscle lies longitudinally in the middle of the plantar side of the foot.

Flexor digitorum longus A muscle resides in the posterior crural region longitudinally over the tibia.

Flexor digitorum profundus An extrinsic hand muscle that acts on the hand while its muscle belly located in the forearm.

Flexor digitorum superficialis Muscle that provides flexion of the middle phalanges of the fingers at the proximal interphalangeal joints.

Flexor hallucis brevis Muscle that originates from the plantar side of the cuboid and lateral cuneiform and inserts on the plantar side of the base of the first proximal phalange.

Flexor hallucis longus Muscle that originates on the posterior side of the distal two-thirds of the fibular diaphysis.

Flexor pollicis brevis A muscle in the hand that flexes the thumb.

Flexor pollicis longus A muscle in the forearm and hand that flexes the thumb.

Fourth and fifth TM joints The articulations between the fourth and fifth metatarsal and the cuboid bone.

Fourth metacarpal Shorter and smaller than the third metacarpal, this has a base that is small and quadrilateral.

Fourth metatarsal A long bone in the foot. Third longest of the five metatarsal bones.

Frontal Bone of the skull that forms the forehead.

Frontal axis An imaginary line that runs from left to right through the center of the body.

Frontal lobe The part of the brain that controls important cognitive skills in humans.

Frontal plane A vertical plane that divides the body into ventral and dorsal sections.

Fundamental position The anatomical position but with the forearms rotated so that the palms face the sides of the body.

Fusiform Spindle-like shape that is wide in the middle and tapers at both ends.

G

Gastrocnemius A superficial two-headed muscle that is in the back part of the lower leg of humans.

Gemellus inferior An important muscle that connects the upper legs to the pelvic region.

Gemellus superior A deep gluteal muscle that acts on the hip joint and stabilizes the pelvis.

General layers of tissue The four layers of tissue in the body are (from superficial to deep): skin, hypodermis, muscular, and skeletal.

Glenohumeral (GH) joint A ball and socket joint between the scapula and the humerus.

Gluteus maximus One of the largest and strongest muscles in the body; it extends from the lower spine to the femur and forms the buttocks.

Gluteus medius A broad, thick, radiating muscle, situated on the outer surface of the pelvis.

Gluteus minimus The primary internal rotator of the hip joint.

Gracilis Muscle responsible for hip adduction and which assists in knee flexion.

H

Hamstrings Three main muscles primarily responsible for flexing the knee that reside in the posterior thigh.

Head The distal end of something or the upper portion of the body.

Hierarchy The structure of the body in terms of fundamental levels of organization that increases in complexity.

Hinge joints Bone joints in which the articular surfaces are molded to each other to permit motion.

Hip adduction and abduction Movement of the leg toward and away from, respectively, the midline of the body.

Hip extension and flexion To straighten and bend, respectively, the hip in the frontal plane about the sagittal axis.

Hip external rotation and internal rotation Rotation of the hip in the transverse plane about the longitudinal axis outward and inward, respectively.

Horizontal abduction Movement of limb away from the midline of body.

Horizontal adduction Movement of the arm or thigh in the transverse plane from a lateral position to an anterior position.

Humeroradial joint The joint between the head of radius and the capitulum of the humerus.

Humeroulnar joint The joint between the humerus and ulna.

Humerus The long bone in the upper arm located between the elbow joint and the shoulder.

Hyperextension Extension beyond 0 degrees.

Hypodermis An epidermal layer of cells that secretes the chitinous cuticle.

I

Iliacus A flat, triangular muscle which fills the iliac fossa.

Iliofemoral ligament A ligament of the hip joint which extends from the ilium to the femur in front of the joint.

Iliopsoas The inner hip muscle whose function is to flex the thigh at the hip joint.

Ilium The uppermost and largest part of the hip bone.

Inferior Away from the head.

Inferior nasal conchae One of the pairs of conchae in the nose.

Infraspinatus A thick triangular muscle, which occupies the chief part of the infraspinatus fossa.

Innervation The process of supplying nerves to an organ or part of the body.

Insertion A muscle's attachment site that is usually more distal and is the more dynamic or mobile one.

Intercostal nerves Pairs of nerves that innervate structures in the thoracic and abdominal regions.

Internal abdominal oblique A muscle that is found at each side of the body, just lateral to the abdomen.

Internal rotation Rotating a joint toward the midline.

Interneurons Neurons that are found exclusively in the central nervous system and connect the sensory neurons and the motor neurons

Interosseous membrane A thick, dense fibrous sheet of connective tissue.

Interphalangeal (IP) joints The hinge joints between the phalanges of the fingers.

Interspinous ligaments Thin and membranous ligaments that connect adjoining spinous processes of the vertebra in the spine.

Intertarsal joints The joints of the tarsal bones in the foot.

Intertransverse ligaments Ligaments that stabilize the transverse processes of the spine.

Intervertebral disc A shock-absorbing structure located between each vertebra; the outer ring is tough fibrocartilage while the inner tissue is a more gelatinous form of cartilage.

Intervertebral joints Joints between each adjacent vertebra from the axis to the sacrum.

Intrinsic Situated within the organ on which it acts; describes a muscle that is entirely (origin, belly, and insertion) located within the region that it acts upon.

Intrinsic muscles Smaller muscles located within the hand itself.

Inversion Rotation of the ankle in the frontal plane around the sagittal axis toward the midline.

Ipsilateral Affecting the same side of the body.

Irregular bones The group of bones having complex forms to provide support and protection.

Ischiofemoral ligament A band of very strong fibers that connect the pelvis and the femur.

Isometric muscular action The production of muscle tension without a change in muscle length or joint angle.

J

Joint capsule An envelope surrounding a synovial joint.

Joint cavity Space enclosed by the synovial membrane and articular cartilages.

Joint motion The displacement of a single body segment.

K

Kinesiology The study of the mechanics of body movements.

Knee extension and flexion Two independent motions that straighten and bend, respectively, the knee in the sagittal plane rotating about the frontal axis.

L

Labrum A piece of fibrocartilage attached to the rim of the shoulder socket and the hip socket.

Lacrimal Facial bone that underlies the area of the proximal edge of the eye.

Large motor unit A unit in which one motor neuron innervates a large number, up to 1,000s, of muscle fibers.

Lateral (fibular) collateral ligament A thin band of tissue running along the outside of the knee.

Lateral compartment There are three muscles residing in the lateral compartment that evert the ankle. They are the fibularis longus, the fibularis brevis, and the fibularis tertius.

Lateral cuneiform Wedge-shaped foot bone, intermediate in size to the other cuneiform bones.

Lateral head The head of the triceps brachii that originates from the posterolateral humerus just inferior to the greater tuberosity.

Lateral meniscus A fibrocartilaginous band that spans the lateral side of the interior of the knee joint.

Latissimus dorsi A large, flat muscle covering the width of the middle and lower back.

Levator scapula Muscle situated at the back and side of the neck that originates from the transverse processes of the first four cervical vertebrae and inserts on the upper scapula.

Ligaments Cord-like tissue that connects bones to other bones to form joints.

Ligamentum flavum One of a series of bands of elastic tissue that runs between the lamina from the axis to the sacrum.

Linea alba A white fibrous structure that runs down the midline of the abdomen from the xiphoid process to the symphysis pubis.

Long bones Hard and dense cylindrical bones that provide strength and mobility.

Long head In the leg, the head of the biceps femoris that originates from the ischial tuberosity. In the arm, the longest of the three heads of the triceps muscle that arises from the infraglenoid tubercle of the scapula.

Longitudinal axis A direction of orientation, which passes from superior to inferior through the transverse plane.

Lower trapezius A shoulder girdle depressor that passes from the vertebral column relatively obliquely through the mid back to the shoulder.

Lumbar Pertaining to the abdominal segment of the torso, between the diaphragm and the sacrum.

Lumbar plexus A network of nerve fibers that supplies the skin and musculature of the lower limb.

Lumbar vertebrae The five vertebrae between the rib cage and the pelvis.

Lumbosacral joint A joint of the body, between the last lumbar vertebra and the first sacral segment of the vertebral column.

Lumbricals Intrinsic muscles of the hand that flex the metacarpophalangeal joints. In the foot, a group of four separate muscles that reside in the plantar foot between the metatarsals.

M

Mandible The largest and strongest bone of the face, it forms the jaw.

Marrow cavity The central cavity of a bone shaft where red bone marrow and yellow bone marrow is stored.

Mastoid process A pyramidal bony projection from the posterior section of the temporal bone.

Maxillae The upper fixed bone of the jaw.

Medial (tibial) collateral ligament The ligament that connects the top of the tibia to the bottom of the femur.

Medial cuneiform The first cuneiform situated at the medial side of the foot.

Medial head The head of the triceps brachii that originates from the entire posterior side of the humeral diaphysis.

Medial meniscus A fibrocartilage semicircular band that spans the knee joint medially.

Median An imaginary line that bisects the body vertically through the midline marked by the naval, dividing the body into left and right sides.

Medulla oblongata A portion of the hindbrain that controls autonomic functions.

Medullary cavity The central cavity of a bone shaft; also known as the marrow cavity.

Metacarpophalangeal (MP) joints These joints are formed by the reception of the rounded heads of the metacarpal bones.

Metatarsals Long bones, convex in shape, that give the foot its arch.

Metatarsophalangeal (MP) joints The joints between the metatarsal bones of the foot and the proximal bones of the toes.

Midbrain Along with the pons, this region mostly sends information to and from various parts of the brain.

Middle (or intermediate) phalange The bone located in the fingers in between the proximal and distal phalange.

Middle cuneiform The intermediate of the three wedge-like bones in the tarsus.

Middle deltoid Muscle that covers the lateral shoulder and is in between the anterior and posterior sections of the deltoid muscle.

Middle phalange The bone in between the proximal and distal phalange.

Middle trapezius A shoulder girdle retractor that brings the shoulder blades back and also provides stabilization for the shoulder during some arm movements.

Midline The median line or median plane of the body or some part of the body.

Midsagittal A vertical cut down the exact center line of the specimen.

Mixed nerve A nerve that contains the axons of both sensory and motor neurons.

Motion The movement of organs, joints, limbs, and specific sections of the body.

Motor division Division of the peripheral nervous system.

Motor neurons A specialized type of brain cell located within the spinal cord and the brain.

Motor unit A unit that is made up of a motor neuron and skeletal muscle fibers.

Motors Organs that can cause an effect or change; this includes all the various muscles and glands.

Multipennate Muscles in which the fibers are arranged in curved bundles in one or more plane.

Muscle fibers Another name for muscle cells; they are cylindrical and have more than one nucleus.

Muscular Relating to muscle or the muscles.

Muscular action The action or particular movement of a muscle relative to the joint or the body part moved.

Muscular contraction Tightening of muscle, commonly in reference to uterine contractions during childbirth.

Muscular system An organ system consisting of skeletal, smooth, and cardiac muscles.

Muscular tissue A soft tissue that composes muscles and gives rise to muscle's ability to contract.

N

Nasal The bone which opens exteriorly at the nostrils.

Navicular It is a boat-shaped bone located in the top inner side of the foot.

Neck extension and flexion Movement of the neck as in looking upward and downward, respectively.

Neck lateral flexion Movement that bends the head to each side through the frontal plane about the sagittal axis.

Neck rotation Movement that turns the head from side to side through the transverse plane about the longitudinal axis.

Nerve fiber Another name for axon, this is a long, slender projection of a nerve cell.

Nerve root The initial segment of a nerve leaving the central nervous system.

Nerves The conduits ("pipes") filled with axons that extend from the spinal cord and mark the beginning of the PNS.

Nervous tissue The main tissue component of the nervous system.

Neuromuscular junction A chemical synapse formed by the contact between a motor neuron and a muscle fiber.

Neurons The fundamental units of the brain and nervous system.

Neurotransmitters Chemical messengers that transmit signals across a synapse.

Nucleus The central and most important part of an object, movement, or group, forming the basis for its activity and growth.

O

Obturator externus A flat, triangular muscle, which covers the outer surface of the anterior wall of the pelvis.

Obturator internus A hip muscle that originates deep within the pelvis, wraps out and inserts on the posterior aspect of the head of the femur.

Occipital The bone that overlies the occipital lobes of the cerebrum.

Occipital lobe The visual processing center of the mammalian brain containing most of the anatomical region of the visual cortex.

Opponens digiti minimi A triangular muscle located immediately beneath the palmaris brevis.

Opponens pollicis A small, triangular muscle in the hand, which functions to oppose the thumb.

Organ systems A group of organs that work together as a biological system to perform one or more functions.

Organelles A tiny cellular structure that performs specific functions within a cell.

Organs The body's recognizable structures (for example, the heart, lungs, liver, eyes, and stomach) that perform specific functions.

Origin The more fixed, central, or larger attachment of a muscle.

Osseous tissue Type of bone tissue with a honeycomb-like internal matrix, which helps provide rigidity.

P

Palatine A bone behind the maxilla that enters into the formation of the hard palate.

Palmar interossei Three small, unipennate muscles in the hand that lie between the metacarpal bones and are attached to the index, ring, and little fingers.

Palmar radiocarpal ligament A broad membranous band, attached above to the distal end of the radius.

Palmaris longus A slender, elongated, spindle shaped muscle, lying on the medial side of the flexor carpi radialis.

Parallel Describes muscles that are used for fast movements and whose fascicles are parallel to the tendon.

Parietal Large cranial bone that forms the upper rear of the skull.

Parietal lobe The brain area important to the function and processing of sensory information.

Passageways Tunnel-like features that extend into the inner parts of bone or all the way through to the other side.

Patella Also called the kneecap, this is a flat, circular-triangular bone which articulates with the femur (thigh bone) and covers and protects the anterior articular surface of the knee joint.

Patella tendon The ligament that connects the patella and the tibia.

Patellofemoral joint The joint formed by the patella and the femur.

Pectineus A flat, quadrangular muscle, situated at the anterior (front) part of the upper and medial (inner) aspect of the thigh.

Pectoralis major It consists of a clavicular part and a sternal part, both converging to a flat tendon that inserts on the humerus.

Pectoralis minor A thick, fan-shaped muscle, situated at the chest of the human body.

Perimysium The sheath of connective tissue surrounding a bundle of muscle fibers.

Periosteum A dense layer of vascular connective tissue enveloping the bones except at the surfaces of the joints.

Peripheral nervous system The division of the nervous system containing all the nerves that lie outside of the central nervous system (CNS). Its primary role is to connect the CNS to the organs, limbs, and skin.

Phalanges The bones that make up the fingers of the hand and the toes of the foot.

Physiology The branch of biology that deals with the normal functions of living organisms and their parts.

Piriformis A small muscle located deep in the buttock, behind the gluteus maximus.

Pivot joints Freely moveable joints (diarthroses) that allows only rotary movement around a single axis.

Plane joints Type of structure in the body formed between two bones in which the articular, or free, surfaces of the bones are flat or nearly flat, enabling the bones to slide over each other.

Plantar flexion A movement in which the top of your foot points away from your leg.

Plantar interossei A group of three muscles lying between metatarsals II–V.

Plantaris One of the superficial muscles of the superficial posterior compartment of the leg.

Plexus A network of nerves or vessels in the body.

Pons Part of the brainstem, and in humans and other bipeds lies inferior to the midbrain, superior to the medulla oblongata and anterior to the cerebellum.

Popliteus The one muscle in the posterior compartment of the lower leg that only acts on the knee and not on the ankle.

Posterior Further back in position; of or nearer the rear or hind end.

Posterior atlanto-occipital membrane A membranous ligament that is connected above to the posterior margin of the foramen magnum and below to the upper border of the posterior arch of the atlas.

Posterior compartment Contains the muscles that plantar flex the ankle.

Posterior cruciate ligament The ligament that connects the posterior intercondylar area of the tibia to the medial condyle of the femur.

Posterior deltoid The muscle forming the rounded contour of the human shoulder.

Posterior longitudinal ligament A long and important ligament located immediately posterior to the vertebral bodies (to which it attaches loosely) and intervertebral discs (to which it is firmly attached).

Posterior ramus The posterior division of a spinal nerve.

Posterior scalene The smallest and deepest of the scalene muscles, which inserts on the second rib.

Posterior talofibular ligament A ligament that runs almost horizontally from the malleolar fossa of the lateral malleolus of the fibula to a prominent tubercle on the posterior surface of the talus immediately lateral to the groove for the tendon of the flexor hallucis longus.

Projection tracts Extend vertically between higher and lower brain areas and spinal cord centers, and carry information between the cerebrum and the rest of the body.

Projections Convex "bumps" that extend from the main surface of a bone.

Pronator quadratus A square-shaped muscle on the distal forearm that acts to pronate (turn so the palm faces downwards) the hand.

Pronator teres A narrow muscle on the distal forearm that serves to pronate the forearm.

Prone A body position in which the person lies flat with the chest down and the back up.

Protraction The action of moving the shoulders forward (rounding the shoulders).

Proximal Nearer to the midline.

Proximal anterior tibiofibular ligament The ligament that travels horizontally over the front surface of the tibiofibular syndesmosis (the meeting area of the fibula and the tibia).

Proximal interphalangeal (PIP) joints Joints located between the first (also called proximal) and second (intermediate) phalanges.

Proximal phalange The proximal bones that join with the metacarpals of the hand or metatarsals of the foot at the metacarpophalangeal joint or metatarsophalangeal joint.

Proximal posterior tibiofibular ligament The ligament that travels horizontally over the rear surface of the tibiofibular syndesmosis (the meeting area of the fibula and the tibia).

Proximal radioulnar joint A synovial pivot joint between the circumference of the head of the radius and the ring formed by the radial notch of the ulna and the annular ligament.

Proximal tibiofibular joint An arthrodial joint between the lateral condyle of the tibia and the head of the fibula.

Pubiofemoral ligament A ligament in the inferior side of hip joint. It extends from the pubic portion of the acetabular rim and passes below the neck of the femur.

Pubis Also called the pubic bone, this is one of the bones that make up the pelvis.

Q

Quadratus femoris A flat, quadrilateral skeletal muscle located on the posterior side of the hip joint.

Quadratus lumborum This muscle supports good posture and helps stabilize the spine when bending to the side or extending the lower back.

Quadratus plantae Is a muscle in the foot that extends from the anterior (front) of the calcaneus (heel bone) to the tendons of the digitorum longus muscle in the leg.

Quadriceps tendon The tendon that connects the knee extensors to the superior pole of the patella.

R

Radial collateral ligament Ligament in the elbow on the side of the radius.

Radial deviation Lateral flexion of the wrist toward the radius; or, movement of bending the wrist to the thumb (radial) side.

Radiocarpal joint The articulation between the distal end of the radius and the proximal carpal bones.

Radioulnar-carpal joint A synovial joint formed by the articulation between the distal radius and the scaphoid, lunate, and triquetrum as well as the soft tissue structures that hold the joint together.

Radius One of the two large bones of the forearm, the other being the ulna.

Rectus femoris A superficial muscle that passes longitudinally through the center of the anterior thigh. It is a double jointed muscle in that it crosses both the hip and the knee and therefore acts upon both. It originates from the anterior inferior iliac spine (AIIS) and inserts on the tibial tuberosity via the patellar tendon.

Red bone marrow Type of marrow in which the stroma primarily contain the developmental stages of erythrocytes, leukocytes, and megakaryocytes; it is present throughout the skeleton during fetal life and at birth. After the fifth postnatal year, it is gradually replaced in the long bones by yellow marrow.

Reflex A rapid, involuntary reaction to an "emergency."

Retraction The motion of pinching the scapulae together in the back.

Rhomboideus major A shoulder girdle retractor that lies obliquely just inferior to the rhomboideus minor.

Rhomboideus minor A shoulder girdle retractor that is the upper of two muscles beneath the trapezius muscle.

Right and left lateral flexion Movements that bend the trunk to each side in the frontal plane rotating around the sagittal axis.

Right and left rotation Right and left turning of the trunk in the transverse plane around the longitudinal axis.

Rotation Turning of a body segment that resides on the midline (the head and trunk) to the right or left side.

Round ligament Ligament that is found on the interior of the coxal joint.

S

Sacral Relating to or laying near the sacrum.

Sacral plexus A nerve plexus formed by the 4th and 5th lumbar and 1st, 2nd, 3rd sacral nerves; supplies the pelvic region and lower limbs.

Sacrum A triangular bone made up of five fused vertebrae and forming the posterior section of the pelvis.

Saddle joints Joints that have such sharp curvatures to their bone ends that they resemble a saddle.

Sagittal axis The axis that passes from anterior to posterior through the frontal plane.

Sagittal plane An anatomical plane that divides the body into right and left parts; this plane may be in the center of the body and split it into two halves or away from the midline and split it into unequal parts.

Sartorius muscle The longest muscle in the body, it originates from the anterior superior iliac spine (ASIS), crosses the hip, then obliquely crosses the thigh, and inserts just below the medial tibial condyle on the medial side.

Scapulothoracic articulation Formed by the convex surface of the posterior thoracic cage and the concave surface of the anterior scapula.

Second metacarpal The longest metacarpal with the largest base.

Second metatarsal The longest of the metatarsal bones, being prolonged backward and held firmly into the recess formed by the three cuneiform bones.

Second TM joint Joint of the second metatarsal and intermediate (or middle) cuneiform.

Semimembranosus A flat, long muscle situated in the posterior region of the thigh, so called because of its membranous tendon of origin.

Semispinalis capitis A long, thin muscle that is located at the back of the neck, on both sides of the spinal column.

Semispinalis cervicis Transversospinal group of muscles, formed of muscles between a spinous process and the transverse process of the upper five or six thoracic vertebrae.

Semitendinosus A rod-shaped muscle residing in the medial side of the posterior thigh superficial to the semimembranosus.

Sensory division Also called the afferent division, this transmits impulses from peripheral organs to the CNS.

Sensory neurons Nerve cells within the nervous system responsible for converting external stimuli from the organism's environment into internal electrical impulses.

Sensory receptors A structure that reacts to a physical stimulus in the environment, whether internal or external.

Serratus anterior A muscle that originates on the surface of the first to eighth ribs at the side of the chest and inserts along the entire anterior length of the medial border of the scapula.

Sesamoid bones Bones that are embedded within a tendon or a muscle.

Short bones These bones provide support and stability, and their length, width, and depth are similar.

Short head In the arm, a flexor and supinator of the elbow joint. In the leg, the head of the biceps femoris that originates from the linea aspera.

Shoulder extension and flexion Opposing shoulder motions in the sagittal plane, which occur about the frontal axis.

Simple body motion The simultaneous displacement of more than one body segment without the entire body being displaced.

Simple joint motion The displacement of a single body segment in one direction within one body plane and around one axis of rotation.

Skeletal Relating to or functioning as a skeleton.

Skeletal muscles Striated muscle tissues that are under the voluntary control of the somatic nervous system; they apply force to and move the bones of the skeleton and hence the body.

Skin The thin layer of tissue forming the natural outer covering of the body of a person or animal.

Small motor units A unit in which one motor neuron innervates a small number of muscle fibers.

Smooth muscles Also called involuntary muscles, they show no cross stripes under microscopic magnification.

Soleus A powerful muscle in the back part of the lower leg (the calf). It runs from just below the knee to the heel, and is involved in standing and walking.

Sphenoid An unpaired bone of the neurocranium. It is situated in the middle of the skull toward the front, in front of the basilar part of the occipital bone.

Spinal cord A long, thin, tubular structure made up of nervous tissue, which extends from the medulla oblongata in the brainstem to the lumbar region of the vertebral column.

Spinal nerves Mixed nerves, which carry motor, sensory, and autonomic signals between the spinal cord and the body.

Splenius capitis Muscle that pulls on the base of the skull from the vertebrae in the neck and upper thorax. It is involved in movements such as shaking the head.

Splenius cervicis "Muscle lower on the spine than the splenius capitis that originates from the spinous processes of T3–T6 vertebrae and inserts on the transverse processes of C1–C3 vertebrae."

Spongy osseous tissue In bone, this tissue is arranged into trabeculae, which are the interconnected columns that create the sponge-like grid of bone.

Stabilizers Muscles that work to stabilize the body and its extremities during multiplane movement.

Sternoclavicular (SC) joint The linkage between the clavicle (collarbone) and the sternum (breastbone).

Sternoclavicular ligament A ligament that extends from the posterior aspect of the sternal end of the clavicle to the posterosuperior manubrium.

Sternocleidomastoid Muscle that runs through the anterolateral neck originating from the manubrium and proximal clavicle. It inserts into the mastoid process of the temporal bone.

Structural joint classifications This division divides joints into fibrous, cartilaginous, and synovial joints depending on the material composing the joint and the presence or absence of a cavity in the joint.

Subcostal nerve This nerve (anterior division of the twelfth thoracic nerve) is larger, and it runs along the lower border of the twelfth rib, often gives a communicating branch to the first lumbar nerve, and passes under the lateral lumbocostal arch.

Subscapularis A large triangular muscle that fills the subscapular fossa and inserts into the lesser tubercle of the humerus and the front of the capsule of the shoulder joint.

Subtalar joint Also known as the talocalcaneal joint, this is a joint of the foot between the talus and the calcaneus.

Superficial A relative term to describe that which is closer to the body's surface.

Superior In a hierarchy or tree structure of any kind, this represents an individual or position at a higher level than another.

Superior, middle, and inferior glenohumeral ligaments Provides anterior stability at 45 degrees and 60 degrees abduction.

Supination Rotation of the forearm and hand so that the palm faces upward.

Supinator A broad muscle in the posterior compartment of the forearm, curved around the upper third of the radius.

Supine The supine position means lying horizontally with the face and torso facing up.

Supraspinatus A relatively small muscle of the upper back that runs from the supraspinous fossa superior portion of the scapula (shoulder blade) to the greater tubercle of the humerus.

Supraspinous ligament A strong fibrous cord, which connects together the apices of the spinous processes from the seventh cervical vertebra to the sacrum.

Sutural bone An irregular isolated bone that can appear in addition to the usual centers of ossification of the skull.

Synapse In the nervous system, a synapse is a structure that permits a neuron (or nerve cell) to pass an electrical or chemical signal to another neuron or to the target.

Synaptic bulb Structure that contains synaptic vesicles, which are filled with neurotransmitters.

Synaptic cleft The space between neurons at a nerve synapse across which a nerve impulse is transmitted by a neurotransmitter.

Synaptic knob Also known as a synaptic bulb, this is one of many terminal endpoints on a neuron's axon.

Synarthroses Joints that permits very little or no movement under normal conditions.

Synergists Muscles that assist the agonists in a motion.

Synovial fluid A slippery lubricant that nourishes structures within the joint cavity, rinses waste products from them, and reduces friction which can wear out the joint.

Synovial joints Also known as diarthroses, these allow for great movement and join bones with a fibrous capsule that is continuous with the periosteum of the joined bones.

Synovial membrane A specialized connective tissue that lines the inner surface of the capsules of synovial joints and the tendon sheath.

T

Talocrural joint The articulation between the talus and the bones in the crural region (the tibia and fibula).

Talus The bone that makes up the lower part of the ankle joint (the tibia and fibula bones of the lower leg make up the upper part of the ankle joint).

Tarsals These form a group of seven articulating bones in the foot located between the bones of the lower leg (tibia, fibula) and the metatarsal bones.

Tarsometatarsal (TM) joints Are gliding joints located in the foot between the bones of the second row of the tarsus and the metatarsal bones.

Temporal Cranial bones that form the side of the skull and attach to the mandible.

Temporal lobe The brain region where sound is processed; it is also a region where auditory language and speech comprehension systems are located.

Tendons Tough bands of fibrous connective tissue that connect muscle to bone and are capable of withstanding tension.

Tensor fasciae latae One of two muscles that abduct the hip, it resides in the anterolateral thigh. It originates from the anterior superior iliac spine (ASIS) of the coxal bone. Its insertion is via the iliotibial (IT) tract (or band).

Teres major Muscle that resides lateral to the scapula and runs through the axilla.

Teres minor Muscle that originates from the lateral border of the scapula just superior to the origin site of the teres major muscle.

Terminal arborization A group of short branches from the end of the nerve fiber.

Terminal end fiber Each branch of the terminal arborization.

Thalamus This organ is referred to as the "sensory gateway" because it receives and relays sensory input to its proper location in the brain.

Third metacarpal The third metacarpal bone (metacarpal bone of the middle finger) is a little smaller than the second.

Third metatarsal The second longest metatarsal.

Thoracic Anterior chest.

Thoracic vertebrae The group of 12 small bones that form the vertebral spine in the upper trunk.

Tibia The main bone of the lower leg, forming what is more commonly known as the shin.

Tibialis anterior This muscle is passes longitudinally through the anterior crural region from the proximal tibia across the ankle to the foot.

Tibialis posterior Most central of all the leg muscles, this muscle is located in the deep posterior compartment of the leg.

Tibiofemoral joint The largest of joint in the body, the articulation between the proximal tibia and the distal femur.

Tissue Any of the distinct types of material of which animals or plants are made, consisting of specialized cells and their products.

Toe abduction and adduction To splay the toes in the frontal plane about the sagittal axis.

Toe extension and flexion To straighten and bend the toes in the sagittal plane about the frontal axis.

Tracts Systems of body parts or organs that act together to perform some function.

Transverse ligament This ligament, unlike the others, does not connect two bones. Rather, it attaches on both sides of the intertubercular sulcus (bicipital groove).

Transverse plane An imaginary plane that divides the body into superior and inferior parts. It is perpendicular to the coronal plane and sagittal plane.

Trapezius This lateral neck flexor is commonly divided into three sections (upper, middle, and lower) because each section causes a different motion.

Triangular Describes muscles that are broad at one end and narrow at the other.

Triceps brachii Large muscle on the back of the upper limb It is the muscle principally responsible for extension of the elbow joint.

Triceps surae A pair of muscles located at the calf—the two-headed gastrocnemius and the soleus.

True ribs The first seven ribs attach to the sternum (the breast bone) in the front.

Trunk extension and flexion Movements that straighten and bend the trunk in the sagittal plane about the frontal axis.

U

Ulna A long bone found in the forearm that stretches from the elbow to the smallest finger, and is found on the medial side of the forearm. It runs parallel to the radius, the other long bone in the forearm, and is the larger and longer of the two.

Ulnar collateral ligament The ligament that stabilizes the medial side of the elbow. It attaches from medial epicondyle of the humerus and fans out to attach on medial parts of the ulna.

Ulnar deviation Bending sideways, but toward the ulna (medially) or toward midline.

Unilateral Relating to or affecting only one side of an organ, the body, or another structure.

Unipennate Having the fibers arranged obliquely and inserting into a tendon only on one side in the manner of a feather barbed on one side.

Upper trapezius A large paired surface muscle that extends longitudinally from the occipital bone to the lower thoracic vertebrae of the spine and laterally to the spine of the scapula. It moves the scapula and supports the arm.

Upward scapular rotation A turning of the scapula in a counterclockwise direction.

V

Vastus intermedius The largest and most powerful part of the quadriceps femoris, it is located on the outside of the thigh.

Vastus lateralis One of the muscles forming the quadriceps femoris, it is located on the inside of the thigh and extends to the knee.

Vastus medialis An extensor muscle located medially in the thigh that extends the knee.

Ventral Relating to the underside of an animal or plant; abdominal.

Ventral root Also called the anterior root, this is the efferent motor root of a spinal nerve. At its distal end, it joins with the dorsal root to form a mixed spinal nerve.

Vertical axis A direction of anatomical orientation, which passes from superior to inferior through the transverse plane.

Vesicles A structure within or outside a cell, consisting of liquid or cytoplasm enclosed by a lipid bilayer.

Vomer Small, thin bone separating the left and right nasal cavities.

W

Wrist flexion and extension Bending of the wrist toward the anterior and straightening toward the posterior, respectively.

Y

Yellow bone marrow Fatty tissue that is no longer is involved in producing red blood cells, it can revert back into red bone marrow if some severe condition creates a greater than normal need for production.

Z

Zygomatic In the human skull, this bone (cheekbone or malar bone) is a paired irregular bone which articulates with the maxilla, the temporal bone, the sphenoid bone, and the frontal bone.

Index

G

H

I

J

K

L

M

N

O

P

Q

R

S